Studies in Mechanobiology, Tissue Engineering and Biomaterials

Volume 2

Series Editor
Dr. Amit Gefen
Department of Biomedical Engineering
The Iby and Aladar Fleischman Faculty of Engineering
Tel Aviv University
69978 Ramat Aviv
Israel
e-mail: gefen@eng.tau.ac.il

Further volumes of this series can be found on our homepage:
http://www.springer.com/series/8415

Krishnendu Roy
Editor

Biomaterials as Stem Cell Niche

Springer

Editor
Prof. Krishnendu Roy, PhD
Cockrell School of Engineering,
Department of Biomedical Engineering
The University of Texas at Austin
1 University Station, C0800
Austin, TX 78712
USA
e-mail: kroy@mail.utexas.edu

ISSN 1868-2006 e-ISSN 1868-2014

ISBN 978-3-642-13892-8 e-ISBN 978-3-642-13893-5

DOI 10.1007/978-3-642-13893-5

Springer Heidelberg Dordrecht London New York

Library of Congress Control Number: 2010935018

Cover design: WMXDesign GmbH

Printed on acid-free paper

Springer is part of Springer Science+Business Media (www.springer.com)

Preface

Recent advances in stem cell biology have opened up new therapeutic strategies for a variety of incurable diseases. But the ultimate clinical success of such therapies lies entirely on our ability to efficiently control and manipulate stem cell fate and produce therapeutic cells in large, pharmaceutically relevant scales. This book presents a treatise on the state-of-the-art developments in biomaterials as synthetic niches for engineering stem cells, both for understanding their behavior under 3D biomimetic conditions as well as to develop new strategies for long term maintenance and lineage-specific differentiation into therapeutic cells. Animal and human stem cells of both embryonic and adult origin are discussed with applications ranging from nerve regeneration, orthopedics, cardiovascular therapy, blood cell production and muscle regeneration. Both synthetic and natural biomaterials have been discussed with specific focus on how material–stem cell interactions direct specific cell fate.

Chapter 1 introduces the reader to matrix biology and how dynamic elements of the stem cell extracellular matrix (ECM) can be engineered through biomaterials.

Chapter 2 and 3 provide a detailed understanding of how material properties and three-dimensional (3D) biomaterial scaffolds can be used to alter the stem cell niche with particular focus on surface energy, mechanical properties as well as controlled release of factors from the niche material.

The concept of niche topography and its effect on stem cell fate is thoroughly discussed in Chaps. 4 and 5. Chapter 4 provides particular emphasis on mechanotransduction while Chap. 5 emphasizes on how nanoscale topography of the stem cell niche influences cell proliferation and differentiation. Specifically the authors present ways to mimic these nanoscale features using three-dimensional (3D) nanofiber matrices and how such matrices influence a variety of stem cell behavior.

The use of micropatterning technologies to create synthetic, soft 3D niches is discussed in Chap. 6. Specifically, hydrogel-type biomaterials are presented as a conducive-microenvironment for stem cell growth and differentiation.

Chapter 7 further elaborates on the use of microfabrication technologies along with high throughput approaches on embryonic stem cell research. These are critical not only in controlling stem cell fate in a precise and deliberate manner but to also study cell–niche interactions in a quantitative and high-throughput way.

Chapters 8–12 focus on use of biomaterial-based niches for lineage-specific differentiation of stem cells. Specifically, Chaps. 8 and 9 discuss cardiovascular differentiation of stem cells while Chap. 10 introduces hematopoietic lineage, specifically T cell differentiation from embryonic and adult stem cells. Chapter 11 provides a review of neural regeneration under hypoxic conditions with specific emphasis on the role of biomaterials. Finally, Chap. 12 provides a unique treatise on the aging stem cell niche in the context of biomaterial-based muscle regeneration.

Austin, Texas Krishnendu Roy

Contents

Engineering ECM Complexity into Biomaterials for Directing Cell Fate 1
Sarah E. Stabenfeldt, Ashley Carson Brown, and Thomas H. Barker

Functional Biomaterials for Controlling Stem Cell Differentiation 19
Ameya Phadke, Chien-Wen Chang, and Shyni Varghese

Integration of Biomaterials into 3D Stem Cell Microenvironments 45
Andres Bratt-Leal, Richard Carpenedo, and Todd McDevitt

Stem Cell Interaction with Topography 61
Benjamin K. K. Teo, Soneela Ankam, and Evelyn K. F. Yim

The Nanofiber Matrix as an Artificial Stem Cell Niche 89
Hai-Quan Mao, Shawn H. Lim, Shuming Zhang, Gregory Christopherson, Korey Kam, and Stephen Fischer

Micropatterned Hydrogels for Stem Cell Culture 119
Sharon K. Hamilton, Hang Lu, and Johnna S. Temenoff

Microengineering Approach for Directing Embryonic Stem Cell Differentiation 153
Hojae Bae, Jason W. Nichol, Amir Foudeh, Behnam Zamanian, Cheong Hoon Kwon, and Ali Khademhosseini

Biomaterials as Stem Cell Niche: Cardiovascular Stem Cells 173
Ge Zhang and Laura J. Suggs

The Integrated Role of Biomaterials and Stem Cells in Vascular Regeneration 195
Guoming Sun, Sravanti Kusuma, and Sharon Gerecht

Synthetic Niches for Stem Cell Differentiation into T cells 225
Ankur Singh and Krishnendu Roy

Understanding Hypoxic Environments: Biomaterials Approaches to Neural Stabilization and Regeneration after Ischemia 247
Jennie B. Leach and Elizabeth M. Powell

Biomaterial Applications in the Adult Skeletal Muscle Satellite Cell Niche: Deliberate Control of Muscle Stem Cells and Muscle Regeneration in the Aged Niche 275
Eric Jabart and Irina Conboy

Author Index 309

Engineering ECM Complexity into Biomaterials for Directing Cell Fate

Sarah E. Stabenfeldt, Ashley Carson Brown and Thomas H. Barker

Abstract By definition, biomaterials and tissue engineering materials aim to replace, restore, and/or regenerate tissue where disease or injury has caused irreparable damage to native tissue. However, mimicking the complexity of the in vivo milieu has proved to be a serious challenge for current biomaterials. For example, one common strategy involves incorporating a short known biologically active motifs into the biomaterial to encourage cellular attachment, migration, and differentiation. In doing so, the intricacy of the dynamic in vivo environment is greatly diminished. This chapter highlights dynamic elements of the extracellular matrix (ECM) that provide critical signaling cues in vivo, the state of current biomaterial approaches, and potential approaches to incorporate more complex components into future biomaterials. The goal of this section is to introduce emerging concepts in matrix biology that have the potential to be exploited in biomaterials design, rather than being an exhaustive review of ECM dynamics.

1 Cell–ECM Interactions

Cell signaling involves a multitude of interactions with the extracellular environment that includes cell–cell, cell–ECM, and cell–growth factor interactions. Early reports of the ECM function described it as merely a 3D matrix required for mechanically supporting cells. However, we now know that the ECM

S. E. Stabenfeldt, A. C. Brown and T. H. Barker (✉)
W.H. Coulter Department of Biomedical Engineering,
Georgia Institute of Technology/Emory University, Atlanta GA, USA
e-mail: thomas.barker@bme.gatech.edu

Stud Mechanobiol Tissue Eng Biomater (2010) 2: 1–18
DOI: 10.1007/8415_2010_1

Published online: 12 June 2010

serves more functions than simply holding cells together to form tissue. The ECM is composed of an array of tissue-type dependent fibrous proteins many of which also exhibit developmental and pathology-specific splice variations, proteoglycans, and matricellular proteins. Moreover, structural fibril proteins such as laminin (Ln), fibronectin (Fn), and collagen also have many binding sites for growth factors and proteases as well as its primary role in the engagement of cell-surface integrins. Therefore, the ECM serves not only as a mechanical support, but also a reservoir for cellular signaling molecules. This chapter highlights the composition, signaling, and dynamic regulation through proteolytic degradation and remodeling of the ECM. Using the neural stem cell niche as a model system we address how these ECM dynamics influence stem cell differentiation and phenotype. Finally, this chapter will outline future additional areas of research in cell–ECM interactions and incorporation of complex ECM motifs into biomaterials for enhanced control of stem cell biology.

1.1 ECM Composition and Signaling

The ECM is comprised of a highly organized cellular microenvironment that is critical to cellular survival and development [14, 34, 74]. The fibril proteins (i.e. Fn, Ln, and collagen) contain many binding domains that interact with cells and other proteins. Without these proteins, development can be significantly altered or even halted. For example, during embryogenesis, Ln-1 ($\alpha 1\beta 1\gamma 1$) and Ln-10 ($\alpha 5\beta 1\gamma 1$) are essential for proper basement membrane assembly [46, 66]. Ablation of specific Ln subunits is lethal to developing murine embryos by day 5.5 ($\beta 1$ and $\gamma 1$) or day 7 ($\alpha 1$) due an absence of a basement membrane and an undifferentiated endoderm [34, 46, 66].

Cell–ECM interactions occur predominately through the receptor class known as integrins. Integrins are heterodimeric cell surface receptors composed of two transmembrane polypeptides, so-called α and β subunits. These $\alpha\beta$-integrin complexes are intracellularly linked to the cytoskeletal associated proteins (e.g. talin, vinculin, and/or paxillin) and signaling enzymes (e.g. focal adhesion kinase; FAK). The functional role of integrin–ECM interactions ranges from directing ECM assembly to stimulating intracellular signaling pathways for migration, differentiation, and survival. For example, β_1 integrins are capable of interacting with Ln-1 ($\alpha 1$-chain) to aid in proper assembly of the basement membrane during embryogenesis [1]. Disruption of Ln-1 assembly has been shown to lead to significant apoptosis in the basal lamina of embryoid bodies [34, 46]. Additionally, the β_1 integrin family has been linked to pro-survival signaling. In particular, Leone et al. [33] reported increased apoptosis in vitro of neural stem cells derived from β_1 integrin conditional knock-out mice. Therefore, ablation of either the ECM or critical ECM receptors eliminates inherent survival mechanisms in developing cells.

1.2 ECM Regulation

Integrin specific engagement has implications in cell development, survival, and differentiation. In development, the ECM environment changes dramatically, from mostly Fn to a predominantly Ln environment. Changes in ECM environment after development, such as in wound healing situations, can be less dramatic but still have serious effects on cell fate. Two ways in which integrin specificity can be regulated are through proteolytic processing of ECM molecules, and through mechanochemical translation of cell-binding ECM domains. These processes are thought to reveal "cryptic" binding sites and/or disturb conformation specific interaction between integrins and their ECM ligands, resulting in changes in integrin binding and potentially leading to an integrin "switch". These so called integrin switches result in the loss of binding of one integrin and the preferential binding of another integrin to an ECM ligand as the result of an alteration in the ligand itself. An emerging concept is that the inclusion of cryptic binding sites within ECM molecules provide a way to include structural cues for cells during development, tissue organization, wound healing, and remodeling processes [62]. This concept is evidenced by findings that reveal that unveiling of cryptic binding sites can govern many cell–ECM interactions, including migration, invasion, adhesion, and differentiation. This type of signaling provides an elegant system for regulating ECM cues in an energy efficient manner, such that the time-sensitive cues are created at the same time and are spatially oriented in the same location but remain cloaked until needed. Proteolytic processing of the ECM must be strictly controlled during wound healing because of the dramatic differences in cell phenotype that result; a requirement which is highlighted by evidence that many disorders, such as metastatic cancer, arise from over active proteases.

1.2.1 Proteolytic Processing of the ECM

Proteolysis of ECM proteins can significantly alter ECM conformation, integrin specificity, and ultimately ECM-mediated cell phenotype. A primary example of this is seen in proteolytic processing of Ln during wound healing of the epidermis [47]. Ln-5 is a known ligand of three main integrins, $\alpha3\beta1$, $\alpha6\beta1$, and $\alpha6\beta4$ [37]. Ln-5 is comprised of $\alpha3$, $\beta2$, and $\gamma2$ Ln chains. The integrin binding domain of Ln is located within the Ln globular (LG) domains found at the C-terminus of the $\alpha3$ chain of the molecule. Within the LN-$\alpha3$ module either $\alpha3\beta1$ or $\alpha6\beta4$ can bind, depending on the processing of the molecule, which can promote either migration or hemidesmosome formation, respectively. During in vivo dermal wound healing, unprocessed Ln-5 is synthesized and assembled into a fibrous network at the wound edge. In this unperturbed state, $\alpha3\beta1$ integrin binds via the LG3 domain, specifically to the PPFLMLLKGSTR motif [28]. This interaction leads to epithelial cellular motility and facilitates wound closure. Upon wound closure, cellular motility must be downregulated and this is elegantly regulated through

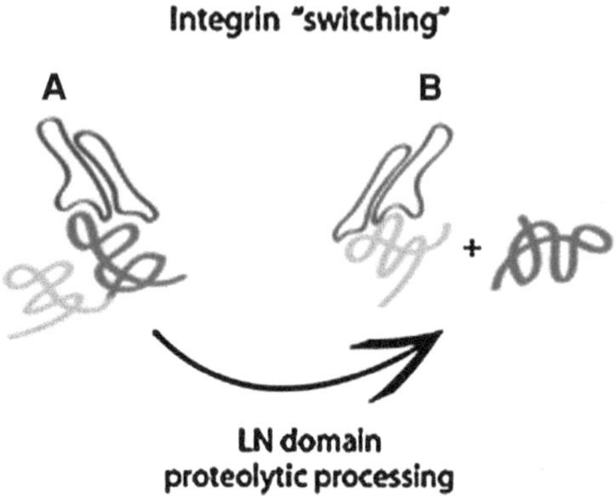

Fig. 1 Proteolytic processing and integrin-binding cryptic site exposure in Ln. *A* In newly synthesized Ln the α3 chain is expressed as the full-length polypeptide containing five globular domains. The LG3 domain supports α3β1 integrin binding. *B* Upon proteolysis the LG4 domain is exposed and drives primarily α6β4 integrin engagement and subsequent hemidesmosome formation

proteolytic processing of the LG3 domain, resulting in an integrin switch (Fig. 1). Specifically, proteolysis occurs between the LG3 and LG4 domains by serine proteases cleaving at Gln 1337-Asp1338, which results in exposure of a preferential ligand for α6β4 [50]. This integrin switch leads to a downregulation of cell motility, induction of hemidesmosome formation, and ultimately stabilization epithelial cell–cell adhesion [15, 22, 51, 52]. The parallels between the wound healing physiology studies and the process of epithelial to mesenchymal transition in the developing embryo, including gastrulation and development of complex epithelial organs like lung, kidney, and glands are striking (Table 1).

Proteolytic processing of matrix molecules can lead to an integrin switch, but the converse has also been observed. Integrin specific interactions with ECM molecules can also induce proteolysis through upregulation of MMPs. Once again in the context of dermal wound healing, keratinocytes located at the migrating epithelial front, where there is no basement membrane, upregulate expression of collagenase (MMP-1). In culture, this upregulation of MMP-1 has been shown to only occur when cells are in contact with native type I collagen [68], and is the result of integrin α2β1 binding. The specific nature of MMP-1 induction through α2β1 integrin binding is evidenced by the observations that α2 integrin blocking

Table 1 Role of specific integrins in cell fate determination

Integrin	Role in cell fate	References
α2β1	Leads to increases in MMP-1 expression on collagen I; facilitates keratinocyte migration on collagen	[68]
α3β1	Facilitates migration on Ln during wound closure	[50, 51]
α5β1	Facilitates wound healing, angiogenesis, enhanced differentation of precursor cells (osteoblasts, myoblasts, NSCs, and MSCs), and migration of Fn	[13, 21, 29, 69, 42]
α6β1	Increased levels are a sign of "stemness"; Facilitates migration on Ln and NSC adhesion and proliferation	[38, 50, 51, 58, 64]
α6β4	Results in hemidesmosome formation	[50–52]
αvβ3	Increased cell addhesion, increased proliferation and decreased differentiation of precursor cells	[8, 69, 13, 42]

antibodies prevent MMP-1 expression. The induction of MMP-1 expression facilitates keratinocyte migration on collagen, by cleaving the molecule. It has been suggested that cleavage of the collagen matrix by MMP-1 may expose cryptic sites on collagen for other receptors, such as αvβ5 integrin [45, 48]. Expression of MMP-1 after initial wounding peaks on the first day and gradually tapers for approximately 14 days until the levels of MMP-1 are undetectable, which corresponds with complete reepitheilization. Interestingly, Ln-1 (only deposited after reepithelilization), but not Ln-5 (expressed by migrating keratinocytes), seems to down regulate MMP-1 expression in keratinocytes that are in contact with collagen-I. This elegant coordination of integrin specific responses to collagen further highlights the importance of proteolytic processing of ECM molecules in regulating cell fate, particularly in the realm of wound healing. Whether these concepts are fully recapitulated in stem cell niches has yet to be fully established, but the principle holds promise for the design of ECM–mimetic materials that are truly interactive with and responsive to cells that are actively differentiating into various cell lineages.

1.2.2 Mechanochemical Translation of Cell-binding ECM Domains

In addition to proteolytic processing, exposure of cryptic binding sites can be regulated by force-mediated conformational changes in ECM molecules. Force-mediated conformational changes are, as the name implies, the direct result of cell-applied forces on ECM molecules. As in proteolytic processing of ECM molecules, mechanochemical translation results in a specific alteration in the molecule, in this case in the conformation, within critical ligand binding sequences. Two types of mechanochemical translation have been observed, first the activation and release of growth factors from the ECM and second, the deactivation/activation of integrin binding sites, both of which ultimately result in changes in local cellular responses. Though force-mediated conformational changes are documented in many molecules including von Willebrand factor [72], titan [36], transforming growth factor β (TGFβ), and the classical example of Fn, specifically in the type III repeats of the molecule.

Fn is a high-molecular weight glycoprotein, comprised of two nearly identical 230–270 kDa monomers cross-linked by a pair of disulfide bonds near their C-termini [41, 53]. Fn exists as a single gene, but alternative splicing of its mRNA results in multiple, tissue-specific and developmental/pathology-specific isoforms. Structurally, each Fn monomer is comprised of three repeating units known as type I, type II, and type III repeats. While type I and type II repeats contain stabilizing disulfide bridges, the type III repeats, composed of two anti-parallel β-sheets, do not, leaving them highly sensitive to force mediated unfolding. The central cell-binding region of Fn is found within the type III repeats, specifically the seventh through tenth type III repeats. The well-described Arg-Gly-Asp (RGD) site, which is a well-known and often employed peptide ligand for many integrins, is found within the tenth type III repeat. The best-defined demonstration of the effect of Fn domain unfolding at the secondary-level on integrin-binding is seen in domain unfolding of the tenth type III repeat.

In addition to the RGD-containing tenth type III repeat, a second integrin binding site, the so-called "synergy" site comprised of the sequence PHSRN, is found on the adjacent ninth type III repeat. The synergy site is located approximately 32 Å from the RGD loop on the tenth type III repeat. The type III repeats show great elasticity in the loops between their F- and G-β strands, known as the FG loop, which allows the ninth and tenth type III repeats to present multiple conformations. Under small applied forces (on the order of 10 pN) Fn's tenth type III repeat is susceptible to partial unfolding, resulting in an intermediate state in which the RGD loop within the tenth type III repeats begins to translocate away from the ninth type III repeat, resulting in an increase in the distance between the RGD and synergy sites from approximately 32 to approximately 55 Å [30]. This capacity to present multiple spatial orientations of the ninth and tenth type III repeats has great implication on cell binding, because the relative positioning of these two domains has been shown to influence integrin α5β1 binding. Integrin α5β1 binds by simultaneously engaging the RGD and synergy sites, as these sites translocate away from one another, they can no longer be bound simultaneously and integrin α5β1 no longer binds (Fig. 2). These findings suggest that an integrin "switch" can be mechanically induced by stretching and partially unfolding these two domains. Further evidence suggesting that a mechanically induced integrin switch exists within the ninth and tenth type III repeats of Fn was shown by studies in which conformational stability was conferred to the ninth or tenth type III repeats, resulted in modulation of integrin specificity. In these studies, the ninth type III repeat was stabilized via a Leu-Pro mutation at amino acid 1,408 [42] or the tenth type III repeat [27] was stabilized through stabilization of the hydrogen bonding within the repeat. In both cases, stabilization of the relative positions of the two repeats resulted in increased affinity for integrin α5β1 over integrin αvβ3. In addition, studies in which the linker region between the ninth and tenth type III

Fig. 2 Mechano-regulation of integrin binding to Fn. *A* The RGD and PHSRN sites of Fn are in close proximity and can be simultaneously bound by integrin α5β1. *B* After the application of force, the Fn domains begin to partially unfold and subsequently the RGD and PHSRN sites are moved away from one another. As a result of this translocation, α5β1 integrin can no longer bind

repeats was increased in length showed reduction in α5β1 binding [17]. These studies have significant implications in biomaterials design. For over a decade the integrin/adhesion ligand of choice in biomaterials has been the RGD peptide, either in linear or cyclic form. It has become increasingly clear that RGD is significantly less capable of directing integrin specificity, and thus cellular responses are mostly uncontrolled in these biomaterials. More recently several groups have begun to explore approaches for incorporating both RGD and synergy peptides as individual entities within biomaterials [5]. However, all such approaches have relied on statistical methods for ensuring sufficient coupling of both RGD and synergy peptides so as to promote the appropriate distances between the motifs. This approach has a major flaw in that the organization of these motifs with the Fn ECM is not homogeneous, nor is it random. The motifs are structurally positioned with respect to one another and the elegance of the Fn system ensures their appropriate position at energy levels below a particular threshold. Thus far the attempts to engineer an RGD + synergy biomaterial have lacked this level of control.

In addition to mediating integrin-specificity, force mediated unfolding of Fn is also required for polymerization of the molecule into a fibrillar form [2, 4, 40]. Cellular forces expose cryptic binding sites required for polymerization. Recent studies have highlighted the importance of cell-mediated unfolding of Fn in formation and maintenance of a polymerized Fn matrix. In vitro, inhibition of actin polymerization resulting in loss of cellular contractility results in relaxation of the Fn matrix, such that Fn exists in a more compact form [3]. Fn molecular collapse has also been observed in cell derived Fn-rich ECMs, in which fibroblasts were extracted from the matrix. Semi-quantitative analysis of Fn molecular unfolding indicates that cells are capable of inducing strain levels of approximately 200–700% into Fn fibers, resulting in domain unfolding of the type III repeats [65]. More recently, it has been shown that force-mediated unfolding of Fn fibers may lead to the association of nontraditional binding partners with Fn. While still not fully understood, there is growing support for the concept that partial unfolding of Fn and perhaps specifically the type III repeats leads to radical shifts in the function of Fn. An underexplored area of this research relates to the release of growth factors from Fn upon mechanical stimulation, although there is evidence that upon force application, Fn-bound TGFβ latency complex is "activated" and releases active TGFβ [73].

As with integrin-switching via proteolytic processing, integrin-switching as the result of mechanically induced translation can have great consequences on cell fate determination. Specifically in the case of Fn unfolding, there is typically a switch between α5β1 and αvβ3 integrin binding, resulting in differences in differentiation and proliferation processes. For example, engagement of the β1 integrin subunit has been shown to enhance differentiation of precursor cells [18]. In contrast, engagement of αv integrins, including αvβ3, contributes to increased cell adhesion and proliferation and decreased differentiation [8]. Specifically, systems resulting in an increase of α5β1 binding over αvβ3 binding to Fn has been shown to induce a switch between proliferation to differentiation of MC3T3-E1 osteoblast-like cells [69] and mouse C2C12 myoblasts to myotubes [13]. Controlling these integrin switches by engineering constructs that incorporate Fn type III repeats in specific conformations offer a potential method for controlling integrin specificity and ultimately cell fate. Highlighting this approach, Martino et al. [42] engineered Fn fragments presenting RGD and synergy in their native conformation as either a wild-type product or a hyperstabilized fragment. Stabilization of the domains lead to significantly enhanced specificity for $\alpha5\beta1$ integrin and enhanced differentiation of mesenchymal stem cells. In addition, $\alpha5\beta1$ integrin has been shown to regulate angiogenesis [29] and wound repair processes [21]. The ability to control cell fate, particularly in the case of stem cell differentiation, has many uses in the field of regenerative medicine.

2 ECM and the Stem Cell Niche

During neural development, ECM proteins are critical players. Ln-1, Ln-10, Fn, and collagen-IV are all ECM proteins that are important to neural signaling and have been associated with cellular survival, adhesion, migration, and differentiation [9, 12, 57]. Moreover, recent analysis of the stem cell niche in the adult subependymal layer (also known as the subventricular zone; SVZ) has identified the co-localization of LN subunits with basic fibroblast growth factor (bFGF) [26, 43, 44]. These findings indicate that the ECM not only provides direct cell support through cell–ECM interactions, but it also sequesters growth factors for neural stem cells (NSC).

2.1 Integrins: A Sign of "Stemness"

Certain integrin subunits have been related to properties of stem cells or "stemness". For example, murine embryonic stem cells (ESC), hematopoietic stem cells (HSC), and neural stem cells (NSC) all express elevated levels of $\alpha_6\beta_1$ integrins [19, 58]. Looking closer at NSCs, both murine and human derived NSCs have been separated from more mature differentiated cells based on the levels of β_1 integrin expression [19, 49]. Nagato et al. [49] demonstrated that sorting of a heterogeneous neural cell population positive for β_1 integrins yielded a population of pluripotent stem cells capable of differentiating into all three neural lineages (neurons, oligodendrocytes, and astrocytes); whereas, cells negative for β_1 integrins yielded a population incapable of neurosphere formation, an identifying quality of NSCs. Similarly, examining the genetic expression of immature and mature neural cell markers from human NSCs indicate that NSC populations with high levels of β_1 integrins also expressed immature NSC markers (i.e. prominin-1, nestin, sox2, sox3, musahi-1, and bmi-1); whereas, populations with low levels of β_1 integrins had increased expression of more mature neuronal markers (i.e. βIII-tubulin) [19].

2.2 Neural Stem Cells and Integrin/ECM Alterations

NSCs within the developing and mature adult brain engage in ECM interactions via integrins. Research has shown that the ECM microenvironment surrounding NSCs and subsequent integrin engagement is critical to maintaining self-renewal properties and then initiating differentiation into the various neural phenotypes. This section highlights the temporal and spatial profiles of ECM components and integrins surrounding NSCs during neural development and within the adult NSC niche.

2.2.1 Integrin and ECM Profile During Neural Development

Developing neural tissue provides an excellent example of integrin alterations that are both spatially and temporally dependent (review by [63]). Throughout development, β_1 and β_5 integrins have been identified as unique in the developing cortex and subcortical tissue. However, integrin distribution shifts and includes a broader range of β integrins in post-natal brains. For example, β_6 integrins only appear in the adult cortex namely on neurons and oligodendrocytes [63]. The neural stem cell niche of the developing ventricular zone (VZ) is dominated with high levels of β_1 integrins on NSCs and Ln (Fig. 3) [10, 38, 56]. At embryonic day 13.5 (E13.5; murine model) the β_1 integrin positive NSCs are present in the VZ, subventricular zone (SVZ), and intermediate cortical regions; these NSCs also co-localize with markers of proliferation [38]. However, by E16 the β_1 integrin positive cells are confined to the VZ/SVZ regions as the cortical neuronal layers become more differentiated (Fig. 3) [38]. More specifically, the VZ is populated with NSCs expressing high levels of $\alpha_6\beta_1$ integrins. Interrogating neural tissue radial out into the intermediate zone and developing cortical regions has revealed that levels of α_6 integrin subunit decrease and levels of $\alpha_5\beta_1$ integrins increase [6, 7]. Corresponding to integrin expression, levels of known β_1 integrin ECM-based ligands (i.e. Ln and Fn) are spatially and temporally regulated in the VZ/SVZ and

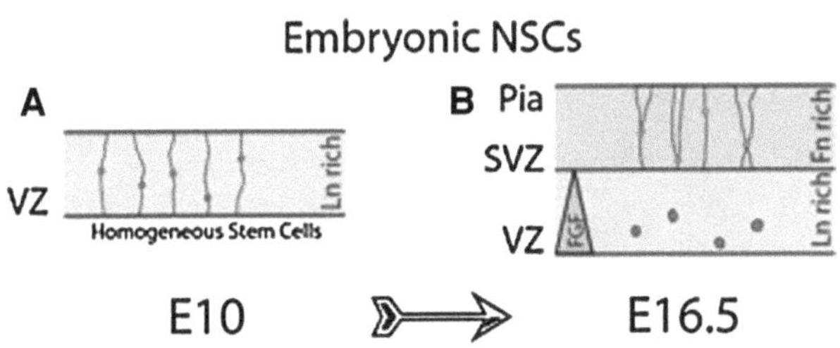

Fig. 3 Integrin switch during neural development. *A* The NSC microenvironment during development consists predominately of bFGF soluble factors and laminin chains with corresponding $\alpha_6\beta_1$ integrins that maintains a relatively homogenous stem cell population. *B* As development continues, NSCs begin to differentiate and migrate out to cortical layers potentially due to integrin switching (e.g. $\alpha_6\beta_1$ to $\alpha_5\beta_1$ integrins), ECM modifications (Ln to fibronectin), and/or decreasing levels of bFGF

developing cortex. Specifically, the levels of the known $\alpha_6\beta_1$ integrin ligand Ln-α2 chain (Ln 2, 4, and 12) peak in the VZ at E12.5 and then decrease throughout the remainder of development [7]. In contrast, Fn, a ligand for $\alpha_5\beta_1$ integrins, is predominately present in the developing cortex [70]. Assessing the potential sources for these ECM components, recent studies suggest that NSCs could participate in producing certain Ln chains. For example, in vitro NSC cultures have been shown to produce and secrete high levels of Ln γ1 chains (ligand for $\alpha_6\beta_1$ integrins; Ln 1-4, 6-11); however, the production of the γ1 chains is significantly reduced after differentiation [67]. This localized ECM production may be a type self-regulation that maintains the NSCs in an undifferentiated state. However, the Ln γ1 chains may be further modified by extracellular proteolytic enzymes or other means to shift the self-maintenance signal to differentiation signal, as was demonstrated in the case of keratinocyte wound repair in the previous section. Therefore, the signaling pathways that stimulate NSC migration and differentiation during development may involve integrin switching from $\alpha_6\beta_1$ integrins to $\alpha_5\beta_1$ integrins. The suggested role of $\alpha5\beta1$ integrin in the differentiation of NSCs echoes the demonstrated role of this integrin in the differentiation of a number of other stem and precursor cells including MC3T3 preosteoblast, MC12 myoblasts, and our own work looking at human MSCs.

2.2.2 ECM and Integrin Profile in Adult Neural Stem Cell Niche

In the adult brain, regions of neurogenic potential have been identified in both the SVZ region and the hippocampus [11, 16]. However for this chapter, only the SVZ will be highlighted. In SVZ, the NSC niches are localized to a subregion between the ependymal layer of the apical ventricle surface and neighboring vasculature (Fig. 4) [26, 31]. Interestingly, a percentage of the NSCs actually penetrate the ependymal cell layer positioning the NSC in contact with the ventricle and span out to the local vasculature (5–10 μm away from the ventricle surface) [64].

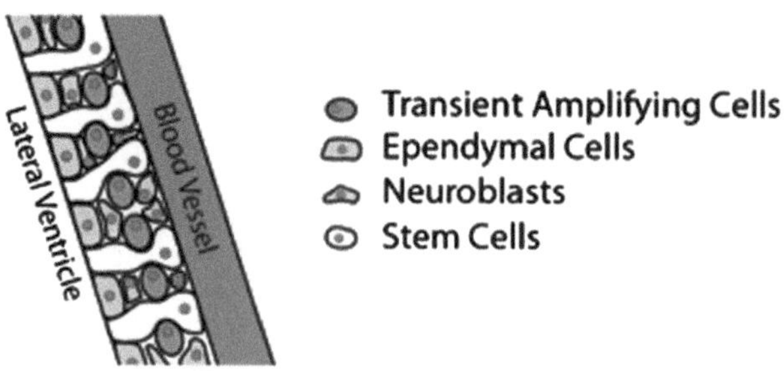

Fig. 4 Adult neural stem cell niche. Located in SVZ, the NSC niches are localized to a subregion between the ependymal layer of the apical ventricle surface and neighboring vasculature. NSCs have been shown to integrate in between ependymal cells while maintaining contact with the blood vessel basal lamina rich in Ln. More committed transient amplifying cells as reside in the SVZ and continue to differentiate into neuroblasts that migrate out of the SVZ. Comparing the integrin profiles of NSCs to neuroblasts there is a significant loss of $\alpha_6\beta_1$-integrin expression

Similar to the NSCs of the VZ/SVZ region in the developmental phases, NSCs of the adult SVZ express high levels of $\alpha_6\beta_1$ integrins; however, the more differentiated migratory neuroblasts have lower levels of $\alpha_6\beta_1$ integrins [64]. Adult vascular endothelial cells are known to secrete a substantial amount of Ln-10, a well-known ligand for $\alpha_6\beta_1$ integrins. So, it is not surprising that the co-localization of NSC $\alpha_6\beta_1$ integrins and endothelial-derived Ln-10 have been observed in vivo [26, 43, 64]. Moreover, in vitro SVZ derived neurospheres adhere and spread on a monolayer of confluent endothelial cells; adhesion is blocked with either $\alpha_6\beta_1$ integrin blocking antibodies or a competitive soluble Ln peptide. By anchoring to both the vasculature and ventricle, adult NSCs have direct access to two sources of soluble signals that may influence intracellular signaling in combination with ECM/integrin signaling.

2.2.3 Functional Role of ECM/Integrin Interactions

Recent publications have investigated the functional role of β_1 integrins within the embryonic and adult neural stem cell niche [38, 64]. Both groups utilized transient function-blocking antibodies delivered intraventricularly in vivo to provide localized β_1 or $\alpha_6\beta_1$ integrin interference. In both the embryonic VZ and adult SVZ NSCs, blocking β_1 integrins predominately affected the location of the NSCs in relation to ventricular surface. NSCs mainly lie within 10 μm of the apical ventricle surface. However, localized and transient blocking of the β_1 integrins resulted in NSC detachment from the apical ventricle surface and NSCs were found at distances greater than 10 μm from the boundary [38, 64]. Additionally, an increased number of proliferating NSCs were also observed. These studies suggest that NSC $\alpha_6\beta_1$ integrins function to maintain NSC adhesion and also regulate the proliferative capacity of the NSCs within the niche. In vitro studies with SVZ derived NSCs revealed that $\alpha_6\beta_1$ integrin signaling in NSCs involves the mitogen-activated protein (MAP) kinase pathway for self-renewal maintenance [7]. In addition to β_1 integrin signaling, the MAPK pathway is also influenced by growth factor signaling such as epidermal growth factor (EGF) and basic fibroblast growth factor basic (bFGF or FGF-2) [7, 32]. The convergence onto MAPK signaling is significant since during neural development, levels of EGF and bFGF fluctuate in the developing neuroepithelium. The initial mitogenic factor, bFGF, peaks at E11 at which point EGF levels begin to increase. Interestingly, NSCs exposed to bFGF upregulate EGF receptors [35]. Therefore, the wave of bFGF may act as a primer for the upcoming wave of EGF. These soluble signals shift in conjunction with modifications to the ECM and integrin expression profiles within the VZ and developing cortical regions play a major role in neural development and differentiation (Fig. 2).

Migration and differentiation patterns of NSCs have been evaluated using controlled in vitro experiments. Specifically, NSCs adhere with greater affinity to Fn and Ln-1 than collagen-I, collagen-IV, and non-treated tissue cultured plastic [71]. Additionally, migration on Ln-1 was significantly higher than Fn. On both Ln-1 and Fn, migration was controlled by β_1 integrins. More specifically, migration on Ln-1 appeared to be regulated by $\alpha_6\beta_1$ integrin signaling, whereas $\alpha_5\beta_1$ integrins modulated migration on Fn [25, 71]. Astrocytic differentiation on both Ln-1 and Fn dominated the differentiation profile; however, modest increases in neuronal phenotypes was observed on Ln-1 compared to Fn [71]. Collectively, characterization of in vivo neural development and in vitro NSC cultures indicate that both spatial and temporal changes in both integrin expression and the ECM are instrumental in NSC maintenance, migration, and differentiation. Taken together with dynamic nature of the ECM, that is, not only is the ECM regulated at the translational and transcription level but also posttranscriptionally in the extracellular space through proteolytic processing and mechanochemical translation events, suggest that the maintenance as well as the differentiation of stems cells, like NSCs, can be partially controlled by the design of ECM–mimetic biomaterials.

3 Current Biomaterials Approaches

Cell–ECM interactions play a critical role in directing cell fate. Cell binding to ECM molecules through integrins is determined by specific sequences within the ECM molecules. In the field of regenerative medicine, the ability to specifically control integrin binding in turn offers an attractive method for controlling cell fate. As it becomes clearer how specific sequences within ECM molecules and how proteolytic processing directs integrin binding, and how this in turn leads to markedly different cellular responses, biomedical approaches are arising to harness these responses. Two approaches have been taken for developing biomaterials to control cell fate as a result of ECM interactions: biomimetic approaches and engineering of protein variants. Both approaches have been used to increase cell adhesion, direct proliferation, and direct differentiation.

3.1 Biomimetic Approaches

Biomimetic approaches typically focus on developing simple environments to facilitate cell adhesion through the addition of cell adhesive peptides, such as RGD, into scaffolds. This bottom-up approach is attractive from the standpoint of simplicity, ease of reproducibility, and cost effectiveness. In general this approach has demonstrated utility in controlling cellular adhesion, but thus far lacks the biological specificity for controlling complex cellular processes such as stem cell differentiation. The widely used RGD sequence for example, is a promiscuous integrin binding sequence and is known to bind many integrins including $\alpha2\beta1$, $\alpha3\beta1$, $\alpha4\beta1$, $\alpha5\beta1$, $\alpha8\beta1$, $\alpha v\beta1$, $\alpha v\beta3$, $\alpha v\beta5$, $\alpha v\beta6$, and $\alpha v\beta8$ [61]. Further highlighting the lack of specificity in the RGD sequence is seen in the number of

ECM proteins that contain the sequence, including Fn, vitronectin, LN, entactin, osteopontin vitronectin, fibrinogen, thrombospondin, von Willebrand factor, tenascin, and several others including some growth factors like TGFβ [61]. Despite the general failure of the RGD peptide to direct cell fate, several other peptide motifs have begun to demonstrate greater control over integrin specificity and thus cell behavior. Of note are two increasingly used peptides, GFOGER and YIGSR, derived from collagen and Ln, respectively. These adhesive ligands direct greater specificity to particular integrins and thus will likely yield greater results as these and peptides are developed. However, what remains lacking in these systems is the temporal control over their presentation to cells as is seen in the native ECM. Several approaches have been explored toward this goal. The discovery and development of enzyme-specific proteolytic peptides has enabled greater biomimicry by allowing biomaterials to be processed in ways similar to the native ECM [39]. In addition, synergistic peptides may be patterned in ways that enable simple peptides to exert greater control over integrin specificity [5]. Despite these efforts, directing specific cell behavior, in particular differentiation, in response to biomaterials for regenerative medicine applications is thus far a mostly unrealized goal in the field of biomaterials and regenerative medicine. [23, 24] Current technological advances inspired by the reductionist point of view have focused on developing simple and merely adequate environments that facilitate simple cell processes, such as cellular adhesion. However, even if ECM-derived peptides have largely demonstrated their utility, such as supporting cell adhesion, the lack of biological specificity of many of these peptide motifs is simply not optimal to control more integrated processes such as differentiation [54]. In an attempt to elicit greater control over integrin binding, efforts have been made to engineer ECM variants, through the creation of recombinant protein fragments, that incorporate additional levels of complexity to direct cell fate.

3.2 Engineering Protein Variants

Engineering protein fragments has several distinct advantages for directing integrin specificity; first extracellular matrix molecules are large, complex molecules that contain not only integrin binding sites, but also many other sites that are involved in other protein/protein interactions as well as numerous other processes that can influence cell fate. By creating protein fragments, these other interactions can be either engaged or avoided and specific integrin interactions can be elicited. Second, as previously mentioned, protein fragments are more specific than simply using peptide sequences and have evolved molecular mechanisms that refine their activity, e.g. the unfolding of Fn's ninth and tenth type III repeats at a threshold force. Integrin specificity can be conferred by adjacent binding sequences, such as the PHSRN sequence in Fn, [30, 41] as well as through conformation induced specificity, which is seen in the requirement of a helical conformation within the integrin sequence of collagen for $\alpha 2\beta 1$ integrin binding [27, 59].

In order to induce integrin specific cellular responses, several groups have created recombinant Fn protein variants that display the ninth and tenth type III repeats displaying both the RGD and PHSRN sites, these protein variants therefore can be used to induce $\alpha5\beta1$ integrin binding and the resulting cell phenotypes. For example, Fn fragments displaying the 7–10 type III repeats were shown to enhance osteoblastic differentiation of bone marrow stromal cells when coated on titanium implants [54, 55], a response that was shown to be the result of an increase in integrin $\alpha5\beta1$ binding compared to RGD functionalized surfaces. Engineered Fn type III repeats have also been used to direct mesenchymal stem cell (MSC) fate in 2D and 3D systems through the presentation of the ninth and tenth type III together (RGD and PHSRN) or the tenth type III repeat alone (RGD only) [42]. Once again the presentation of RGD and PHSRN together resulted in the increase of integrin $\alpha5\beta1$ binding compared to MSCs cultured on RGD alone, which predominantly engaged integrin $\alpha v\beta3$. This observed integrin switch was accompanied by an enhanced osteoblastic differentiation.

While these examples focus on conferring integrin specificity to cells through Fn type III repeats, there are a number of other proteins that contain specific, known sequences that direct integrin binding, including the GFOGER sequence found in type I collagen. The GFOGER sequence, which has also been used as a single polypeptide adhesion ligand, is found in residues 502–507 within the molecule and direct integrin $\alpha2\beta1$ binding [27, 59]. Interestingly, the tertiary structure of the binding sequence is just as important as the amino acid sequence for integrin binding, such that the native triple-helical structure is critical for cell binding [27]. Several groups have created collagen-1 fragments that present the GFOGER sequence in its native triple helical sequence and have shown the fragments utility in directing cell fate in various systems, including liver and bone [20, 27]. By combining several fragments that direct integrin specific engagement, one could potentially elicit complex cellular responses and explore synergistic effects. An example of this is seen through the combination of triple helical GFOGER fragments for directing $\alpha2\beta1$ engagement with Fn type III repeat fragments for directing $\alpha5\beta1$ engagement. By eliciting $\alpha2\beta1$ and $\alpha5\beta1$ binding, Reyes and Garcia [60] were able to enhance osteoblastic differentiation and matrix mineralization.

These examples highlight the utility of engineering protein variants in directing integrin specificity and cell fate. The possibilities for engineering protein variants are ever increasing as biologists uncover previously unknown binding sites within ECM molecules and as more becomes known about the signaling networks involved.

3.3 Future Directions for Biomaterials as Stem Cell Niches

As new biomaterials and regenerative medicine approaches are developed there will be a significant need to acknowledge and account for the inherent complexity presented in the native stem cell niche and in developmental programs. From an engineering perspective, increasing complexity can be incorporated into emerging systems through modern approaches such as 2D and 3D patterning. Several groups have successfully employed micro and nano-scale patterning technologies in 2D that are capable of controlling integrin cluster size. Whether this level of control is physiologically relevant and/or is capable of exerting the level of control over cell fate remains to be determined. Using 2D systems to understand how we can use the spatial orientation of the ECM to regulate cell fate is just the beginning. The current efforts to pattern 3D hydrogels via non-cell destructive means (i.e. nonUV/radical) must continue to progress. The benefit of developing patternable systems is that one can take both the reductionist approach of employing peptides, as well as employ the more physiologically relevant ECM systems (e.g. recombinant ECM fragments). Related to the idea of patterning nano and micro-scale "spots" is the concept of integrin clustering via the design and creation of tandem repeats of ECM fragments and/or peptides. Native ECM protein fibers are assembled such that adhesion motifs are presented in a tandem repeated fashion. Several groups have begun to explore the creation of recombinant tandem repeating structures presenting multiple adhesion factors. This approach potentially allows the time sensitive, via enzymatic degradation, loss of motif that enhances integrin clustering and downstream signaling. One could then take this concept a step further in the development of systems that enables one to control not only the spatial orientation in a single modification step, but rather, through the design of gel–sol materials, control for the rates of clustering by modification of the fluidic properties of such complex materials. Through these and other future developments that continually strive to emulate the complexity of the ECM microenvironment, the possibility of engineering biomaterials as stem cells niches will become reality.

References

1. Aumailley, M., Pesch, M., et al.: Altered synthesis of laminin 1 and absence of basement membrane component deposition in (beta)1 integrin-deficient embryoid bodies. J. Cell Sci. **113**(Pt 2), 259–268 (2000)
2. Baneyx, G., Baugh, L., et al.: Coexisting conformations of fibronectin in cell culture imaged using fluorescence resonance energy transfer. Proc. Natl Acad. Sci. USA **98**(25), 14464–144648 (2001)
3. Baneyx, G., Baugh, L., et al.: Fibronectin extension and unfolding within cell matrix fibrils controlled by cytoskeletal tension. Proc. Natl Acad. Sci. USA **99**(8), 5139–5143 (2002)
4. Barker, T.H., Baneyx, G., et al.: SPARC regulates extracellular matrix organization through its modulation of integrin-linked kinase activity. J Biol Chem **280**(43), 36483–36493 (2005)
5. Benoit, D.S., Anseth, K.S.: The effect on osteoblast function of colocalized RGD and PHSRN epitopes on PEG surfaces. Biomaterials **26**(25), 5209–5220 (2005)
6. Campos, L.S.: Beta1 integrins and neural stem cells: making sense of the extracellular environment. Bioessays **27**(7), 698–707 (2005)
7. Campos, L.S., Leone, D.P., et al.: Beta1 integrins activate a MAPK signalling pathway in neural stem cells that contributes to their maintenance. Development **131**(14), 3433–3444 (2004)
8. Cheng, S.L., Lai, C.F. et al.: Bone mineralization and osteoblast differentiation are negatively modulated by integrin alpha(v)beta3. J. Bone Miner. Res. **16**(2), 277–288 (2001)

9. Chothia, C., Jones, E.Y.: The molecular structure of cell adhesion molecules. Ann. Rev. Biochem. **66**, 823–862 (1997)
10. Cousin, B., Leloup, C., et al.: Developmental changes in integrin beta-subunits in rat cerebral cortex. Neurosci. Lett. **234**(2–3), 161–165 (1997)
11. Doetsch, F., Garcia-Verdugo, J.M. et al.: Cellular composition and three-dimensional organization of the subventricular germinal zone in the adult mammalian brain. J. Neurosci. **17**(13), 5046–5061 (1997)
12. Freire, E., Gomes, F.C.A., et al.: Structure of laminin substrate modulates cellular signaling for neuritogenesis. J. Cell Sci. **115**(24), 4867–4876 (2002)
13. Garcia, A.J., Vega, M.D., et al.: Modulation of cell proliferation and differentiation through substrate-dependent changes in fibronectin conformation. Mol. Biol. Cell. **10**(3), 785–798 (1999)
14. Gersdorff, N., Muller, M., et al.: Basement membrane composition in the early mouse embryo day 7. Dev. Dyn. **233**(3), 1140–1148 (2005)
15. Goldfinger, L.E., Hopkinson, S.B., et al.: The alpha3 laminin subunit, alpha6beta4 and alpha3beta1 integrin coordinately regulate wound healing in cultured epithelial cells and in the skin. J. Cell Sci. **112**(Pt 16), 2615–2629 (1999)
16. Gould, E., Reeves, A.J., et al.: Hippocampal neurogenesis in adult old world primates. Proc. Natl Acad. Sci. USA **96**(9), 5263–5267 (1999)
17. Grant, R.P., Spitzfaden, C., et al.: Structural requirements for biological activity of the ninth and tenth FIII domains of human fibronectin. J. Biol. Chem. **272**(10), 6159–6166 (1997)
18. Gronthos, S., Simmons, P.J., et al.: Integrin-mediated interactions between human bone marrow stromal precursor cells and the extracellular matrix. Bone **28**(2), 174–181 (2001)
19. Hall, P.E., Lathia, J.D., et al.: Integrins are markers of human neural stem cells. Stem Cells **24**(9), 2078–2084 (2006)
20. Hennessy, K.M., Pollot, B.E., et al.: The effect of collagen I mimetic peptides on mesenchymal stem cell adhesion and differentiation, and on bone formation at hydroxyapatite surfaces. Biomaterials **30**(10), 1898–1909 (2009)
21. Herard, A.L., Pierrot, D., et al.: Fibronectin and its alpha 5 beta 1-integrin receptor are involved in the wound-repair process of airway epithelium. Am. J. Physiol. **271**(5 Pt 1), L726–L733 (1996)
22. Hintermann, E., Bilban, M., et al.: Inhibitory role of alpha 6 beta 4-associated erbB-2 and phosphoinositide 3-kinase in keratinocyte haptotactic migration dependent on alpha 3 beta 1 integrin. J. Cell Biol. **153**(3), 465–478 (2001)
23. Hubbell, J.A.: Bioactive biomaterials. Curr. Opin. Biotechnol. **10**(2), 123–129 (1999)
24. Hubbell, J.A.: Materials as morphogenetic guides in tissue engineering. Curr. Opin. Biotechnol. **14**(5), 551–558 (2003)
25. Jacques, T.S., Relvas, J.B., et al.: Neural precursor cell chain migration and division are regulated through different beta 1 integrins. Development **125**(16), 3167–3177 (1998)
26. Kerever, A., Schnack, J., et al.: Novel extracellular matrix structures in the neural stem cell niche capture the neurogenic factor FGF-2 from the extracellular milieu. Stem Cells **25**(9), 2146–2157 (2007)
27. Khew, S.T., Zhu, X.H., et al.: An integrin-specific collagen-mimetic peptide approach for optimizing Hep3B liver cell adhesion, proliferation, and cellular functions. Tissue Eng. **13**(10), 2451–2463 (2007)
28. Kim, J.M., Park, W.H., et al.: The PPFLMLLKGSTR motif in globular domain 3 of the human laminin-5 alpha3 chain is crucial for integrin alpha3beta1 binding and cell adhesion. Exp. Cell Res. **304**(1), 317–327 (2005)
29. Kim, S., Bell, K., et al.: Regulation of angiogenesis in vivo by ligation of integrin alpha5beta1 with the central cell-binding domain of fibronectin. Am. J. Pathol. **156**(4), 1345–1362 (2000)
30. Krammer, A., Craig, D., et al.: A structural model for force regulated integrin binding to fibronectin's RGD-synergy site. Matrix Biol. **21**(2), 139–147 (2002)
31. Lathia, J.D., Rao, M.S., et al.: The microenvironment of the embryonic neural stem cell: lessons from adult niches? Dev. Dyn. **236**(12), 3267–3282 (2007)

32. Learish, R.D., Bruss, M.D., et al.: Inhibition of mitogen-activated protein kinase kinase blocks proliferation of neural progenitor cells. Brain Res. Dev. Brain Res. **122**(1), 97–109 (2000)
33. Leone, D.P., Relvas, J.B., et al.: Regulation of neural progenitor proliferation and survival by β1 integrins. J. Cell Sci. **118**, 2589–2599 (2005)
34. Li, S., Harrison, D., et al.: Matrix assembly, regulation, and survival functions of laminin and its receptors in embryonic stem cell differentiation. J. Cell Biol. **157**(7), 1279–1290 (2002)
35. Lillien, L., Raphael, H.: BMP and FGF regulate the development of EGF-responsive neural progenitor cells. Development **127**(22), 4993–5005 (2000)
36. Linke, W.A., Grutzner, A.: Pulling single molecules of titin by AFM—recent advances and physiological implications. Pflugers Arch. **456**(1), 101–115 (2008)
37. Litjens, S.H., de Pereda, J.M., et al.: Current insights into the formation and breakdown of hemidesmosomes. Trends Cell Biol. **16**(7), 376–383 (2006)
38. Loulier, K., Lathia, J.D., et al.: beta1 integrin maintains integrity of the embryonic neocortical stem cell niche. PLoS Biol **7**(8), e1000176 (2009)
39. Lutolf, M.P., Lauer-Fields, J.L. et al.: Synthetic matrix metalloproteinase-sensitive hydrogels for the conduction of tissue regeneration: engineering cell-invasion characteristics. Proc Natl. Acad. Sci. USA **100**(9), 5413–5418 (2003)
40. Mao, Y., Schwarzbauer, J.E.: Fibronectin fibrillogenesis, a cell-mediated matrix assembly process. Matrix Biol. **24**(6), 389–399 (2005)
41. Mao, Y., Schwarzbauer, J.E.: Accessibility to the fibronectin synergy site in a 3D matrix regulates engagement of alpha5beta1 versus alphavbeta3 integrin receptors. Cell Commun. Adhes. **13**(5–6), 267–277 (2006)
42. Martino, M.M., Mochizuki, M., et al.: Controlling integrin specificity and stem cell differentiation in 2D and 3D environments through regulation of fibronectin domain stability. Biomaterials **30**(6), 1089–1097 (2009)
43. Mercier, F., Kitasako, J.T., et al.: Anatomy of the brain neurogenic zones revisited: fractones and the fibroblast/macrophage network. J. Comp. Neurol. **451**(2), 170–188 (2002)
44. Mercier, F., Kitasako, J.T., et al.: Fractones and other basal laminae in the hypothalamus. J. Comp. Neurol. **455**(3), 324–340 (2003)
45. Messent, A.J., Tuckwell, D.S., et al.: Effects of collagenase-cleavage of type I collagen on alpha2beta1 integrin-mediated cell adhesion. J. Cell Sci. **111** (Pt 8), 1127–1135
46. Miner, J.H., Li, C., et al.: Compositional and structural requirements for laminin and basement membranes during mouse embryo implantation and gastrulation. Development **131**(10), 2247–2256 (2004)
47. Miner, J.H., Yurchenco, P.D.: Laminin functions in tissue morphogenesis. Annu. Rev. Cell. Dev. Biol. **20**, 255–284 (2004)
48. Murphy, G., Gavrilovic, J.: Proteolysis and cell migration: creating a path? Curr. Opin. Cell Biol. **11**(5), 614–621 (1999)
49. Nagato, M., Heike, T., et al.: Prospective characterization of neural stem cells by flow cytometry analysis using a combination of surface markers. J. Neurosci. Res. **80**(4), 456–466 (2005)
50. Nguyen, B.P., Gil, S.G., et al.: Deposition of laminin 5 by keratinocytes regulates integrin adhesion and signaling. J. Biol. Chem. **275**(41), 31896–31907 (2000)
51. Nguyen, B.P., Ryan, M.C., et al.: Deposition of laminin 5 in epidermal wounds regulates integrin signaling and adhesion. Curr. Opin. Cell Biol. **12**(5), 554–562 (2000)
52. O'Toole, E.A., Marinkovich, M.P., et al.: Laminin-5 inhibits human keratinocyte migration. Exp. Cell Res. **233**(2), 330–339 (1997)
53. Pankov, R.Y.K. Fibronectin at a glance. J. Cell Sci. **115**(20), 3861–3863 (2002)
54. Petrie, T.A., Capadona, J.R., et al.: Integrin specificity and enhanced cellular activities associated with surfaces presenting a recombinant fibronectin fragment compared to RGD supports. Biomaterials **27**(31), 5459–5470 (2006)
55. Petrie, T.A., Raynor, J.E., et al.: The effect of integrin-specific bioactive coatings on tissue healing and implant osseointegration. Biomaterials **29**(19), 2849–2857 (2008)

56. Pinkstaff, J.K., Detterich, J., et al. Integrin subunit gene expression is regionally differentiated in adult brain. J Neurosci **19**(5), 1541–1556 (1999)
57. Powell, S.K., Kleinman, H.K. Neuronal laminins and their cellular receptors. Int. J. Biochem. Cell Biol. **29**(3), 401–414 (1997)
58. Ramalho-Santos, M., Yoon, S., et al.: "Stemness": transcriptional profiling of embryonic and adult stem cells. Science **298**(5593), 597–600 (2002)
59. Reyes, C.D., Garcia, A.J.: Engineering integrin-specific surfaces with a triple-helical collagen-mimetic peptide. J. Biomed. Mater. Res. A **65**(4), 511–523 (2003)
60. Reyes, C.D., Garcia, A.J.: Alpha2beta1 integrin-specific collagen-mimetic surfaces supporting osteoblastic differentiation. J. Biomed. Mater. Res. A **69**(4), 591–600
61. Ruoslahti, E.: RGD and other recognition sequences for integrins. Annu. Rev. Cell Dev. Biol. **12**, 697–715 (1996)
62. Schenk, S., Quaranta, V.: Tales from the crypt[ic] sites of the extracellular matrix. Trends Cell Biol. **13**(7), 366–375 (2003)
63. Schmid, R.S., Anton, E.S.: Role of integrins in the development of the cerebral cortex. Cereb. Cortex **13**(3), 219–224 (2003)
64. Shen, Q., Wang, Y., et al.: Adult SVZ stem cells lie in a vascular niche: a quantitative analysis of niche cell–cell interactions. Cell Stem Cell **3**(3), 289–300 (2008)
65. Smith, M.L., Gourdon, D., et al.: Force-induced unfolding of fibronectin in the extracellular matrix of living cells. PLoS Biol. **5**(10), e268 (2007)
66. Smyth, N., Vatansever, H.S., et al.: Absence of basement membranes after targeting the LAMC1 gene results in embryonic lethality due to failure of endoderm differentiation. J. Cell Biol. **144**(1), 151–160 (1999)
67. Staquicini, F.I., Dias-Neto, E., et al.: Discovery of a functional protein complex of netrin-4, laminin gamma1 chain, and integrin alpha6beta1 in mouse neural stem cells. Proc. Natl Acad. Sci. USA **106**(8), 2903–2908 (2009)
68. Steffensen, B., Hakkinen, L., et al.: Proteolytic events of wound-healing—coordinated interactions among matrix metalloproteinases (MMPs), integrins, and extracellular matrix molecules. Crit. Rev. Oral Biol. Med. **12**(5), 373–398 (2001)
69. Stephansson, S.N., Byers, B.A., et al.: Enhanced expression of the osteoblastic phenotype on substrates that modulate fibronectin conformation and integrin receptor binding. Biomaterials **23**(12), 2527–2534 (2002)
70. Stettler, E.M., Galileo, D.S.: Radial glia produce and align the ligand fibronectin during neuronal migration in the developing chick brain. J. Comp. Neurol. **468**(3), 441–451 (2004)
71. Tate, M.C., Garcia, A.J., et al.: Specific beta-1 integrins mediate adhesion, migration, and differentiation of neural progenitors derived from the embryonic striatum. Mol. Cell. Neurosci. **27**(1), 22–31 (2004)
72. Tsai, H.M.: Shear stress and von Willebrand factor in health and disease. Semin. Thromb. Hemost. **29**(5), 479–488 (2003)
73. Wipff, P.J., Rifkin, D.B., et al.: Myofibroblast contraction activates latent TGF-beta1 from the extracellular matrix. J. Cell. Biol. **179**(6), 1311–1323 (2007)
74. Yurchenco, P.D., Wadsworth, W.G.: Assembly and tissue functions of early embryonic laminins and netrins. Curr. Opin. Cell Biol. **16**(5), 572–579 (2004)

Functional Biomaterials for Controlling Stem Cell Differentiation

Ameya Phadke, Chien-Wen Chang and Shyni Varghese

Abstract Differentiation of stem cells has shown to be strongly influenced through several cues provided by reciprocal interactions with the extracellular microenvironment, consisting of soluble bioactive agents and the extracellular matrix. While the dynamic extracellular matrix is difficult to mimic in its entirety, recent research has successfully mimicked individual matrix-centric cues using synthetic polymeric systems to influence differentiation of stem cells into tissue-specific lineages. Material properties that have been shown to direct this differentiation include chemical functionality, mechanical properties, as well as tissue-mimetic modifications such as mineralization. Another aspect of the extracellular microenvironment that has been mimicked in the controlled differentiation of stem cells is the presence of specific bioactive agents. Material-based delivery of these agents allows for the spatiotemporal variation in their presentation to stem cells, allowing for precise control over their terminally differentiated phenotype. Thus, the delivery of bioactive agents to cells via synthetic materials has also been an effective method to influence stem cell differentiation to various tissue-specific lineages. In this chapter, we discuss the use of synthetic materials to direct stem cell differentiation through both, capitulation of matrix-specific biochemical, mechanical and physical cues, as well as the controlled delivery of specific bioactive agents.

A. Phadke, C.-W. Chang and S. Varghese (✉)
Department of Bioengineering, MC 0412, University of California, San Diego USA
e-mail: svarghese@ucsd.edu

Stud Mechanobiol Tissue Eng Biomater (2010) 2: 19–44
DOI: 10.1007/8415_2010_3

Published online: 12 June 2010

1 Introduction

1.1 Emergence of Stem Cell Engineering in Regenerative Medicine

Stem cells are proving to be an extremely invaluable tool in understanding developmental processes, disease progression, epigenetics, pathophysiology, drug screening and cell based therapies. Among these, cell therapies represent the most challenging yet potentially most fruitful applications for stem cells. Upon differentiation into a suitable phenotype, stem cells can be introduced at a damaged site in a tissue in order to facilitate its regeneration, halting any further tissue damage and even possibly reversing it. Approaches combining the use of stem cells and appropriate materials have thus shown great promise in treating several conditions emerging from the degeneration of tissues. When utilizing such a strategy however, it is important to understand the interaction between stem cells and materials and the effect of these interactions on the efficacy of the desired therapy in regenerating the desired tissue. A comprehensive understanding of these interactions allows for the effective design and development of materials capable of influencing stem cell adhesion as well as the lineage into which these cells differentiate. This requires a multidisciplinary approach integrating concepts in material science, chemistry, cell biology and physiology. In this chapter, we present such an approach capable of aiding in the design of suitable materials and subsequently efficient regenerative therapies.

1.2 Stem Cell Sources

Multipotent and pluripotent cells capable of differentiating into several lineages have been obtained from a variety of sources and are often classified based on the source from which they are obtained. Embryonic stem cells (ESCs) are obtained from embryonic sources and were first isolated through the in vitro fertilization of preimplantation blastocysts [1]. They are characterized by their high telomerase activity and pluripotent differentiation potential. Mesenchymal stem cells are multipotent progenitor cells and are typically isolated from bone marrow, although they have been isolated from a variety of adult tissues such as bone, cartilage, skin, fat and muscle [2]. They are characterized by a spindle-like morphology and have been shown to differentiate into adipocytes, chondrocytes and osteoblasts [3]. A more recent advance in sourcing stem cells has been the development of induced pluripotent stem cells (iPS) [4–7]. First reported by Takahashi and Yamanaka [6], these stem cells are obtained by genetic reprogramming of differentiated cells into a de-differentiated state resembling embryonic stem cells. These cells represent a promising method of obtaining autologous pluripotent stem cells sourced from adult tissues.

2 Stem Cell Expansion and Differentiation Using Biomaterials

2.1 Roles of ECM in Stem Cell Differentiation

The extracellular environment provides essential structural support and regulates signaling to cells [8]. Cells are organized in extracellular matrix (ECM), a hydrated extracellular environment specifically for supporting cell–cell and cell–ECM interactions. The interactions between cells and ECM are crucial in embryogenesis, tissue differentiation, wound healing and tumorigenesis [9]. ECM components regulate stem cell differentiation mainly by providing two-way biophysical and biochemical communications to the cells. The ECM is a 3D hydrophilic network comprising of fibrous structural proteins (collagens, fibronectin, laminins, elastin and vitronectin) and glycosaminoglycan (GAG) network. Among these structural components, collagen and elastin networks provide tissue with mechanical resistance to shear and tensile stress. Osmotic pressure created by the negatively charged GAGs results in the highly swollen viscous matrix, thereby providing compressive strength to the tissue. In addition to provide mechanical protection to cells, ECM contains various cell adhesion molecules to support cell attachment and proliferation. Figure 1 details cues influencing cell behavior in the extracellular environment.

ECM has a profound effect on stem cell differentiation. During the process of tissue development and morphogenesis, the dynamic remodeling of ECM components is required to direct differentiation of uncommitted progenitor cells into a specific lineage. It is generally believed that the interactions between ECM and cells initiate various signal transduction pathways [10] thereby regulating lineage of differentiation. This was shown by Datta et al. [11] in a study that demonstrated the ability of bonelike ECM to promote osteogenesis of human mesenchymal stem cells (hMSCs). An interesting study by Hoshiba et al. [12]

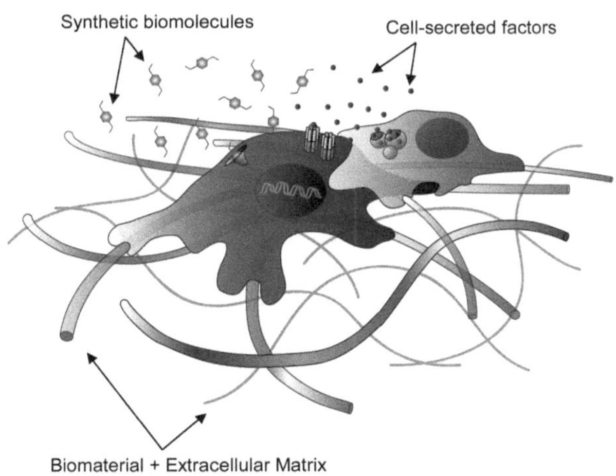

Fig. 1 Schematic demonstrating the reciprocal molecular interactions between the cells and their microenvironment comprising of extracellular matrix components, soluble factors and the surrounding cells

showed that enhanced osteogenesis of MSCs was observed when these cells were cultured in matrices produced by mesenchymal stem cells in the early stage of ostegenesis when compared to MSCs cultured on matrices on obtained from late-stage osteogenesis or from undifferentiated MSCs. This suggests that the ECM may show different structure and composition in different stages of differentiation.

Additionally, a number of studies have demonstrated the potential of using single ECM components to tailor materials to achieve desired stem cell lineage. Chung et al. [13] demonstrated the ability of hydrogels of hyaluronic acid (a component found abundantly within the ECM of cartilage) to promote chondrogenic differentiation of encapsulated hMSCs. Previous studies have demonstrated the capability of collagen gels to induce chondrogenic differentiation of MSCs [14, 15]. Brännvall et al. [16] reported the efficient neuronal differentiation of neural stem/progenitor cells (NS/PC) upon encapsulation in collagen–hyaluronan composite hydrogels, while Awad et al. [17] demonstrated that encapsulation within scaffolds of gelatin (a denatured form of collagen) promoted chondrogenic differentiation of human adipose derived stem cells. These examples illustrate the importance of utilizing ECM components in order to regulate the differentiation of stem cells into tissue specific cells. Although these naturally derived materials provide the necessary biological cues for cell–matrix interactions, they often suffer from batch-to-batch variations and challenges associated with modifications. In contrast, synthetic materials offer great control over structural and mechanical properties but lack biological cues. Hybrid scaffolds containing both naturally derived materials and synthetic materials often offer a "best of both worlds" approach and hence are extremely promising prospects as materials for cell culture matrices.

2.2 Mimicking ECM with Synthetic Biomaterials

2.2.1 Mimicking the Biophysical and Biochemical Properties of ECM

The initial goal of studies involving synthetic biomaterial based scaffolds was to provide a 3D architectural/structural support to cells [18]. Of late, there has been an emphasis on the development of synthetic biomaterials eliciting various interactions observed between cells and ECM in native tissues by mimicking several well studied extracellular biochemical cues and biophysical cues. By utilizing several inherent properties of matrix materials, cell–matrix interactions can be harnessed to modulate stem cell differentiation. These properties include: matrix functional groups, mechanical properties, matrix degradability, surface geometry and microarchitecture [11, 19–27]. Additionally, scaffolds may also be modified through processes such as mineralization in order to stimulate differentiation into osteogenic lineage. A detailed discussion on the role of biomineralized scaffolds on osteogenic differentiation of stem cells is presented in Sect. 2.2.3.

Functionalization of Synthetic Substrates with ECM Derived Ligands

In addition to incorporating entire ECM components, biomaterials can be functionalized using specific ligands representing the ECM binding sites to modulate cell attachment, proliferation and differentiation [28]. This functionalization can be achieved by a variety of methods such as blending [29], copolymerization [30], and immobilization using techniques such as *N*-hydroxysuccinamide (NHS) chemistry [31].

Several studies have made use of well studied cell-binding peptide sequences such as RGD, YISR and IKVAV to improve cell adhesion to synthetic biomaterials [32–34]. However, it is important to consider that orientation of these ligands within the scaffold material may affect their ability to promote cell adhesion [35]. It is interesting to note that these RGD based peptide ligands promote cell adhesion and migration more effectively when clustered in scaffolds rather than when sparsely dispersed within the scaffold [36].

Modification of polymers with ECM derived ligands has also been reported to affect differentiation of stem cells. Silva et al. [37] demonstrated the differentiation of neural precursor cells into neurons and astrocytes by incorporating IKVAV moieties in self assembled ampiphilic nanofibrous matrices. A recent study indicated that the presence of decorin moieties tethered to PEG hydrogels stimulated chondrogenesis of encapsulated hMSCs [38]. Hwang et al. [39] reported that encapsulation of human embryonic stem cells within RGD modified polyethylene diacrylate (PEGDA) based hydrogels promoted increased chondrogenic differentiation, when compared to unmodified PEGDA hydrogels as well PEGDA hydrogels incorporating ECM molecules such as hyaluronic acid, collagen type I and collagen type II. This was attributed to reports from other studies indicating that RGD binding to integrin $\alpha v\beta_1$ stimulated release of TGF-β1, thereby stimulating chondrogenic differentiation.

Interestingly, several studies have reported enhancement of osteogenic differentiation by RGD incorporation into biomaterial scaffolds [40, 41]. Shin et al. [42] report that incorporation of RGD peptide into oligo-poly(ethylene glycol) fumerate stimulated osteogenic differentiation of rat bone marrow stromal cells even in the absence of β-glycerolphosphate and dexamethasone (DEX), typically used as supplements in medium to trigger osteogenic differentiation. It was suggested that the interaction between RGD peptide and surface integrins in these cells activated intracellular pathways triggering osteogenic differentiation in a manner similar to that seen when such cells are exposed to dexamethasone. This is supported by other studies indicating that selective activation of integrins can trigger osteogenic differentiation of progenitor cells [43, 44].

2.2.2 Effects of the Cell–Matrix Interface

Surface Chemistry and Interfacial Energy

The ability of cells to respond to differences in surface chemistry of synthetic biomaterials has been well demonstrated [45]. By altering the chemical structure of the surface of synthetic materials, the binding of proteins to these surfaces and their orientation (and hence the binding of cells to the surfaces) can be affected. In other words, cellular response to biomaterials may be controlled by altering interaction of material with serum components. By tailoring biomaterial surfaces with specific surface properties providing specific extracellular microenvironments, desired degrees of attachment, proliferation and differentiation can be achieved [32, 46]. This is especially important in mimicking cell–matrix interactions to obtain materials with desired capacity to promote cell adhesion and tissue specific differentiation. Although recreating a synthetic mimic of the dynamic extracellular environment is fairly challenging, there have recently been rapid advances made in developing synthetic analogs incorporating various chemical functionalities typically observed within the ECM. For example, modification of polymeric surfaces with anionic groups causes formation of a negatively charged material; this is one potential method to obtain a highly water-swollen matrix with the ability to resist compression, thereby mimicking the role of glycosaminoglycans (GAGs) in load bearing tissues.

Plasma grafting has been explored as a potential method to alter surface chemistry of biomaterials. It is important to note that plasma grafting can be used to obtain highly localized modified domains (from several hundred angstroms to 10 mm) leaving the bulk properties of the materials unaffected. Mwale et al. [26] observed that altering surface chemistry through glow discharge plasma using ammonia affected the differentiation of hMSCs. This process, when applied to nylon 6-polyamide and biaxially oriented polypropylene, led to enrichment of the surface with nitrogen atoms, thereby promoting cell adhesion. Interestingly, application of this treatment to nylon 6-polyamide promoted osteogenic differentiation of MSCs while plasma-treated biaxially oriented polypropylene (BOPP) was found to suppress osteogenic differentiation. The authors attributed this suppression to the possibility that BOPP inhibited the formation of collagenous extracellular matrix by the seeded stem cells, thereby inhibiting further differentiation. Mwale et al. [47] also reported in another study, that doping of BOPP with nitrogen rich plasma polymerized ethylene (referred to as PPE:N) suppressed not only expression of collagen type X but also several osteogenic markers such as alkaline phosphatase, osteocalcin and bone sialoprotein in differentiating hMSCs. This indicates a potential application for this technique in promoting chondrogenic differentiation of MSCs while suppressing/delaying their endochondral ossification. Of interest is the fact that hMSCs for this study were sourced from patients aged 60–80 years undergoing treatment for osteoarthritis. Although these MSCs inherently expressed hypertrophic markers such as collagen type X and osteogenic markers under control conditions, culturing them on plasma treated BOPP down regulated these markers without affecting the markers which are characteristic of hMSC-derived chondrocytes. Plasma grafting was also used to great effect by Wan et al. [48] to alter the surface chemistry, surface energy and surface topology of poly(L-lactic acid) (PLLA) films, thereby increasing their ability to support cell retention.

Jiao and Cui [49] summarized several well-studied methods to modify surface chemistry of polyester biomaterials thereby improving their ability to support cell growth. Hydrolytic degradation is often used in polyesters, thereby cleaving surface ester bonds and leading to formation of carboxyl and hydroxyl residues at the surface. This serves to increase the hydrophilicity and decrease the interfacial surface energy of the material, thereby altering its cellular response. Croll et al. [50] proposed two methods i.e. base hydrolysis and aminolysis to lead to formation of carboxyl or primary and secondary amines respectively on the surface of poly(L-lactic-co-glycolic acid) (PLGA), while minimizing collateral surface degradation.

Surfaces may also be modified through anchoring of monomers such as vinyl acetate, acrylic acid and acrylamide onto PLLA films by means of photo-induced grafting. Additionally, functional groups such as carboxyl, hydroxyl and amide groups were incorporated by means of photo-grafting of hydroxyethyl methacrylate, methacrylic acid and acrylamide, respectively, onto the surface. Such a modification was shown to improve the ability of the biomaterial (film) to support chondrocyte growth [51].

While there have been several well studied methods of modifying surface chemistry of polymers to improve cell adhesion, effects of chemistry of polymeric matrices on differentiation of multipotent cells have only recently been reported [20, 22]. Chastain et al. [52] used two different materials viz. PLGA and polycaprolactone (PCL) to modulate the preferential adsorption of ECM proteins from serum and demonstrated that depending upon the adsorbed protein the material showed differential effect on osteogenic differentiation of hMSCs. Photochemical modification of polystyrene surfaces with azodiphenyl derivatives of hydrophilic polymers such as polyacrylic acid (PAAc), polyacrylamide (PAAm) and polyethylene glycol (PEG) was shown by Guo et al. [22] to affect chondrogenic differentiation of hMSCs. While modification with PAAc and PAAm were found to improve cell adhesion, modification with PEG was found to inhibit the cell adhesion. Additionally, surfaces modified with PAAm showed more rapid adhesion of cells than PAAc. The authors attributed this difference to the electrostatic attraction between the positively charged surface formed from photochemical modification with PAAm and the negatively charged cells. Modification with PEG would form a neutral surface which would not exhibit this attractive force with cells and as a result, cells cultured on PEG modified surfaces showed aggregation into pellets, indicating dominance of cell–cell interactions over cell–material interactions. PEG-modified and PAAm-modified surfaces were found to promote chondrogenic differentiation of these cells upon culturing in chondrogenic medium. Curran et al. [53, 54] reported that chondrogenesis is promoted on glass slides by the presence of surface hydroxyl and carboxyl groups whereas surface amine and thiol groups were found to stimulate osteogenesis. Methylated and untreated glass surfaces were found to maintain undifferentiated phenotype of MSCs. Similar results were also reported by Lee et al. [55] with Chinese hamster ovary cells. This study reported that low density polyethylene sheets functionalized with amine groups promoted cell adhesion to the greatest degree among charged groups (carboxyl and amine), while hydrophilic neutral groups (such as OH) promoted cell adhesion to a greater degree than hydrophobic neutral groups (amide groups).

An effect of material chemistry on differentiation of cells under 2D and 3D culture conditions was recently reported by Benoit et al. [20]. Human mesenchymal stem cells were plated on PEG hydrogel surfaces functionalized with carboxyl, phosphate and *t*-butyl groups, under 2D culture conditions. Morphological observations indicated that cells cultured on carboxyl-modified PEG surfaces showed a rounded morphology similar to that seen in chondrocytes cultured in 2D conditions, cells cultured on phosphate modified PEG surfaces assumed a spread morphology similar to that observed in osteoblasts and cells cultured on *t*-butyl modified PEG surfaces showed adipocyte-like morphology (see Fig. 2), along with the presence of intracellular lipid droplets (not shown).

This was confirmed through FISH analysis (see Fig. 3) wherein cells cultured on surfaces modified with carboxyl, phosphate and *t*-butyl groups showed elevated expression of aggrecan (a chondrogenic marker), core binding factor α-1 (an osteogenic marker) and peroxisome proliferator-activated receptor γ (PPARγ) (an adipogenic marker) respectively when compared to a control, unmodified PEG surface.

It has been reported that lineage into which stem cells differentiate can be influenced by cell shape, spreading and matrix stiffness [21, 25] and that these can be influenced through material chemistry. To evaluate whether material chemistry affected cell lineage in a manner independent of the aforementioned properties, MSCs were encapsulated by Benoit et al. in three-dimensional scaffolds functionalized with tertiary butyl and phosphate groups, respectively; due to small

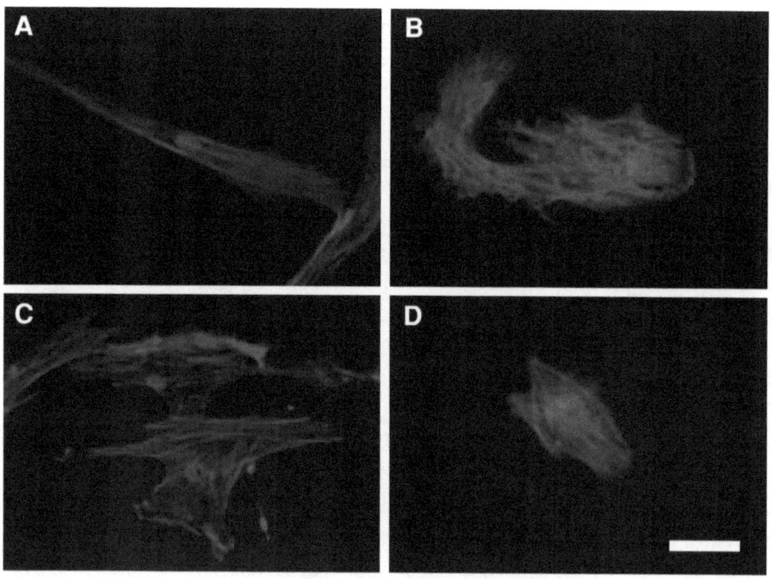

Fig. 2 Images from immunostaining (F-actin and nuclei) of hMSCs seeded on (**a**) unmodified PEG (**b**) carboxyl-modified PEG (**c**) phosphate-modified PEG and (**d**) *t*-butyl modified PEG. (Adapted from [20] with permission, copyright Nature Publishing Group, 2008)

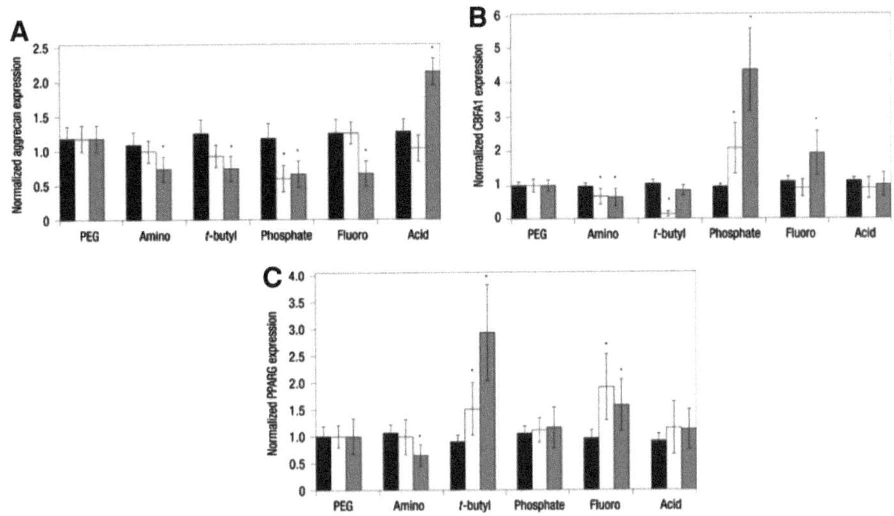

Fig. 3 Gene expression profiles of (**a**) aggrecan (**b**) CBFα-1 (**c**) PPARγ for hMSCs seeded on surfaces modified with various functional groups, normalized to expression on control surface of unmodified PEG. *Black bars* represent expression after 4 days, *white bars* represent expression after 7 days and *gray bars* represent expression after 14 days (adapted from [20] with permission, copyright Nature publishing group, 2008)

mesh size, encapsulated cells were restricted to a rounded morphology independent of material chemistry. Additionally, sufficiently small concentrations of the functional groups were used so as not to affect the bulk mechanical properties of the polymeric scaffold material to a significant degree. In this manner, effects of cell–material interaction on differentiation were effectively decoupled from effects on differentiation of cell morphology and matrix elasticity. Immunoblotting revealed that cells encapsulated in *t*-butyl-modified-PEG showed an upregulation of the adipogenic marker PPARγ after 14 days which remained constant at 21 days suggesting adipogenic differentiation. Cells encapsulated in phosphate-modified PEG showed an upregulation of the osteogenic marker CBFα-1 after 14 days and showed increased levels of expression after 21 days suggesting a differentiation into osteogenic phenotype. The authors proposed that these difference in matrix functionality triggered differentiation down different pathways through two potential mechanisms. Firstly, through interaction with surface receptors, it is possible that different functional groups triggered different intracellular signaling pathways promoting differentiation into varying lineages. Another mechanism is the selective sequestration of cell secreted factors by the functional groups; the sequestered factors may then influence the differentiation of the cells down a particular pathway.

It is also important to consider the effect of scaffold material chemistry on the adsorption of serum components such as fibronectin and their resulting interaction with cellular receptors such as integrins. Recently, chemistry of matrix materials has been used to vary the quantity and conformation of adsorbed fibronectin which in turn was shown to influence the adhesion and differentiation of cells [24, 56]. Michael et al. [57] demonstrated an effect of varying surface chemistry of materials on the conformation of fibronectin in a coated layer; alterations in surface energy by varying surface functional groups (surface chemistry) affected the quantity of adsorbed fibronectin; neutral hydrophilic (OH) and hydrophobic (CH_3) groups promoted fibronectin binding to a greater extent than charged functional groups (-NH_2 and -COOH). Additionally, it was shown that surfaces modified with methyl, carboxyl, hydroxyl and amine groups respectively produced markedly different conformational changes in fibronectin coated on these surfaces, thereby allowing variation in exposure of sites capable of binding to specific integrins. This conformational variation was also shown to affect osteogenic differentiation in self assembled monolayers [56]. It was observed that surfaces containing hydroxyl and amine functionalities respectively promoted osteogenic differentiation to a greater degree than surfaces functionalized with carboxyl and methyl groups, respectively. Through immunological studies, it was determined that based on the surface chemistry (and hence conformation of fibronectin in the coated layer), different integrin binding sites (matrix ligands) were exposed and different cell surface integrins were activated. Based on the surface chemistry of the material, activity of binding sites for integrins $\alpha 5\beta 1$ and $\alpha v\beta 3$ were reported. It was observed that surfaces modified with carboxyl, amine and hydroxyl groups respectively promoted binding of integrin $\alpha 5\beta 1$ to fibronectin; carboxyl modified surface additionally showed binding of integrin $\alpha v\beta 3$ to fibronectin. Interestingly, mineralization (indicating osteogenic differentiation) was observed only on amine and hydroxyl modified surfaces. Treatment of carboxyl modified surfaces with $\beta 3$ blocking antibody promoted mineralization on this surface; additionally treatment of amine modified surfaces with the same antibody led to an increase in mineralization. It is important to note that amine modified surfaces were expected to show greater $\beta 3$ affinity than hydroxyl modified surfaces. These observations led the authors to suggest that binding of integrin $\alpha 5\beta 1$ promotes osteogenic differentiation while $\alpha v\beta 3$ suppresses it. Moursi et al. [44] reported that activity of integrin $\alpha 5\beta 1$ is essential for triggering of upregulation of factors representing osteoblast activity. It has also been shown that over expression of $\alpha v\beta 3$ in MC3T3-E1 cells suppresses osteoblastic activity [58].

Changes in surface chemistry also affect material hydrophobicity/hydrophilicty, thereby affecting protein adsorption on biomaterial surfaces. This in turn can potentially affect interaction between cells and materials [45] and has been characterized as interfacial energy [59]. Surfaces with net positive or negative charges tend to be more hydrophilic than neutral surfaces. Several studies have reported the effect of surface energy on binding of ECM proteins such as fibronectin, vitronectin, albumin, globulin and fibrinogen to material surfaces [60, 61]. Previous studies showed improved cell adhesion to hydrophilic surfaces [62–64]. Changes in surface energy can lead to conformational changes in adsorbed fibronectin, influencing binding to cellular integrins. Binding to lineage-specific integrins may activate signaling pathways specific to a particular lineage and thereby influence the differentiation of stem cells.

Lieb et al. [24] reported the effect of surface energy on osteogenic differentiation of marrow stromal cells. This study focused on decreasing the hydrophobicity of poly (D,L-lactic acid) by preparing poly (D,L-lactic acid)–poly(ethylene glycol)–monomethyl ether (PLA–PEG–MmE) diblock copolymers. Interestingly, PLA–PEG–MmE diblock copolymers showed lower cell attachment and proliferation than PLA, PLGA and tissue culture polystyrene. This was attributed to the decreased adsorption of serum proteins in the case of PLA–PEG–MmE due to decreased hydrophobicity. However, upon long term culturing on these materials, it was observed that cells cultured on PLA–PEG–MmE showed significantly higher alkaline phosphatase activity and greater degree of mineralization (evaluated through Von Kossa silver staining) than PLA, PLGA and tissue culture polystyrene. These serve as biomarkers for osteogenic differentiation and suggest that a small decrease in the hydrophobicity of the material promoted differentiation into osteogenic lineage. The authors attributed this to specific conformational changes in adsorbed proteins due to changes in surface energy, thereby exposing binding sites specific to integrins active during osteoblast activity. Such conformational changes in ECM proteins have previously been shown to modulate osteogenic differentiation on synthetic materials [56]. It is also mentioned however, that the PEG–PLA diblock copolymers also did show rough surfaces, which previously have indeed been shown to promote osteogenic differentiation [65, 66]. Indeed a study by Dalby et al. [67] demonstrated the influence of nanoscale surface topology on the osteogenic differentiation of hMSCs. In sum, conformational changes in adsorbed proteins and the resulting selective activation of cell adhesion molecules seem to serve as the predominant mechanism through which surface energy and chemistry of biomaterials promotes differentiation of stem cells.

2.2.3 Mineralization of Matrix Materials

The term ‘biomineralization’ refers to the modification of materials by integration with a crystalline/semicrystalline inorganic phase resembling that seen in mineralized tissues, such as bone or tooth enamel. Presence of mineralized coating consisting of an apatite layer mimicking bone mineral has been shown to extensively promote osseointegration of implant materials while also promoting bone healing [68]. Bone mineral is similar to hydroxyapatite ($Ca_5(PO_4)_3OH$), although studies of Raman spectra of bone suggest the substitution of hydroxyl group by carbonate groups in the crystal lattice [69]. Materials such as bioactive glasses and calcium phosphate based ceramics have demonstrated the ability to form a layer of apatite resembling bone mineral upon incubation in simulated body fluid, a solution mimicking the ionic composition and pH of plasma [68, 70]. Of late, there has been an increased interest in generation of mineral–polymer composite materials; one reason for this is the similarity of these materials to the native structure of bone consisting of a composite of an organic phase associated with a crystalline, inorganic phase. The successful syntheses of hydrogel–apatite composites showing strong adhesion between the two phases indicate the vast potential of polymeric matrices in this field.

Mineralization of Polymeric Matrices

A variety of methods have been used to incorporate inorganic apatite phases into polymeric matrices and have been discussed in detail by Kretlow and Mikos [71]. A popular method is the incorporation of anionic polar groups into scaffold materials; upon soaking in simulated body fluid (SBF), these groups serve as potential initiators of apatite nucleation by binding to calcium. For example, incorporation of anionic functional groups was observed to induce mineralization of poly hydroxyl-2-ethyl methacrylate (pHEMA) scaffolds upon soaking in simulated body fluid [72]. Supplementing the simulated body fluid with fetal bovine serum/albumin has been demonstrated to promote mineralization of pHEMA without any chemical modification [73]. In another study, Song et al. [74] achieved mineralization of pHEMA scaffold materials utilizing a pH-mediated templating process from the thermal decomposition of urea. This mineralization process has also been successfully utilized to generate apatite–polymer composites with polycaprolactone [75], PLGA and PLLA [76].

Effect of Mineralization on Cell Adhesion, Proliferation and Differentiation

Several studies have indicated that incorporation of a mineral phase into polymeric scaffold material serves to enhance attachment, proliferation and osteogenic differentiation of multipotent cells. Koç et al. [77] observed that rat MSCs seeded onto mineralized PLGA foams showed excellent adhesion to the scaffold material and underwent osteogenic differentiation. This study suggested that presence of a mineralized layer not only promotes osteoinduction of seeded cells but also promotes cell adhesion as a result of increased surface roughness. A study by Osathanon et al. [78] compared osteoinductive capacity of two kinds of mineral–polymer composites involving electrospun fibrin scaffolds: one wherein varying quantities of nanoparticles of hydroxyapatite were incorporated directly into the polymeric phase and one in which fibrin scaffolds were incubated in a solution containing concentrations of calcium and phosphate ions resembling those seen physiologically, leading to the deposition of a mineral. Upon seeding with murine calvarial cells and incubation with osteogenic medium, both of these materials showed enhanced expression of osteogenic markers (BSP, OCN, COL 1, ALP, CBFA 1 and OSX) as compared to non-composite scaffolds consisting of fibrin alone. However, mineralized fibrin scaffolds showed higher expression of these markers at earlier time points as compared to scaffolds prepared by incorporation of nano-size hydroxyapatite as well as higher alkaline phosphatase activity and a greater increase in calcium content. Additionally, mineralized fibrin scaffolds showed greater calcium phosphate dissolution than fibrin/nanosize hydroxyapatite scaffolds, leading the authors to suggest a link between extracellular calcium and phosphate concentrations and osteogenic differentiation through resultant upregulation of osteogenic markers (see Fig. 4).

This is supported in a study by Dvorak et al. [79] in which murine and rat fetal calvarial cells exposed to higher extracellular ionized calcium levels led to

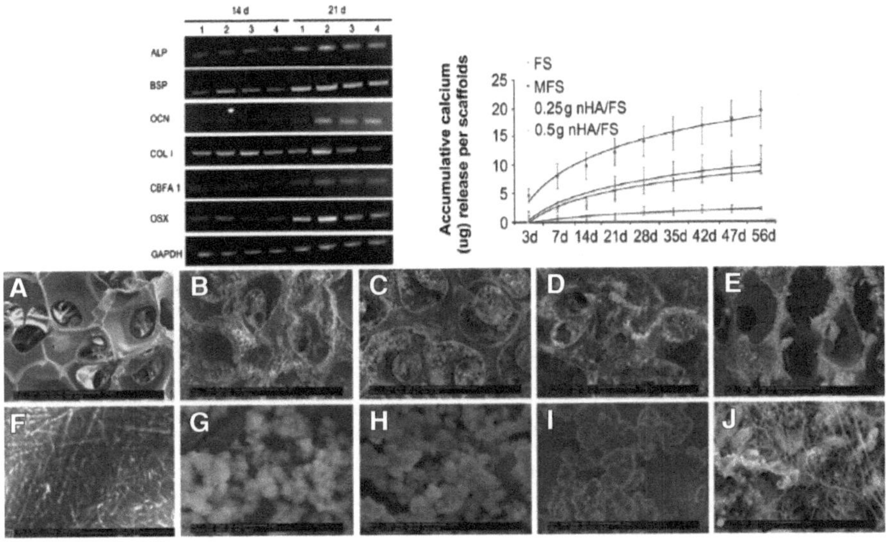

Fig. 4 *Top, left* gene expression for murine calvarial cells seeded on (*1*) fibrin scaffolds (*2*) mineralized fibrin scaffolds (*3*) fibrin scaffolds with 0.25 g nanosize hydroxyapatite and (*4*) fibrin scaffolds seeded with 0.5 g nanosize hydroxyapatite. *Top, right* dissolution profile for fibrin scaffolds (*FS*), mineralized fibrin scaffolds (*MFS*), fibrin scaffolds with 0.25 g nanosize hydroxyapatite (0.25 g *nHA/FS*) and fibrin scaffolds seeded with 0.5 g nanosize hydroxyapatite (0.5 g *nHA/FS*). *Bottom* Scanning electron micrographs of (**a**, **f**) FS in PBS for 14 days (**b**–**d**, **g**–**i**) *MFS* after soaking in simulated body fluid for (**b**, **g**) 24 h (**c**, **h**) 7 days (**d**, **i**) 14 days and (**e**, **j**) 0.5 ng HA/FS (adapted from [78] with permission, copyright Elsevier, 2008)

upregulation of core binding factor α-1, collagen type 1, osteopontin and osteocalcin, all of which are markers indicating osteogenic differentiation of progenitor cells. These findings imply that mineralized scaffolds may be able to promote osteogenic differentiation even in the absence of soluble factors often used to promote it. These mineralized scaffolds promoted osteogenic differentiation by exposing the murine calvarial cells to higher extracellular calcium levels in the microenvironment around the material due to dissolution of the inorganic mineralized phase. It is important however, to note that scaffolds undergoing excessively rapid dissolution may produce excessively high extracellular calcium concentration, thereby inhibiting osteogenisis [80].

While such methods have been used to ascertain the effect of mineralization of matrices formed from polymers of natural origin, similar reasoning can be used to predict the effect of mineralization of synthetic matrices on cell lineage. A study by Yu et al. [81] reports the ability of surface mineralization of nanofibrous ε-polycaprolactone scaffolds to stimulate osteogenesis of seeded rat MSCs. Mineralized scaffolds were found to promote cell adhesion and proliferation to a greater extent than unmineralized scaffolds. Additionally, cells seeded on unmineralized scaffolds showed increased proliferation. It was proposed by the authors based on microscopic evidence that cells seeded on mineralized scaffolds reached confluence before unmineralized scaffolds thereby inhibiting further proliferation; this would promote differentiation of these cells.

As evidenced in the above studies, mineralization of synthetic scaffolds is an effective technique to promote cell adhesion, stimulate differentiation into osteogenic lineage and improve ossteointegration of synthetic implant materials. In other words, this technique shows promise in application to cell based therapies for the efficient healing of mineralized tissues.

2.2.4 Mechanical Properties

In addition to the effects of extracellular biochemical cues, reciprocal mechanical interactions between cells and environment have significant impact on the differentiation of stem cells [21]. The effects of mechanical forces on cells due to the matrix can be observed from single cell level to the development of complex tissues. At the single cell level, adhesion of cell to a material with specific stiffness triggers signaling transduction cascades allowing translation of extracellular mechanical cues into intracellular events [82]. Dynamic interactions between cell and matrix control several cell behaviors such as spreading, migration and cell shape through binding with integrins, the chief mechanotransducers for cells [83]. For example, cell geometry (spreading) is a result of pre-stress between ECM and cellular microtubules [84]. Previous studies have demonstrated that cell spreading controls processes such as proliferation and apoptosis [85]. The ability of various types of cells to respond to mechanical differences in the extracellular environment has been described in detail in a review by Discher et al. [86]. A recent study by Engler et al. [21] showed that stem cell differentiation can be directed by varying elasticity of matrix. In this study, hMSCs were cultured on 2D polyacrylamide hydrogels with different elasticity prepared from using different amount of crosslinker. Neurogenesis, myogenesis and osteogenesis of hMSCs were observed on soft, medium and stiff hydrogel matrices, respectively. Khatiwala et al. [87] have evaluated the effect of ECM compliance on osteogenesis of progenitor cells and the signaling pathways involved. ECM compliance was found to affect activity of extracellular signaling kinase (ERK), with stiffer matrices promoting osteogenic differentiation. Additionally, a potential mechanism suggested by the authors was the downstream ERK–mitogen-activated protein kinase (MAPK) activation of the RhoA–Rho associated protein kinase (ROCK) signaling pathway.

2.3 Biomaterial Based Delivery of Soluble Factors for 3D Cell Culture

2.3.1 Incorporation of Bioactive Agents into Matrix Materials

In addition to the extracellular matrix, soluble bioactive agents (such as growth factors, hormones, proteins, small molecules, cytokines and chemokines) also play many roles in regulating proliferation and differentiation of stem cells. Incorporation of bioactive agents into growth medium represents the simplest way to harness the beneficial effects of such factors on stem cell differentiation [88]. However, this approach suffers from certain limitations. For instance, hydrophobic agents show poor solubility in aqueous media. Hydrophilic bioactive agents can be readily dissolved but their stability is strongly dependent on environmental factors such as ionic strength, pH and enzymatic degradation/inactivation [89]. This approach has major limitations for 3D cell-laden systems of critical size due to limited degree of diffusion across cell-laden matrices. Heterogeneous differentiation may occur due to non uniform distribution of growth factors throughout the cell laden construct upon their delivery through incorporation into medium. Additionally, growth factor signaling during development involves precise concentration of these factors and their spatial and temporal gradients. Figure 5 represents the various methods by which bioactive agents can be delivered to progenitor cells via functional biomaterials.

A variety of biomaterial-based technologies have been developed of late to precisely control delivery of bioactive agents to stem cells in a spatiotemporal manner. One such approach involves the direct incorporation of bioactive agents within the biomaterial; however, this bolus delivery approach does not provide sustained delivery of the desired agents. Studies have shown that manipulation of material microstructure allows retention of growth factors within the scaffold thereby offering their sustained delivery to embedded cells. For instance, β-sheet microstructure was created in one study by treating lyophilized silk with an organic solvent. The steric hindrance effect arising from resultant β-sheet microstructure was thought to contribute to the sustained release of IGF-I from scaffolds constructed with these modified silk materials [90].

Another approach involves nano- and micro- carrier based delivery vehicles in which carriers containing the bioactive agents of interest are dispersed within the 3D scaffold seeded with stem cells to achieve localized controlled delivery of the agents. Such sustained release of soluble factors to differentiating stem cells inside

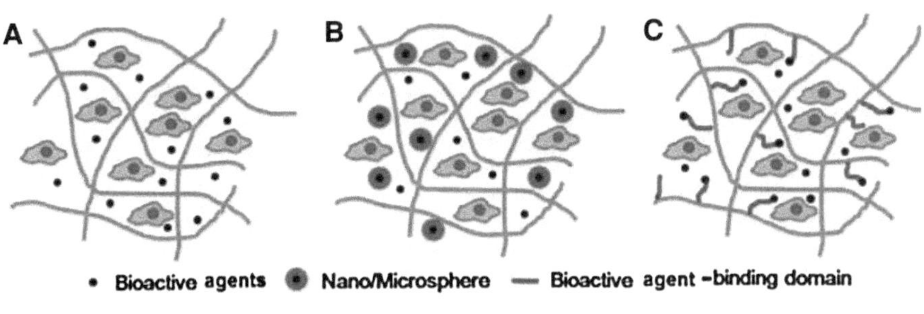

Fig. 5 Schematic drawing of various methods used to deliver bioactive molecules to cells within a biomaterial scaffold. **a** Dispersion of the agent within the scaffold material. **b** Encapsulation of bioactive agents in degradable nano/microspheres. **c** Use of interactions between bioactive agents and binding domains in the scaffold material

biomaterial matrices usually exhibits higher differentiation efficiency. Potential toxicity is also minimized by preventing the effects of administration of soluble factors at excessively high doses, referred to as 'dose dumping'. Thus, the incorporation of delivery vehicles carrying bioactive agents into tissue scaffolds allows for the engineering of stem cells in a 3D defined microenvironment [91]. The release profile of soluble factors from carriers within the scaffold depends on various properties of the biomaterial such as porosity [92–94], composition [92, 93], degradability and microstructure [90]. Generally, biomaterial chemistry determines the strength of interactions between soluble factors and the material [94]. Soluble factors not interacting appreciably with materials can diffuse freely and thus shorten the delivery time to the encapsulated cells. Due to the sensitivity of bioactive molecules it is important to use mild encapsulation conditions so as to avoid their deactivation and/or degradation, again restricting the choice of biomaterial [89, 95]. Commonly used delivery vehicle systems include biodegradable nano/microsphere (beads) [96] and polyelectrolyte complexes, formed by electrostatic interactions between the bioactive molecules and materials. A variety of biodegradable synthetic polymers such as PLA, PGA, PLGA, PEG–PLGA, PEG–PCL have been widely used to prepare drug/protein-loaded nano/microspheres [96]. In addition to synthetic biodegradable polymers, natural materials such as gelatin and chitosan can also be used to prepare protein-loaded nano/micro particles. The application of degradable particle allows the sustained protein release via particle degradation.

2.3.2 Effects of Controlled Delivery of Bioactive Agents on Stem Cell Differentiation

Delivery of Bioactive Agents to Embryonic Stem Cells

Addition of soluble bioactive molecules directly into culture medium represents a convenient approach for directing ESCs differentiation (through embryoid body, EBD, formation) with modest efficiency [88]. Homogenous delivery of soluble bioactive agents into EBD is highly challenging, because of their 3D spheroid nature [97]. As a result, cell differentiation within embryonic bodies is usually heterogenous and disordered due to the inefficient intra-EBD transport of soluble factors. Sachlos et al. [97] have characterized the composition of the shell surrounding EBDs and have reported improving diffusive transport into the EBD interior through the disruption of this membrane. The shell was found to consist of ECM comprised of collagen type I, a squammous cellular layer with tight cell–cell adhesions associated with E-cadherin and a basement membrane, as indicated by the presence of collagen type IV lining. In the latter part of the study, this basement membrane was disrupted either by preventing its formation with noggin or degrading it using collagenase. This treatment was found to increase the diffusive transport of retinoic acid and subsequently promote the neuronal differentiation of the ESCs.

Several carrier-based methods such as peptide-transmembrane domain (PTD)-protein conjugates [98] and microparticles [99] have been developed to further increase the delivery efficiency. Similarly, Carpanedo et al. [100] have adapted in situ release of retinoic acid to the interior region of EBDs using degradable PLGA microspheres. The in-situ release of soluble factors within EBDs has showed promising results by promoting homogenous differentiation of ESCs within EBDs into phenotypes resembling those observed in early mouse streak embryos (see Fig. 6).

Other studies such as those by Newman et al. [101] and Nojehdehian et al. [102] have demonstrated the efficacy of utilizing degradable PLGA microspheres to deliver retinoic acid to pluripotent P19 embryonic carcinoma cells for differentiation into neuronal lineage. As stated by Newman et al. these microspheres served two purposes viz. the delivery of retinoic acid to the cells and as potential transplantation matrices to support cell attachment, growth and differentiation. Nojehdian et al. also demonstrated that the presence of a poly-L-lysine coating on the microspheres further enhanced neuronal differentiation of the embryonic carcinoma cells.

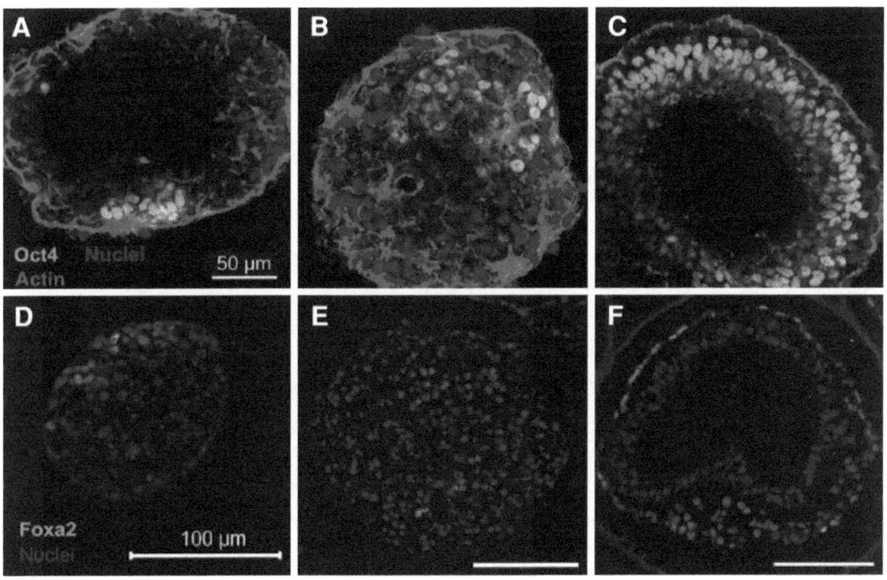

Fig. 6 Immunostaining of embryoid bodies. OCT4 staining was performed on day 10 untreated embryoid bodies (**a**), embryoid bodies with unloaded microspheres (**b**) and embryoid bodies with retinoic acid-loaded microspheres (**c**). Embryoid bodies with untreated and unloaded microspheres contained clusters of OCT4t cells, while OCT4t cells in embryoid bodies with retinoic acid-loaded microspheres were localized to the columnar cell layer. FOXA2, a marker of visceral endoderm, was also expressed in clusters of untreated (**d**) and unloaded MS EBs (**e**), but was localized to the outermost layer of cell in RA MS EBs (**f**). **a–c**, *bar* 50 mm; **d–f**, *bar* 100 mm. (Adapted from [100] with permission, Copyright Elsevier, 2009)

Tissue Specific Differentiation of Stem Cells Using Delivery of Bioactive Agents

Osteogenic differentiation of stem cells can be stimulated with various growth factors including bone morphogenetic proteins (BMPs), dexamethasone, bFGF and platelet rich plasma (PRP) derived growth factors. The sustained release of bioactive agents such as dexamethasone from microparticles has been shown to enhance osteoegenesis of stem cells in a 3D environment [103]. In this study, carboxymethylchitosan/poly(amidoamine) dendrimer loaded with dexamethasone was used to enhance the osteogenesis of rat bone marrow stromal cells (RBMSCs) growing on porous degradable hydroxyapatite and starch–polycaprolactone scaffolds, respectively [103]. Though biodegradability of the materials is a major consideration, it is extremely important to study biomaterials with inherent growth factor binding capability. In studies by Anseth et al. [104, 105], heparan and PEG were used to fabricate scaffolds for sustained growth factor delivery. Heparan, a highly sulfated glycoaminoglycan, contains binding domain for various growth factors such as bFGF and BMP-2. Their work showed that sustained release of bFGF (up to 5 weeks) from heparan domains in the scaffolds increased the expression of alkaline phosphatase (ALP) and other osteogenic markers such as collagen I and osteopontin at the transcriptional level. In another study by the same research group, Benoit et al. made use of fluvastatin release from PEG hydrogel scaffolds in order to activate release of BMP-2 from encapsulated hMSCs, thereby stimulating osteogenic differentiation of these cells [106]. In a study by Basmanav et al. [92], microspheres prepared by complexing poly(4-vinyl pyridine) (P(4)VN) and alginic acid loaded with the growth factors BMP-2 and BMP-7 showed enhanced osteogenic differentiation of bone marrow stem cells (BMSC) in porous scaffolds. In another study [94], the combination of degradable chitosan beads with degradable scaffolds (Porous nano-hydroxyapatite/collagen/poly(L-lactic acid)/chitosan microspheres (nHAC/PLLA/CMs)) supported sustained release of active BMP-2 derived synthetic peptides, the release rate of which was dependent on the degradation rate of both microspheres and scaffold. Enhanced osteogenesis of rabbit marrow mesenchymal stem cells was achieved using this system.

Similar to osteogenesis, bioactive agents are required to control and or enhance chondrogenic differentiation of stem cells. Chondrogenic differentiation of MSCs in vitro is often achieved by culturing them in a three-dimensional (3D) condition in the presence of TGF-β superfamily growth factors [107, 108]. The incorporation of these chondrogenic morphogens into biomaterial scaffolds is advantageous due the ability of this method to provide spatially and temporally moderated delivery of bioactive agents. In one study, a dual delivery system was designed by using double bead microspheres to deliver dexamethasone (DEXA) and dehydroepiandrosterone (DHEA) simultaneously to engineer inflammation-free tissue in the vicinity of the implant [96]. These microspheres contained a PLGA core which hydrolyzed in aqueous environment, releasing these two chondrogenic factors simultaneously. Up-regulation of collagen II, aggrecan, GAGs, cartilage oligomeric matrix protein (COMP) and down-regulation of collagen I were observed from mMSCs. Another example of dual delivery systems is the delivery of TGF-β and dexamethasone from PLGA scaffolds [91]. In another study, increased chondrogenesis was observed upon release of IGF along with TGF-β release from silk scaffolds [90]. In a study by Park et al. [109] gelatin microparticles loaded with insulin-like growth factor-1 (IGF-1) and transforming growth factor β1 (TGF-β1) were incorporated into a degradable PEG fumarate hydrogel. These two growth factors were released from gelatin particles via simple diffusion. The degradation of PEG-fumarate hydrogels further enhances the delivery efficiency by increasing the mobility of encapsulated nano/microparticles which promoted chondrogenesis of rabbit MSCs. Researchers have also utilized the ability to manipulate the spatial- and temporal release of growth factors from biomaterial based delivery vehicles for engineering composite tissues. For instance, Wang et al. [110] have incorporated microspheres containing growth factors, BMP-2 and IGF-1, in a gradient fashion within the alginate scaffold to regulate osteochondral differentiation of MSCs.

2.4 In Vivo Applications

As seen above, the ability of functional materials to influence the terminal phenotype of various stem cells is quite well-illustrated. In addition to this, several studies have illustrated the ability of similar functional materials to recruit progenitor cells and promote the formation of new tissue in vivo, particularly cartilage and bone tissue. A study by Sharma et al. demonstrates in vivo chondrogensis of goat MSCs encapsulated in photopolymerized PEG hydrogels containing hyaluronic acid (HA) along with TGF-β, followed by the generation of cartilage-specific ECM (specifically collagen II and proteoglycan) upon subcutaneous injection into athymic nude mice [111]. Inclusion of HA and TGF-β in the hydrogels was found to promote proteoglycan synthesis; in the absence of HA, there was significant expression of collagen I, while no proteoglycan was produced in the absence of TGF-β. This synergetic effect of TGF-β and HA on promoting chondrogenic differentiation was attributed to the possibility of increased retention of TGF-β within the construct through restriction of its diffusion by inclusion of HA. The technique of cell encapsulation has also been used to create a tissue engineered mandibular condoyle [112, 113]. In these studies, chondrogenic and osteogenic cells derived from rat MSCs were encapsulated in PEG hydrogels; upon implantation, these constructs were found to contain stratified layers of both, osseous components such as mineral nodules and cartilaginous components such as glycosaminoglycans. Several in vivo studies in which functional materials are used to differentiate and/or deliver stem cells are summarized by Chai and Leong [114]. In one such study by Levenberg et al. [115], scaffolds consisting of poly(L-lactic acid) and poly(L-lactic-co-glycolic acid) were used in conjunction with various chemical cues such as retinoic acid, activin-A, TGF-β and Insulin Growth Factor 1 (IGF-1) respectively to commit them to germ layers to develop into tissues such as liver, cartilage and nervous tissue. Upon implantation of these constructs in severe combined immunodeficient (SCID) mice, these constructs were found to maintain their viability for at least 2 weeks. Additionally, evidence of integrating with host vasculature was also observed. In another study, Cho et al. [116] demonstrated in vivo adipogenesis of adipocyte derived stem cells seeded in phospholipase/ poly(glycolide) scaffolds upon implantation in athymic mice, although in this case, the scaffold served predominantly to provide mechanical support to the construct, cells and the resultant extracellular matrix.

Lee et al. [117] used thermosensitive hydrogels consisting of triblock copolymer of poly(ethylene glycol-b(DL-lactic acid-co-glycolic acid)-b-ethylene glycol) (PEG–PLGA–PEG) for the efficient engraftment of muscle derived stem cells (MDSCs) for the efficient healing of diabetic wounds. It was observed that MDSCs differentiated into fibroblasts, myofibroblasts and endothelial cells, thereby accelerating the healing and subsequent closure of the diabetic wound. In another study, Boldrin et al. [118] demonstrated the efficient delivery of human muscle precursor cells (hMPCs) by means of a degradable poly(glycolic acid) scaffold. Upon implantation in CD1 nude mice, these scaffolds showed a human nuclear antigen signal (utilized to evaluate the presence of human-origin cells as opposed to the host mouse cells) nearly threefold of that obtained by mere injection of hMPCs, indicating a much higher efficiency of delivery in the case of the cell-seeded polymer scaffolds. Kim et al. [119] have demonstrated the ability of committed adipose derived stem cells (ADSCs) in conjunction with injectable PLGA microspheres to form muscle tissue in vivo. ADSC-seeded PLGA microspheres were cultured in myogenic medium for 21 days and then injected subcutaneously into the necks of nude athymic mice. Mice injected with PLGA microspheres alone showed the formation of acellular matrix at the injection site while those injected with cell-laden microspheres showed the formation of muscle tissue morphologically resembling native muscle.

2.5 Future Perspectives

Both genomic [120] and proteomic [121] approaches have been intensively studied to elucidate the complex molecular regulation network behind stem cell differentiation. With these advances, more and more molecular markers for stem cell differentiation have been discovered. These differentiation markers in combination with high throughput assay-based methods have facilitated the discovery of biologically active small molecules and new materials for controlling stem cell engineering [122, 123]. The advances in this field could lead to the development of novel biomaterials with higher efficiency in controlling tissue specific differentiation of stem cells. While several advances have been made in developing materials capable of controlling stem cell differentiation, there are a multitude of areas which show great promise in revolutionizing stem cell based therapies. One such area is the synthesis of materials showing anisotropy in their chemical and physical properties, especially across controlled gradients. Such materials would be especially useful in cell-based therapies for the repair of interfaces between different tissues such as the osteochondral interface. Controlled variations in material properties can allow for the simultaneous differentiation of common progenitor cells into multiple phenotypes on a single scaffold. Another potentially promising area is the development of self-healing materials capable of supporting cell adhesion and differentiation. Previous studies have investigated self healing using microvascular networks in materials [124]; self healing materials may also be used in conjunction with cells to mend damaged tissues, providing mechanical support at the damage site while simultaneously directing differentiation of cells to the appropriate phenotype. Native tissue cells are also subjected to dynamic mechanical cues and therefore developing advanced multi-functional scaffolds that can provide dynamic chemical and mechanical cues to encapsulated cells beyond being a structural support and/or static chemical and mechanical cues will have a profound influence on the field of biomaterials and stem cell engineering.

Acknowledgments The authors would like to acknowledge financial support from the California Institute for Regenerative Medicine and the University of California, San Diego. Assistance from Dr. Ramsés Ayala and Dr. Nivedita Sangaj in the preparation and review of this manuscript is also gratefully acknowledged.

References

1. Thomson, J.A. et al.: Embryonic stem cell lines derived from human blastocysts. Science **282**(5391), 1145–1147 (1998)
2. Hwang, N.S., Varghese, S., Elisseeff, J.: Controlled differentiation of stem cells. Adv. Drug Deliv. Rev. **60**(2), 199–214 (2008)
3. Atala, A., et al. (eds): Principles of Regenerative Medicine, vol. 1448. Elsevier, Burlington (2008)
4. Okita, K., Ichisaka, T., Yamanaka, S.: Generation of germline-competent induced pluripotent stem cells. Nature **448**(7151), 313–317 (2007)
5. Takahashi, K., et al.: Induction of pluripotent stem cells from adult human fibroblasts by defined factors. Cell **131**(5), 861–872 (2007)
6. Takahashi, K., Yamanaka, S.: Induction of pluripotent stem cells from mouse embryonic and adult fibroblast cultures by defined factors. Cell **126**(4), 663–676 (2006)
7. Yu, J., et al.: Induced pluripotent stem cell lines derived from human somatic cells. Science **318**(5858), 1917–1920 (2007)
8. Daley, W.P., Peters, S.B., Larsen, M.: Extracellular matrix dynamics in development and regenerative medicine. J. Cell Sci. **121**(Pt 3), 255–264 (2008)
9. Engler, A.J., et al.: Multiscale modeling of form and function. Science **324**(5924), 208–212 (2009)
10. Page-McCaw, A., Ewald, A.J., Werb, Z.: Matrix metalloproteinases and the regulation of tissue remodelling. Nat. Rev. Mol. Cell. Biol. **8**(3), 221–233 (2007)
11. Datta, N.H., et al.: In vitro generated extracellular matrix and fluid shear stress synergistically enhance 3D osteoblastic differentiation. Proc. Natl Acad. Sci. USA **103**(8), 2488–2493 (2006)
12. Hoshiba, T., et al.: Development of stepwise osteogenesis-mimicking matrices for the regulation of mesenchymal stem cell functions. J. Biol. Chem. (2009)

13. Chung, C., Burdick, J.A.: Influence of three-dimensional hyaluronic acid microenvironments on mesenchymal stem cell chondrogenesis. Tissue Eng. Part A **15**(2), 243–254 (2008)
14. Nöth, U., et al.: Chondrogenic differentiation of human mesenchymal stem cells in collagen type I hydrogels. J. Biomed. Mater. Res. Part A **83A**(3), 626–635 (2007)
15. Yokoyama, A., et al.: In vitro cartilage formation of composites of synovium-derived mesenchymal stem cells with collagen gel. Cell Tissue Res. **322**(2), 289–298 (2005)
16. Brännvall, K., et al.: Enhanced neuronal differentiation in a three-dimensional collagen-hyaluronan matrix. J. Neurosci. Res. **85**(10), 2138–2146 (2007)
17. Awad, H.A., et al.: Chondrogenic differentiation of adipose-derived adult stem cells in agarose, alginate, and gelatin scaffolds. Biomaterials **25**(16), 3211–3222 (2004)
18. Thomson, R., et al.: Biodegradable polymer scaffolds to regenerate organs. Biopolymers II **122**, 245–274 (1995)
19. Banerjee, A., et al.: The influence of hydrogel modulus on the proliferation and differentiation of encapsulated neural stem cells. Biomaterials **30**(27), 4695–4699 (2009)
20. Benoit, D.S.W., et al.: Small functional groups for controlled differentiation of hydrogel-encapsulated human mesenchymal stem cells. Nat Mater. **7**(10), 816–823 (2008)
21. Engler, A.J., et al.: Matrix elasticity directs stem cell lineage specification. Cell **126**(4), 677–689 (2006)
22. Guo, L., et al.: Chondrogenic differentiation of human mesenchymal stem cells on photoreactive polymer-modified surfaces. Biomaterials **29**(1), 23–32 (2008)
23. Huang, S., Ingber, D.E.: Shape-dependent control of cell growth, differentiation, and apoptosis: switching between attractors in cell regulatory networks. Exp. Cell Res. **261**(1), 91–103 (2000)
24. Lieb, E., et al.: Poly(D,L-lactic acid)–poly(ethylene glycol)–monomethyl ether diblock copolymers control adhesion and osteoblastic differentiation of marrow stromal cells. Tissue Eng. **9**(1), 71–84 (2004)
25. McBeath, R., et al.: Cell shape, cytoskeletal tension, and RhoA regulate stem cell lineage commitment. Dev. Cell **6**(4), 483–495 (2004)
26. Mwale, F., et al.: The effect of glow discharge plasma surface modification of polymers on the osteogenic differentiation of committed human mesenchymal stem cells. Biomaterials **27**(10), 2258–2264 (2006)
27. Watt, F.M., Jordan, P.W., O'Neill, C.H.: Cell shape controls terminal differentiation of human epidermal keratinocytes. Proc. Natl Acad. Sci. USA **85**(15), 5576–5580 (1988)
28. Drumheller, P.D., Hubbell, J.A.: Polymer networks with grafted cell adhesion peptides for highly biospecific cell adhesive substrates. Anal. Biochem. **222**(2), 380–388 (1994)
29. Neff, J.A., Tresco, P.A., Caldwell, K.D.: Surface modification for controlled studies of cell–ligand interactions. Biomaterials **20**(23–24), 2377–2393 (1999)
30. Banerjee, P., et al.: Polymer latexes for cell-resistant and cell-interactive surfaces. J. Biomed. Mater. Res. **50**(3), 331–339 (2000)
31. Jo, S., Engel, P.S., Mikos, A.G.: Synthesis of poly(ethylene glycol)-tethered poly(propylene fumarate) and its modification with GRGD peptide. Polymer **41**(21), 7595–7604 (2000)
32. Lutolf, M.P., Hubbell, J.A.: Synthetic biomaterials as instructive extracellular micro-environments for morphogenesis in tissue engineering. Nat. Biotech. **23**(1), 47–55 (2005)
33. Massia, S.P., Hubbell, J.A.: Convalent surface immobilization of Arg-Gly-Asp- and Tyr-Ile-Gly-Ser-Arg-containing peptides to obtain well-defined cell-adhesive substrates. Anal. Biochem. **187**(2), 292–301 (1990)
34. Santiago, L.Y., et al.: Peptide-surface modification of poly(caprolactone) with laminin-derived sequences for adipose-derived stem cell applications. Biomaterials **27**(15), 2962–2969 (2006)
35. Pierschbacher, M.D., Ruoslahti, E.: Influence of stereochemistry of the sequence Arg-Gly-Asp-Xaa on binding specificity in cell adhesion. J. Biol. Chem. **262**(36), 17294–17298 (1987)

36. Irvine, D.J., Mayes, A.M., Griffith, L.G.: Nanoscale clustering of RGD peptides at surfaces using comb polymers. 1. Synthesis and characterization of comb thin films. Biomacromolecules **2**(1), 85–94 (2000)
37. Silva, G.A., et al.: Selective differentiation of neural progenitor cells by high-epitope density nanofibers. Science **303**(5662), 1352–1355 (2004)
38. Salinas, C.N., Anseth, K.S.: Decorin moieties tethered into PEG networks induce chondrogenesis of human mesenchymal stem cells. J. Biomed. Mater. Res. Part A **90A**(2), 456–464 (2009)
39. Hwang, N.S., et al.: Chondrogenic differentiation of human embryonic stem cell-derived cells in arginine-glycine-aspartate-modified hydrogels. Tissue Eng. **12**(9), 2695–2706 (2006)
40. Hu, Y., et al.: Porous polymer scaffolds surface-modified with arginine-glycine-aspartic acid enhance bone cell attachment and differentiation in vitro. J. Biomed. Mater. Res. **64A**(3), 583–590 (2003)
41. Huang, H., et al.: Enhanced osteoblast functions on RGD immobilized surface. J. Oral Implantol. **29**(2), 73–79 (2003)
42. Shin, H., et al.: Osteogenic differentiation of rat bone marrow stromal cells cultured on Arg-Gly-Asp modified hydrogels without dexamethasone and [beta]-glycerol phosphate. Biomaterials **26**(17), 3645–3654 (2005)
43. Kundu, A.K., Khatiwala, C.B., Putnam, A.J.: Extracellular matrix remodeling, integrin expression, and downstream signaling pathways influence the osteogenic differentiation of mesenchymal stem cells on poly(lactide-co-glycolide) substrates. Tissue Eng. Part A **15**(2), 273–283 (2008)
44. Moursi, A.M., Globus, R.K., Damsky, C.H.: Interactions between integrin receptors and fibronectin are required for calvarial osteoblast differentiation in vitro. J. Cell Sci. **110**(18), 2187–2196 (1997)
45. Boyan, B.D., et al., Role of material surfaces in regulating bone and cartilage cell response. Biomaterials **17**(2), 137–146 (1996)
46. Burdick, J.A., Vunjak-Novakovic, G.: Engineered microenvironments for controlled stem cell differentiation. Tissue Eng. Part A **15**(2), 205–219 (2008)
47. Mwale, F., et al.: Suppression of genes related to hypertrophy and osteogenesis in committed human mesenchymal stem cells cultured on novel nitrogen-rich plasma polymer coatings. Tissue Eng. **12**(9), 2639–2647 (2006)
48. Wan, Y., et al.: Cell adhesion on gaseous plasma modified poly-(-lactide) surface under shear stress field. Biomaterials **24**(21), 3757–3764 (2003)
49. Jiao, Y.-P., Cui, F.-Z.: Surface modification of polyester biomaterials for tissue engineering. Biomed. Mater. **2**(R), 24–37 (2007)
50. Croll, T.I., et al.: Controllable surface modification of poly(lactic-co-glycolic acid) (PLGA) by hydrolysis or aminolysis i: physical, chemical, and theoretical aspects. Biomacromolecules **5**(2), 463–473 (2004)
51. Chu, P.K., et al.: Plasma-surface modification of biomaterials. Mater. Sci. Eng. R Reports **36**(5–6), 143–206 (2002)
52. Chastain, S.R., et al.: Adhesion of mesenchymal stem cells to polymer scaffolds occurs via distinct ECM ligands and controls their osteogenic differentiation. J. Biomed. Mater. Res. Part A **78A**(1), 73–85 (2006)
53. Curran, J.M., Chen, R., Hunt, J.A.: Controlling the phenotype and function of mesenchymal stem cells in vitro by adhesion to silane-modified clean glass surfaces. Biomaterials **26**(34), 7057–7067 (2005)
54. Curran, J.M., Chen, R., Hunt, J.A.: The guidance of human mesenchymal stem cell differentiation in vitro by controlled modifications to the cell substrate. Biomaterials **27**(27), 4783–4793 (2006)
55. Lee, J.H., et al.: Cell behaviour on polymer surfaces with different functional groups. Biomaterials **15**(9), 705–711 (1994)

56. Keselowsky, B.G., Collard, D.M., Garcia, A.J.: Integrin binding specificity regulates biomaterial surface chemistry effects on cell differentiation. PNAS **102**(17), 5953–5957 (2005)
57. Michael, K.E., et al.: Adsorption-induced conformational changes in fibronectin due to interactions with well-defined surface chemistries. Langmuir **19**(19), 8033–8040 (2003)
58. Cheng, S.-L., et al.: Bone mineralization and osteoblast differentiation are negatively modulated by integrin $\alpha_v\beta_3$. J. Bone Miner. Res. **16**(2), 277–288 (2006)
59. Andrade, J.D.: Interfacial phenomena and biomaterials. Med. Instrumen. **7**(2), 110–119 (1973)
60. Baszkin, A., Lyman, D.J.: The interaction of plasma proteins with polymers. I. Relationship between polymer surface energy and protein adsorption/desorption. J. Biomed. Mater. Res. **14**(4), 393–403 (1980)
61. Kennedy, S.B., et al.: Combinatorial screen of the effect of surface energy on fibronectin-mediated osteoblast adhesion, spreading and proliferation. Biomaterials **27**(20), 3817–3824 (2006)
62. Jansen, E.J.P., et al.: Hydrophobicity as a design criterion for polymer scaffolds in bone tissue engineering. Biomaterials **26**(21), 4423–4431 (2005)
63. Mahmood, T.A., et al.: Adhesion-mediated signal transduction in human articular chondrocytes: the influence of biomaterial chemistry and tenascin-C. Exp. Cell Res. **301**(2), 179–188 (2004)
64. Schneider, G.B., et al.: The effect of hydrogel charge density on cell attachment. Biomaterials **25**(15), 3023–3028 (2004)
65. Boyan, B.D., et al.: Surface roughness mediates its effects on osteoblasts via protein kinase A and phospholipase A2. Biomaterials **20**(23–24), 2305–2310 (1999)
66. Kieswetter, K., et al.: Surface roughness modulates the local production of growth factors and cytokines by osteoblast-like MG-63 cells. J. Biomed. Mater. Res. **32**(1), 55–63 (1996)
67. Dalby, M.J., et al.: The control of human mesenchymal cell differentiation using nanoscale symmetry and disorder. Nat. Mater. **6**(12), 997–1003 (2007)
68. Ducheyne, P., Qiu, Q.: Bioactive ceramics: the effect of surface reactivity on bone formation and bone cell function. Biomaterials **20**(23–24), 2287–2303 (1999)
69. Gamsjäger, S., et al.: In: Maher, S.A. (ed.) Raman application in bone imaging, in Raman spectroscopy for soft matter applications, pp. 225–267. Wiley, New York (2009)
70. Oyane, A., et al.: Preparation and assessment of revised simulated body fluids. J. Biomed. Mater. Res. Part A **65A**(2), 188–195 (2003)
71. Kretlow, J.D., Mikos, A.G.: Review: mineralization of synthetic polymer scaffolds for bone tissue engineering. Tissue Eng. **13**(5), 927–938 (2007)
72. Filmon, R., et al.: Effects of negatively charged groups (carboxymethyl) on the calcification of poly(2-hydroxyethyl methacrylate). Biomaterials **23**, 3053–3059 (2002)
73. Zainuddin, et al.: In-vitro study of the spontaneous calcification of PHEMA-based hydrogels in simulated body fluid. J. Mater. Sci. Mater. Med. **17**(12), 1245–1254 (2006)
74. Song, J., Saiz, E., Bertozzi, C.R.: A new approach to mineralization of biocompatible hydrogel scaffolds: an efficient process toward 3-dimensional bonelike composites. J. Am. Chem. Soc. **125**(5), 1236–1243 (2003)
75. Oyane, A., et al.: Simple surface modification of poly([epsilon]-caprolactone) for apatite deposition from simulated body fluid. Biomaterials **26**(15), 2407–2413 (2005)
76. Murphy, W.L., Mooney, D.J.: Bioinspired growth of crystalline carbonate apatite on biodegradable polymer substrata. J. Am. Chem. Soc. **124**(9), 1910–1917 (2002)
77. Koç, A., et al.: In vitro osteogenic differentiation of rat mesenchymal stem cells in a microgravity bioreactor. J. Bioactive Comp. Polym. **23**(3), 244–261 (2008)
78. Osathanon, T., et al.: Microporous nanofibrous fibrin-based scaffolds for bone tissue engineering. Biomaterials **29**(30), 4091–4099 (2008)
79. Dvorak, M.M., et al.: Physiological changes in extracellular calcium concentration directly control osteoblast function in the absence of calciotropic hormones. Proc. Natl Acad. Sci. USA **101**(14), 5140–5145 (2004)

80. Yuan, H., et al.: Osteoinduction by calcium phosphate biomaterials. J. Mater. Sci. Mater. Med. **9**(12), 723–726 (1998)
81. Yu, H.-S., Hong, S.-J., Kim, H.-W.: Surface-mineralized polymeric nanofiber for the population and osteogenic stimulation of rat bone-marrow stromal cells. Mater. Chem. Phys. **113**(2–3), 873–877 (2009)
82. Ingber, D.E.: Cellular mechanotransduction: putting all the pieces together again. FASEB J. **20**(7), 811–827 (2006)
83. Schwartz, M.A., DeSimone, D.W.: Cell adhesion receptors in mechanotransduction. Curr. Opin. Cell Biol. **20**(5), 551–556 (2008)
84. Hu, S., Chen, J., Wang, N.: Cell spreading controls balance of prestress by microtubules and extracellular matrix. Front Biosci. **9**, 2177–2182 (2004)
85. Chen, C.S., et al.: Geometric control of cell life and death. Science **276**(5317), 1425–1428 (1997)
86. Discher, D.E., Janmey, P., Wang, Y.-L.: Tissue cells feel and respond to the stiffness of their substrate. Science **310**(5751), 1139–1143 (2005)
87. Khatiwala, C.B., et al.: ECM compliance regulates osteogenesis by influencing MAPK signaling downstream of RhoA and ROCK. J. Bone Miner. Res. **24**(5), 886–898 (2008)
88. Schuldiner, M., et al.: Effects of eight growth factors on the differentiation of cells derived from human embryonic stem cells. Proc. Natl Acad. Sci. USA **97**(21), 11307–11312 (2000)
89. Park, T.G., Lu, W., Crotts, G.: Importance of in vitro experimental conditions on protein release kinetics, stability and polymer degradation in protein encapsulated poly (-lactic acid-co-glycolic acid) microspheres. J. Control. Release **33**(2), 211–222 (1995)
90. Uebersax, L., Merkle, H.P., Meinel, L.: Insulin-like growth factor I releasing silk fibroin scaffolds induce chondrogenic differentiation of human mesenchymal stem cells. J. Control. Release **127**(1), 12–21 (2008)
91. Park, K., et al.: Functional PLGA scaffolds for chondrogenesis of bone-marrow-derived mesenchymal stem cells. Macromol. Biosci. **9**(3), 221–229 (2009)
92. Basmanav, F.B., Kose, G.T., Hasirci, V.: Sequential growth factor delivery from complexed microspheres for bone tissue engineering. Biomaterials **29**(31), 4195–4204 (2008)
93. Takemoto, S., et al.: Preparation of collagen/gelatin sponge scaffold for sustained release of bFGF. Tissue Eng. Part A **14**(10), 1629–1638 (2008)
94. Niu, X., et al.: Porous nano-HA/collagen/PLLA scaffold containing chitosan microspheres for controlled delivery of synthetic peptide derived from BMP-2. J. Control. Release **134**(2), 111–117 (2009)
95. van de Weert, M., Hennink, W.E., Jiskoot, W.: Protein Instability in Poly(Lactic-co-Glycolic Acid) Microparticles. Pharm. Res. **17**(10), 1159–1167 (2000)
96. Park, K., et al.: The use of chondrogenic differentiation drugs to induce stem cell differentiation using double bead microsphere structure. Biomaterials **29**(16), 2490–2500 (2008)
97. Sachlos, E., Auguste, D.T.: Embryoid body morphology influences diffusive transport of inductive biochemicals: a strategy for stem cell differentiation. Biomaterials **29**(34), 4471–480 (2008)
98. Joliot, A., Prochiantz, A.: Transduction peptides: from technology to physiology. Nat Cell. Biol. **6**(3), 189–196 (2004)
99. Tsakiridis, A., et al.: Microsphere-based tracing and molecular delivery in embryonic stem cells. Biomaterials **30**(29), 5853–5861 (2009)
100. Carpenedo, R.L., et al.: Homogeneous and organized differentiation within embryoid bodies induced by microsphere-mediated delivery of small molecules. Biomaterials **30**(13), 2507–2515 (2009)
101. Newman, K.D., McBurney, M.W.: Poly(D,Llactic-co-glycolic acid) microspheres as biodegradable microcarriers for pluripotent stem cells. Biomaterials **25**(26), 5763–5771 (2004)

102. Nojehdehian, H., et al.: Preparation and surface characterization of poly-L-lysine-coated PLGA microsphere scaffolds containing retinoic acid for nerve tissue engineering: in vitro study. Colloids Surf. B Biointerfaces **73**(1), 23–29 (2009)
103. Oliveira, J.M., et al.: The osteogenic differentiation of rat bone marrow stromal cells cultured with dexamethasone-loaded carboxymethylchitosan/poly(amidoamine) dendrimer nanoparticles. Biomaterials **30**(5), 804–813 (2009)
104. Benoit, D.S., Durney, A.R., Anseth, K.S.: The effect of heparin-functionalized PEG hydrogels on three-dimensional human mesenchymal stem cell osteogenic differentiation. Biomaterials **28**(1), 66–77 (2007)
105. Benoit, D.S.W., Anseth, K.S.: Heparin functionalized PEG gels that modulate protein adsorption for hMSC adhesion and differentiation. Acta Biomater. **1**(4), 461–470 (2005)
106. Benoit, D.S., Collins, S.D., Anseth, K.S.: Multifunctional hydrogels that promote osteogenic human mesenchymal stem cell differentiation through stimulation and sequestering of bone morphogenic protein 2. Adv. Funct. Mater. **17**(13), 2085–2093 (2007)
107. Centrella, M., et al.: Transforming growth factor-{beta} gene family members and bone. Endocr. Rev. **15**(1), 27–39 (1994)
108. Haas, A.R., Tuan, R.S.: Chondrogenic differentiation of murine C3H10T1/2 multipotential mesenchymal cells: II. Stimulation by bone morphogenetic protein-2 requires modulation of N-cadherin expression and function. Differentiation **64**(2), 77–89 (1999)
109. Park, H., et al.: Effect of dual growth factor delivery on chondrogenic differentiation of rabbit marrow mesenchymal stem cells encapsulated in injectable hydrogel composites. J. Biomed. Mater. Res. A **88**(4), 889–897 (2009)
110. Wang, X., et al.: Growth factor gradients via microsphere delivery in biopolymer scaffolds for osteochondral tissue engineering. J. Control. Release **134**(2), 81–90 (2009)
111. Sharma, B., et al.: In vivo chondrogenesis of mesenchymal stem cells in a photopolymerized hydrogel. Plastic Reconstr. Surg. **119**(1), 112–120 (2007) 10.1097/01.prs.0000236896.22479.52
112. Alhadlaq, A., Mao, J.J.: Tissue-engineered neogenesis of human-shaped mandibular condyle from rat mesenchymal stem cells. J. Dent. Res. **82**(12), 951–956 (2003)
113. Alhadlaq, A., et al.: Adult stem cell driven genesis of human-shaped articular condyle. Ann. Biomed. Eng. **32**(7), 911–923 (2004)
114. Chai, C., Leong, K.W.: Biomaterials approach to expand and direct differentiation of stem cells. Mol. Ther. **15**(3), 467–480 (2007)
115. Levenberg, S., et al.: Differentiation of human embryonic stem cells on three-dimensional polymer scaffolds. Proc. Natl Acad. Sci. USA **100**(22), 12741–12746 (2003)
116. Cho, S.-W., et al.: Engineering of volume-stable adipose tissues. Biomaterials **26**(17), 3577–3585 (2005)
117. Lee, P.Y., et al.: Thermosensitive hydrogel PEG–PLGA–PEG enhances engraftment of muscle-derived stem cells and promotes healing in diabetic wound. Mol. Ther. **15**(6), 1189–1194 (2007)
118. Boldrin, L., et al.: Efficient delivery of human single fiber-derived muscle precursor cells via biocompatible scaffold. Cell Transplant. **17**, 577–584 (2008)
119. Kim, M., et al.: Muscle regeneration by adipose tissue-derived adult stem cells attached to injectable PLGA spheres. Biochem. Biophys. Res. Commun. **348**(2), 386–392 (2006)
120. Walker, E., et al.: Prediction and testing of novel transcriptional networks regulating embryonic stem cell self-renewal and commitment. Cell Stem Cell **1**(1), 71–86 (2007)
121. Sampath, P., et al.: A hierarchical network controls protein translation during murine embryonic stem cell self-renewal and differentiation. Cell Stem Cell **2**(5), 448–460 (2008)
122. Yao, S., et al.: Long-term self-renewal and directed differentiation of human embryonic stem cells in chemically defined conditions. Proc Natl Acad Sci USA **103**(18), 6907–6912 (2006)
123. Anderson, D.G., Levenberg, S., Langer, R.: Nanoliter-scale synthesis of arrayed biomaterials and application to human embryonic stem cells. Nat. Biotech. **22**(7), 863–866 (2004)
124. Toohey, K.S., et al.: Self-healing materials with microvascular networks. Nat. Mater. **6**(8), 581–585 (2007)

Integration of Biomaterials into 3D Stem Cell Microenvironments

Andres Bratt-Leal, Richard Carpenedo and Todd McDevitt

Abstract Stem cells receive physical and chemical cues capable of influencing their phenotype from inter-related elements of the microenvironment, such as cell–cell contacts, soluble molecule signals and physical interactions with the ECM. In contrast to conventional 2D culture systems, barriers to diffusion within 3D cultures limit the effectiveness of media manipulation as a method to direct cell behavior. Efforts to engineer stem cell microenvironments in 3D using biomaterials have generally been attempted by either scaffold seeding, cell encapsulation, or microcarrier/microparticle based approaches. These different methods have been applied not only for the propagation of pluri- and multipotent stem cells, but also to direct the differentiation of such stem cells into more differentiated phenotypes. This chapter discusses the unique benefits, as well as associated challenges of integrating biomaterials into 3D stem cell microenvironments.

1 Introduction

Stem cell fate is regulated by transcription factors that control gene and protein expression. The balance between self-renewal and differentiation of stem cells is determined by the sensitivity of gene regulation to changes in the extracellular microenvironment. A complex assembly of soluble factors, cell–cell and cell–extracellular matrix interactions constitute the molecular regulators of the microenvironment through which stem cell fate is determined [1]. Particularly in the case of pluripotent and multipotent stem cells, the microenvironment requires

A. Bratt-Leal, R. Carpenedo and T. McDevitt (✉)
Wallace H. Coulter Department of Biomedical Engineering,
Georgia Institute of Technology, Atlanta GA, 30332-0535, USA
e-mail: todd.mcdevitt@bme.gatech.edu

Stud Mechanobiol Tissue Eng Biomater (2010) 2: 45–59

DOI: 10.1007/8415_2010_7

Published online: 15 June 2010

special consideration because of the vastly divergent array of cell lineages that can be adopted by such stem cells. In the developing embryo, morphogen gradients, cell–cell and cell–extracellular matrix signals are integrated to provide essential spatial and temporal information to direct cell differentiation, migration and tissue formation.

Pluripotent embryonic stem cells (ESCs) can differentiate into cells of all three germ lineages (ecto-, endo- and mesoderm); however, when the microenvironment is not strictly regulated, ESCs differentiate in a spontaneous, uncontrolled manner. Thus, much of the current focus of ESC research is to better understand environmental mechanisms regulating cell differentiation and to develop the essential technologies to direct differentiation more efficiently towards a homogeneous population of cells that could be used for regenerative therapies. Recently, the ability to produce "induced" pluripotent stem (iPS) cells from human somatic cell types has been reported, allowing for patient specific pluripotent cells to be developed, and thereby avoiding the potential immune rejection of allogenic ESCs [2, 3].

Multipotent adult stem cells, present in most tissues of mature organisms, serve to repair and regenerate tissues and thereby enable organisms to live beyond the life-span of individual cells [4]. Multipotent stem cells are also responsive to signals in the microenvironment capable of directing self-renewal or differentiation. For example, in many tissues, contact, or a loss of contact with a specialized basement membrane is believed to influence the fate of stem cell daughter cells [4].

Stem cells are typically expanded in an environment intended to maintain "stemness" before being subjected to a different set of conditions designed to promote differentiation, preferably to a specific phenotype. Biochemical factors such as basic fibroblast growth factor (bFGF) for human ESCs [5, 6] or leukemia inhibitor factor (LIF) for mouse ESCs [7, 8] are routinely added to the culture medium to grow ESCs in an undifferentiated state. Withdrawal of these factors from the culture media allows differentiation to proceed. In contrast, specific factors have yet to be identified to stably maintain mesenchymal stem cells (MSCs) and extended passaging of MSCs in vitro generally results in genetic abnormalities and reduced differentiation potential.

The substrate on which stem cells are cultured also can be used to inhibit or promote differentiation. Feeder layers of cells can be used, such as mouse embryonic fibroblasts (MEFs) [5], to promote ESC self-renewal. In addition, ECM coatings, such as Matrigel™, a complex mixture of basement membrane components isolated from mouse tumor cells, or gelatin can be used as alternative surface coatings to feeder cell layers [9]. Matrigel™ and MEFs contain or produce ECM and soluble factors which favor ESC self-renewal and growth while limiting cell differentiation. The introduction of xenogenic material, either through the substrate, or through animal serum, can lead to the expression of immunogenic animal proteins on human stem cells, limiting their clinical uses [10]. Thus, chemically defined media, which avoid the introduction of animal serum proteins, have been produced in recent years for the purpose of presenting the most basic biochemical factors required for stem cell growth and self-renewal [11]. Defined culture media and ECM substrates can also be advantageous in directed differentiation protocols in order to present defined amounts of molecules and avoid presentation of conflicting signals which could contribute to heterogeneous differentiation.

1.1 Culture in Two or Three Dimensions

Traditionally, two-dimensional culture (2D) is used to maintain cells in an undifferentiated state whereas three-dimensional (3D) culture techniques are more commonly implemented in differentiation protocols. For example, ESCs are commonly differentiated through the formation of 3D multicellular aggregates, referred to as embryoid bodies (EBs) [12]. Spheroid culture provides a platform for scalable culture of cells because they can be grown in suspension and require a lower tissue culture surface area-to-volume ratio compared to cells in monolayer. The 3D spheroid microenvironment is complex and becomes more so as differentiation progresses. Differentiating cells deposit extracellular matrix and are subjected to homo- and heterotypic cell–cell interactions, as well as autocrine and paracrine factors.

Soluble factors added to the culture media are more accessible to cells grown in monolayer and the temporal resolution of these factors can be fairly well controlled by simply exchanging the culture media at specific times [13]. In contrast, 3D culture provides more physiological cues such as increased cell–cell interactions and the potential for increased cell–extracellular matrix interaction; however, diffusion limitations complicate 3D cultures of cells. Concentration gradients of soluble factors added to the media, which result from the diffusion properties of 3D cellular aggregates or constructs, can result in differences in the microenvironment depending on the spatial positioning of cells. Stem cells can be very sensitive to small perturbations in the biochemical composition of their surroundings and the effects of soluble factors often vary in a dose-dependent manner. This limited control of the 3D environment has necessitated the development of biomaterial technologies to engineer the microenvironment of 3D culture systems in order to further develop stem cell differentiation protocols.

1.2 Strategies for Biomaterial Control of the 3D Microenvironment

As introduced above, adhesive protein biomaterials (i.e. Matrigel and gelatin) have been used as culture substrates for stem cell maintenance culture in 2D. Biomaterials can also be integrated within 3D stem cell environments in order to control the increased complexity of the microenvironment. Strategies to control stem cell behavior using biomaterials have largely aimed to deconstruct elements of native biological complexity and integrate defined components into controlled systems to present molecular cues to stem cells.

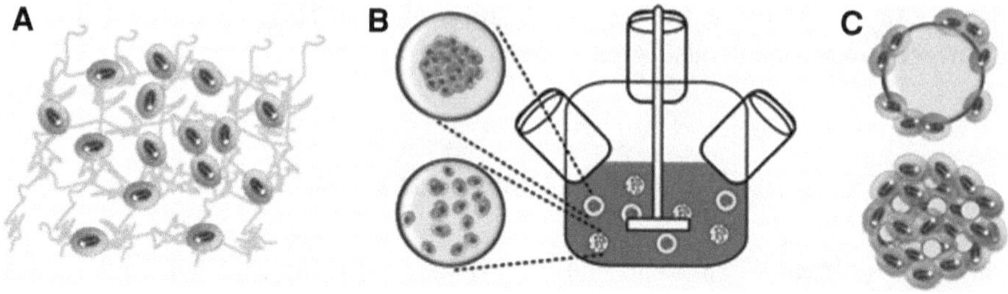

Fig. 1 Biomaterials can be incorporated within 3D stem cell microenvironments to direct cell behavior. Cells can be cultured on or within polymeric scaffolds (**a**) which provide physio-chemical cues. Encapsulation of cell aggregates or single cells (**b**) can be used to increase the surface area to volume ratio for scalable culture and to provide an artificial matrix. "Inside–out" approaches to direct the microenvironment include culture of cells on microcarriers (**c**, *top*) and the incorporation of microparticles within stem cell aggregates (**c**, *bottom*)

Parameters such as hydrophobicity, porosity, degradation kinetics and surface coating can be engineered to create the desired material properties. The ability to engineer biomaterial surface properties can be utilized to present insoluble factors to mimic cell–cell or extracellular matrix interactions. Stem cells and biomaterials can be combined as scaffolds in the classic tissue engineering paradigm, wherein a spongy or fibrous scaffold provides mechanical support for attachment and migration guidance initially and then degrades as the cells produce their own natural scaffolding. In addition to scaffolds, which have been used extensively with other cell types, single cell suspensions or EBs are often completely encapsulated in either a natural ECM matrix or in a polymer designed to provide differentiation cues. Encapsulation typically occurs in the form of small spherical beads (hundreds of microns in diameter) grown in suspension culture whereas scaffolds are much larger and are grown in static or perfused cultures. Another approach is to integrate biomaterials within stem cell spheroids, either using microcarriers or microparticles, to control the microenvironment from the "inside–out." Strategies utilizing scaffolds, encapsulation and microcarrier/microparticles for control of the stem cell microenvironment are depicted in Fig. 1 and are discussed in further detail below.

2 Scaffolds

The application of polymeric scaffolds to support somatic cells was one of the original tenets of tissue engineering strategies [14]. Polymer scaffolds were originally designed primarily as carriers for cell transplantation which provided temporary structural support until cells could adequately synthesize their own matrix to replace the biodegradable synthetic material. However, advancements in the field of biomaterials have led to the development of more sophisticated scaffolds in various forms (e.g. porous, fibrous), capable of responding to environmental changes (e.g. temperature, pH, electrical stimulation, proteases) and directly incorporating biomolecular cues to mediate cell attachment, proliferation and differentiation. Recently, pluripotent and multipotent stem cells have been seeded onto polymer scaffolds as a means to examine self-renewal and differentiation properties in 3D.

Scaffolds have been studied in combination with mesenchymal stem cells (MSCs) for a wide variety of applications including bone and cartilage regeneration. The ability of MSCs, derived from bone marrow or other tissues (e.g. adipose tissue), to differentiate into a variety of cell types, including osteo-, chondro-, and adipogenic lineages, has made MSCs the most common cell source for musculoskeletal tissue engineering strategies. A variety of synthetic and natural polymers have been utilized for both osteogenesis and chondrogenesis of MSCs, including nanofibrous electrospun poly(caprolactone) (PCL) [15–17], PLGA [18–22], and silk [23–26]. The ability of multiple scaffold types with a wide range of chemical and physical properties to support MSC proliferation and differentiation makes this a promising and active area of stem cell research.

Pluripotent stem cells can likewise be cultured on or within scaffolds, and much of the knowledge gained from previous studies with MSCs can be applied to ESC culture. As discussed above, culture of undifferentiated ESCs is typically performed in monolayer, with cells grown on either an inactivated MEF feeder layer or Matrigel. However, the use of synthetic scaffolds for self-renewal culture may circumvent the issues of xenogenic contact and scale-up feasibility associated with MEF and Matrigel substrates. The culture of hESCs on an artificial matrix composed of semi-interpenetrating polymer networks (sIPN) supported short-term maintenance of pluripotency [27]. The sIPN hydrogels were functionalized with the arginine-glycine-aspartic acid (RGD) peptide sequence, and RGD concentration, as well as the mechanical properties of the hydrogel, were varied independently to identify conditions which promoted self-renewal of hESCs. Artificial extracellular matrices for stem cell renewal can be used as a xeno-free alternative to defined media.

Additionally, ESCs have been cultured on nanofibrillar polyamide matrices known as Ultra-Web under self-renewal conditions [28]. ESCs grown on Ultra-Web displayed higher alkaline phosphatase activity, indicative of pluripotency, as well as enhanced proliferation, relative to gelatin-coated glass coverslips. Activation of Rac, a small GTPase, was also enhanced in cells cultured on Ultra-Web, as was activation of the PI3K pathway and Nanog expression. These data indicate that the 3D architecture on which cells are cultured may play an important role in cell fate determination and must be taken into account in future design of cell culture systems for ESC self-renewal.

Synthetic scaffolds have also been applied to differentiation approaches for ESCs. Porous scaffolds composed of PLGA/PLLA have been investigated as substrates for hESC adhesion and differentiation, with the intent of forming complex tissue architectures to be used in transplantation therapies [29]. The combination of seeding human ESCs on porous scaffolds and media supplementation with growth factors was found to induce differentiation into various cell types that expressed markers of neural, chondrogenic and hepatic lineages. Cells remained viable on the scaffolds following implantation in severely immune compromised mice, and continued differentiation and reorganization was observed. Studies focusing specifically on neural differentiation of hESCs seeded onto PLGA scaffolds were performed, with the effect of various media supplements reported [30]. The addition of nerve growth factor and neurotrophin 3 to the scaffold cultures enhanced differentiation to nestin and βIII tubulin positive cells, indicative of neural progenitors and neurons, respectively. However, formation of functional, higher-ordered tissues will likely require greater sophistication in scaffold architecture and differentiation-cue presentation to ESCs.

Scaffolds composed of biomimetic and natural polymers have been used in scaffold fabrication in order to present instructive microenvironments to pluripotent cells. ESCs cultured on the biomimetic material Cytomatrix formed 3D structures similar to EBs, but displayed enhanced ECM production as well as increased efficiency in differentiation to hematopoietic precursor cells [31]. Genes associated with ECM production as well as proliferation and differentiation were found to be enhanced relative to traditional EBs [32]. Incorporation of ESCs into porous alginate scaffolds resulted in efficient, homogeneous EB formation, with EBs spatially restricted within the pores. Agglomeration of EBs appeared to be inhibited, resulting in efficient cell proliferation and differentiation [33].

Fibrin scaffolds have also been investigated for use in directed ESC differentiation to neural cell types [34]. Cells from both dissociated and intact EBs were seeded within fibrin scaffolds, and conditions including cell density and fibrinogen and thrombin concentration were optimized for cell proliferation and differentiation. After 14 days, successful differentiation of ESCs into neurons and astrocytes was observed. In an independent study, cells were seeded onto fibrin scaffolds as well as PEGylated fibrin scaffolds, and proliferation and differentiation were assessed relative to traditional EBs as well as EBs grown in semi-solid methylcellulose [35]. Proliferation in both types of fibrin scaffolds was enhanced relative to EBs and methylcellulose EBs. Culture in non-PEGylated fibrin resulted in differentiation similar to that observed in traditional EBs, with down-regulation of OCT4 and expression of VE-Cadherin, while ESCs growth in PEGylated fibrin were more similar to methylcellulose controls.

Semi-interpenetrating polymer networks composed of the natural polymers collagen, fibronectin and laminin were examined as scaffolds for ESC differentiation [36]. Differentiation was found to be a function of both network composition and concentration, as high collagen concentration inhibited EB cavitation, fibronectin appeared to enhance endothelial differentiation, and laminin enhanced cardiomyogenesis. Use of the self-assembling nanofibrillar peptide scaffold PuraMatrix was investigated for osteogenic differentiation of ESCs [37]. EB-derived cells were seeded onto PuraMatrix scaffolds after 8 days of differentiation, and cells plated onto tissue culture polystyrene served as a 2D control. Both PuraMatrix and 2D substrates supported osteogenic differentiation, although maintenance of OCT4 positive cells was more prevalent in 3D. Scaffolds can be used to control physiochemical elements of the microenvironment, however, their use for large scale production of homogeneous cells may be limited due to diffusional limitations of nutrients in large constructs lacking vasculature.

3 Encapsulation

Encapsulation of stem cells into hydrogels represents a scalable way to present factors locally to cells. Unlike scaffolds, large numbers of capsules can be cultured in suspension culture. Encapsulation of stem cells can affect diffusion, control aggregate size and prevent agglomeration as well as provide an instructive environment depending on the properties of the material chosen for encapsulation. From a bioprocessing perspective, encapsulation provides a method to grow anchorage dependent cells in suspension thereby increasing the surface area to volume ratio and scale up potential. MSCs are difficult to maintain as aggregates in suspension [38] and for this reason they are often studied using encapsulation. In addition to bioprocessing advantages, the creation of an artificial matrix aids in study of cellular response to specific elements of native ECM. Elements such as material stiffness or peptide density can be varied independent of other factors.

Single cell suspensions or cell spheroids can be encapsulated several ways depending on the material properties. Thermosetting hydrogels such as agarose can be used to encapsulate cells by emulsifying a mixture of agarose, cells and oil. Agarose capsules containing EBs are formed in the oil phase after emulsification and can be gelled by lowering the temperature below the gelation point of agarose [39]. Materials such as alginate are useful for encapsulation because gelation occurs in the presence of Ca^{2+} ions and does not require oil phase emulsion or temperature change that can lower cell viability. Precisely sized droplets of alginate and cells can be created and gelled in $CaCl_2$ baths and cells can be retrieved at later time points after transfer to a medium without Ca^{2+} [40–43]. Artificial polymers such as poly(NiPAAm-co-AAC) [44], poly(ethylene glycol) (PEG) [45], and PEG derivatives such as PEG diacrylate [46, 47] and oligo(poly(ethylene glycol)) fumerate [48] have been utilized as well.

Encapsulation also provides a method to investigate interactions between cells and specific signaling sequences in an artificial ECM environment in which the ligand density can be precisely controlled. For example, alginate can be modified to present small peptide sequences such as arginine-glycine-aspartic acid (RGD) [43], found on ECM proteins such as fibronectin, fibrin and vitronectin. Increasing RGD density in alginate gels resulted in dose dependent decrease in encapsulated MSC response to TGF-B1 and dexamethasone, components of chondrogenic media. It was hypothesized that integrin mediated signaling may be responsible for inhibition of chondrogenesis in the cells and control of integrin signaling may be a useful target for directed differentiation strategies. Mimicking ECM interactions using small peptide sequences can aid in understanding the mechanisms by which ECM components contribute to the cellular microenvironment.

In regards to ESCs, encapsulation originated as a method to control the homogeneity of differentiation culture. ESCs express high levels of E-cadherin, a homotypic cell–cell adhesion molecule, which has been shown to be primarily responsible for EB formation in suspension culture [39, 49]. High levels of E-cadherin also can lead to agglomeration which is particularly problematic in static cultures. Agglomeration leads to heterogeneity in EB culture and contributes to the heterogeneity of the resultant differentiated cell population. In addition, the size of the initial ESC aggregate has been implicated in the trajectory of subsequent differentiation, and therefore precise size control of EB formation is considered advantageous [50–52]. In addition, EBs formed using other methods can be later be encapsulated with one EB/capsule to prevent agglomeration and to shield EBs from shear forces experienced in a stirred bioreactor. ESCs can be encapsulated as a single-cell suspension and depending on the material used and the size of the capsule formed, the result can be single EBs or small individual clumps of cells.

ESCs can be encapsulated in natural polymers such as hyaluronic acid [53] or alginate [54] to maintain a pluripotent state useful for production of large amounts of cells. Cells can then either be retrieved from the gels or switched to differentiation conditions for further culture. Retrieval from hyaluronic acid encapsulation requires that the capsules be incubated in hyaluronidase, while alginate capsules can be depolymerized through the removal of divalent cations. Encapsulation can be further used to promote differentiation into hepatocytes [42], chondrocytes [55], cardiomyocytes [56], and definitive endoderm [57]. In some cases, encapsulation is used as a method to support differentiation of ESCs; however, directed differentiation techniques, such as the addition of soluble factors, can also be combined with encapsulation to promote specific directed differentiation.

Bioengineers can apply knowledge of biological processes which are known to occur naturally to direct stem cell behavior, such as the presentation the RGD peptide to promote adhesion, however, another strategy is to screen biomaterials with different surface chemistries to discover new non-physiological interactions which can be useful in directing cell differentiation. Using this strategy, the gene expression of stem cells on different surfaces can be analyzed on an array set-up with high-throughput analysis and materials that promote the desired differentiation can be further analyzed in 3D culture. PEG hydrogels functionalized with small side functional groups of varying charge and hydrophilicity illicit different gene expression profiles of encapsulated hMSCs [58].

4 Microcarriers and Microparticles

Encapsulation of ESCs and other stem cell types is a method to introduce biomaterial control of differentiation cues; however, this method is an "outside–in" strategy. Biomaterial interaction with the microenvironment is directly imparted on cells of the surface of encapsulated aggregates whereas interior cells are not directly affected. Alternative strategies have similarly focused on incorporating biomaterials with cells cultured in suspension; however, they rely either on culture of cells on microcarriers or the incorporation of microparticles within stem cell aggregates.

4.1 Microcarriers

Microcarriers are spherical beads, normally 150–500 microns in diameter, and can be made of a variety of materials such as polystyrene, dextran and glass. Cells can be grown on the surface of microcarriers to increase the available growing surface area per unit volume and have been used to scale up culture of anchorage dependent cells. Microcarriers have been reported to support maintenance culture of human ESCs [59–61], mouse ESCs [49, 62, 63] and MSCs [64, 65] and importantly, population doubling times remain comparable to 2D culture standards. Dextran beads coated in collagen are most often used for ESC attachment. These cells can then be differentiated while adhered to the bead or they can be separated from the beads for use in other differentiation protocols. The choice of coating is important to the cell yield and in the case of polystyrene beads, cells can adhere without a coating through electrostatic interactions. Cell collection from uncoated polystyrene is difficult and results in decreased cell viability, whereas with gelatin coated dextran, trypsin can digest the collagen layer and cells can be recovered with high yield. Matrigel has also been successfully used as a dextran bead coating as it is known that Matrigel will support undifferentiated growth of ESCs [59, 61].

In contrast to scaffold-based approaches, the material properties of microcarriers have not been extensively studied in regards to directed differentiation or microenvironmental control. Microcarrier materials are evaluated on their ability to expand large amounts of undifferentiated cells. This is in part due to the fact that microcarrier culture is analogous to 2D culture where media manipulation using growth factors or small molecules is a potent regulator of cell behavior. Limitations remain with microcarrier culture including agglomeration and low cell viability after collection procedures. Stirred suspension bioreactors are commonly utilized to agitate the culture and prevent agglomeration and a balance must be reached between shear forces experienced by cells and agglomeration at lower agitation rates.

4.2 Microparticles

While microcarriers are used to scale up the culture monolayers of cells, smaller-sized particles, ranging from 250 nm to 10 μm in diameter, can be incorporated within larger cell spheroids to take advantage of increased cell–cell contacts and 3D

ECM contact. The incorporation of materials within stem cell aggregates is a relatively new approach of "inside–out" engineering that can be used to place cells on the interior of the aggregate in direct contact with the biomaterials. Biomaterial microparticles have been widely used in the field of drug delivery as vehicles for controlled release of encapsulated molecules, and their surface can be functionalized with cell specific adhesion ligands for cell-targeted delivery, especially useful in cancer therapies. In addition, microparticle surfaces can be modified to mimic cell–cell interactions, loaded with soluble morphogens for controlled release or can be used to introduce ECM proteins for control of matrix properties [66]. The surfaces of materials such as polystyrene and poly(2-hydroxyethyl methacrylate) have previously been modified in 2D cultures to present LIF to prevent differentiation or Jagged-1 to mimic cell–cell signaling [67, 68]. Microparticles can be combined with other techniques discussed previously to add further control of soluble factor release. Microparticles incorporated in scaffolds [69, 70] can continually release encapsulated factors throughout the construct and this could similarly be used in capsules of encapsulated cells.

Microparticles incorporated within progenitor cell spheroids was first used to improve post-transplantation cell viability of neural cells [71]. Fetal rat brain cells were mixed with poly(lactic-co-glycolic acid) (PLGA) microparticles which released nerve growth factor (NGF) to increase cell viability after transplantation. This concept has also been applied to synthetic microenvironments for ESCs for the purpose of directed differentiation [72, 73]. As introduced in the beginning of the chapter, ESC spheroids present barriers to diffusion and therefore cells in the interior are not completely accessible to molecules in the media as is the case with 2D culture. Cell–cell contacts and deposited ECM can limit the diffusion of even small molecules and the formation of gradients is likely to contribute to the heterogeneous nature of EB differentiation.

Incorporation of biomaterials within EBs circumvents the diffusion barriers to cells on the EB interior and microparticles can act as point sources continuously releasing morphogen within the EB. In this way, gradients of molecules can be minimized to create a more homogeneous environment for the cells within EBs. An example of microparticle-mediated control of the EB microenvironment is the incorporation of PLGA microparticles within mouse EBs [72]. The microparticles were loaded with retinoic acid (RA), a small, hydrophobic morphogen, and were designed to continually release RA throughout EB culture. The resulting EBs upregulated gene expression of genes characteristic of epiblast stage embryos and uniform cavitation was observed in large populations of EBs. This effect could not be matched through any soluble addition of RA to the EB medium, suggesting that the controlled release of the RA by the particles inside the EBs was needed to provide the appropriate microenvironment for epiblast-like EB formation. Evidence that this effect was widespread throughout the entire culture indicates that this strategy could be used to direct differentiation of ESCs in a scalable manner. Controlled release of morphogens is also desired to conserve growth factor for large scale experiments. Biomaterials can preserve the bioactivity of encapsulated growth factor by maintaining the molecule in a bound state and preventing degradation.

Soluble addition of growth factors must be replenished as determined by the half-life of the molecule in order to maintain the desired concentration for cell signaling.

5 Summary and Conclusions

Stem cell differentiation is controlled by a complex system of extracellular signals from the microenvironment. Control over the microenvironment in 3D culture systems is limited when using traditional cell culture methods and therefore engineered biomaterials have been integrated within 3D stem cell environments for directed differentiation. Scaffolds, hydrogel encapsulation, microparticles and microcarriers all are used to control cell–cell interactions, cell–ECM interactions and the presentation of soluble or immobilized factors to cells. These methods have been used to increase the efficiency and scalability of maintenance culture as well as for directed differentiation.

References

1. Watt, F.M., Hogan, B.L.: Out of Eden: stem cells and their niches. Science **287**(5457), 1427–1430 (2000)
2. Takahashi, K., Tanabe, K., Ohnuki, M., Narita, M., Ichisaka, T., Tomoda, K., Yamanaka, S.: Induction of pluripotent stem cells from adult human fibroblasts by defined factors. Cell **131**(5), 861–872 (2007)
3. Yu, J., Vodyanik, M.A., Smuga-Otto, K., Antosiewicz-Bourget, J., Frane, J.L., Tian, S., Nie, J., Jonsdottir, G.A., Ruotti, V., Stewart, R., Slukvin, II, Thomson, J.A.: Induced pluripotent stem cell lines derived from human somatic cells. Science **318**(5858), 1917–1920 (2007)
4. Spradling, A., Drummond-Barbosa, D., Kai, T.: Stem cells find their niche. Nature **414**(6859), 98–104 (2001)
5. Thomson, J.A., Itskovitz-Eldor, J., Shapiro, S.S., Waknitz, M.A., Swiergiel, J.J., Marshall, V.S., Jones, J.M.: Embryonic stem cell lines derived from human blastocysts. Science **282**(5391), 1145–1147 (1998)
6. Amit, M., Carpenter, M.K., Inokuma, M.S., Chiu, C.P., Harris, C.P., Waknitz, M.A., Itskovitz-Eldor, J., Thomson, J.A.: Clonally derived human embryonic stem cell lines maintain pluripotency and proliferative potential for prolonged periods of culture. Dev. Biol. **227**(2), 271–278 (2000)
7. Smith, A.G., Heath, J.K., Donaldson, D.D., Wong, G.G., Moreau, J., Stahl, M., Rogers, D.: Inhibition of pluripotential embryonic stem cell differentiation by purified polypeptides. Nature **336**(6200), 688–690 (1988)
8. Williams, R.L., Hilton, D.J., Pease, S., Willson, T.A., Stewart, C.L., Gearing, D.P., Wagner, E.F., Metcalf, D., Nicola, N.A., Gough, N.M.: Myeloid leukaemia inhibitory factor maintains the developmental potential of embryonic stem cells. Nature **336**(6200), 684–687 (1988)
9. Xu, C., Inokuma, M.S., Denham, J., Golds, K., Kundu, P., Gold, J.D., Carpenter, M.K.: Feeder-free growth of undifferentiated human embryonic stem cells. Nat. Biotechnol. **19**(10), 971–974 (2001)
10. Martin, M.J., Muotri, A., Gage, F., Varki, A.: Human embryonic stem cells express an immunogenic nonhuman sialic acid. Nat. Med. **11**(2), 228–232 (2005)

11. Amit, M., Shariki, C., Margulets, V., Itskovitz-Eldor, J.: Feeder layer- and serum-free culture of human embryonic stem cells. Biol. Reprod. **70**(3), 837–845 (2004)
12. Doetschman, T.C., Eistetter, H., Katz, M., Schmidt, W., Kemler, R.: The in vitro development of blastocyst-derived embryonic stem cell lines: formation of visceral yolk sac, blood islands and myocardium. J. Embryol. Exp. Morphol. **87**, 27–45 (1985)
13. Wada, T., Honda, M., Minami, I., Tooi, N., Amagai, Y., Nakatsuji, N., Aiba, K.: Highly efficient differentiation and enrichment of spinal motor neurons derived from human and monkey embryonic stem cells. PLoS One **4**(8), e6722 (2009)
14. Langer, R., Vacanti, J.P.: Tissue engineering. Science **260**(5110), 920–926 (1993)
15. Li, W.J., Tuli, R., Okafor, C., Derfoul, A., Danielson, K.G., Hall, D.J., Tuan, R.S.: A three-dimensional nanofibrous scaffold for cartilage tissue engineering using human mesenchymal stem cells. Biomaterials **26**(6), 599–609 (2005)
16. Li, W.J., Chiang, H., Kuo, T.F., Lee, H.S., Jiang, C.C., Tuan, R.S.: Evaluation of articular cartilage repair using biodegradable nanofibrous scaffolds in a swine model: a pilot study. J. Tissue Eng. Regen. Med. **3**(1), 1–10 (2009)
17. Li, W.J., Tuli, R., Huang, X., Laquerriere, P., Tuan, R.S.: Multilineage differentiation of human mesenchymal stem cells in a three-dimensional nanofibrous scaffold. Biomaterials **26**(25), 5158–5166 (2005)
18. Yang, J., Cao, C., Wang, W., Tong, X., Shi, D., Wu, F., Zheng, Q., Guo, C., Pan, Z., Gao, C., Wang, J.: Proliferation and osteogenesis of immortalized bone marrow-derived mesenchymal stem cells in porous polylactic glycolic acid scaffolds under perfusion culture. J. Biomed. Mater. Res. A (2009)
19. Xin, X., Hussain, M., Mao, J.J.: Continuing differentiation of human mesenchymal stem cells and induced chondrogenic and osteogenic lineages in electrospun PLGA nanofiber scaffold. Biomaterials **28**(2), 316–325 (2007)
20. Park, K., Cho, K.J., Kim, J.J., Kim, I.H., Han, D.K.: Functional PLGA scaffolds for chondrogenesis of bone-marrow-derived mesenchymal stem cells. Macromol. Biosci. **9**(3), 221–229 (2009)
21. Tanaka, T., Hirose, M., Kotobuki, N., Tadokoro, M., Ohgushi, H., Fukuchi, T., Sato, J., Seto, K.: Bone augmentation by bone marrow mesenchymal stem cells cultured in three-dimensional biodegradable polymer scaffolds. J. Biomed. Mater. Res. A **91**(2), 428–435 (2009)
22. Stiehler, M., Bunger, C., Baatrup, A., Lind, M., Kassem, M., Mygind, T.: Effect of dynamic 3-D culture on proliferation, distribution, and osteogenic differentiation of human mesenchymal stem cells. J. Biomed. Mater. Res. A **89**(1), 96–107 (2009)
23. Hofmann, S., Knecht, S., Langer, R., Kaplan, D.L., Vunjak-Novakovic, G., Merkle, H.P., Meinel, L.: Cartilage-like tissue engineering using silk scaffolds and mesenchymal stem cells. Tissue Eng. **12**(10), 2729–2738 (2006)
24. Meinel, L., Hofmann, S., Karageorgiou, V., Zichner, L., Langer, R., Kaplan, D., Vunjak-Novakovic, G.: Engineering cartilage-like tissue using human mesenchymal stem cells and silk protein scaffolds. Biotechnol. Bioeng. **88**(3), 379–391 (2004)
25. Meinel, L., Karageorgiou, V., Hofmann, S., Fajardo, R., Snyder, B., Li, C., Zichner, L., Langer, R., Vunjak-Novakovic, G., Kaplan, D.L.: Engineering bone-like tissue in vitro using human bone marrow stem cells and silk scaffolds. J. Biomed. Mater. Res. A **71**(1), 25–34 (2004)
26. Kim, H.J., Kim, U.J., Kim, H.S., Li, C., Wada, M., Leisk, G.G., Kaplan, D.L.: Bone tissue engineering with premineralized silk scaffolds. Bone **42**(6), 1226–1234 (2008)
27. Li, Y.J., Chung, E.H., Rodriguez, R.T., Firpo, M.T., Healy, K.E.: Hydrogels as artificial matrices for human embryonic stem cell self-renewal. J. Biomed. Mater. Res. A **79**(1), 1–5 (2006)
28. Nur, E.K.A., Ahmed, I., Kamal, J., Schindler, M., Meiners, S.: Three-dimensional nanofibrillar surfaces promote self-renewal in mouse embryonic stem cells. Stem Cells **24**(2), 426–433 (2006)
29. Levenberg, S., Huang, N.F., Lavik, E., Rogers, A.B., Itskovitz-Eldor, J., Langer, R.: Differentiation of human embryonic stem cells on three-dimensional polymer scaffolds. Proc. Natl Acad. Sci. USA **100**(22), 12741–12746 (2003)

30. Levenberg, S., Burdick, J.A., Kraehenbuehl, T., Langer, R.: Neurotrophin-induced differentiation of human embryonic stem cells on three-dimensional polymeric scaffolds. Tissue Eng. **11**(3–4), 506–512 (2005)
31. Liu, H., Roy, K.: Biomimetic three-dimensional cultures significantly increase hematopoietic differentiation efficacy of embryonic stem cells. Tissue Eng. **11**(1–2), 319–330 (2005)
32. Liu, H., Lin, J., Roy, K.: Effect of 3D scaffold and dynamic culture condition on the global gene expression profile of mouse embryonic stem cells. Biomaterials **27**(36), 5978–5989 (2006)
33. Gerecht-Nir, S., Cohen, S., Itskovitz-Eldor, J.: Bioreactor cultivation enhances the efficiency of human embryoid body (hEB) formation and differentiation. Biotechnol. Bioeng. **86**(5), 493–502 (2004)
34. Willerth, S.M., Arendas, K.J., Gottlieb, D.I., Sakiyama-Elbert, S.E.: Optimization of fibrin scaffolds for differentiation of murine embryonic stem cells into neural lineage cells. Biomaterials **27**(36), 5990–6003 (2006)
35. Liu, H., Collins, S.F., Suggs, L.J.: Three-dimensional culture for expansion and differentiation of mouse embryonic stem cells. Biomaterials **27**(36), 6004–6014 (2006)
36. Battista, S., Guarnieri, D., Borselli, C., Zeppetelli, S., Borzacchiello, A., Mayol, L., Gerbasio, D., Keene, D.R., Ambrosio, L., Netti, P.A.: The effect of matrix composition of 3D constructs on embryonic stem cell differentiation. Biomaterials **26**(31), 6194–6207 (2005)
37. Garreta, E., Genove, E., Borros, S., Semino, C.E.: Osteogenic differentiation of mouse embryonic stem cells and mouse embryonic fibroblasts in a three-dimensional self-assembling peptide scaffold. Tissue Eng. **12**(8), 2215–2227 (2006)
38. Karoubi, G., Ormiston, M.L., Stewart, D.J., Courtman, D.W.: Single-cell hydrogel encapsulation for enhanced survival of human marrow stromal cells. Biomaterials **30**(29), 5445–5455 (2009)
39. Dang, S.M., Gerecht-Nir, S., Chen, J., Itskovitz-Eldor, J., Zandstra, P.W.: Controlled, scalable embryonic stem cell differentiation culture. Stem Cells **22**(3), 275–282 (2004)
40. Markusen, J.F., Mason, C., Hull, D.A., Town, M.A., Tabor, A.B., Clements, M., Boshoff, C.H., Dunnill, P.: Behavior of adult human mesenchymal stem cells entrapped in alginate-GRGDY beads. Tissue Eng. **12**(4), 821–830 (2006)
41. Ma, H.L., Hung, S.C., Lin, S.Y., Chen, Y.L., Lo, W.H.: Chondrogenesis of human mesenchymal stem cells encapsulated in alginate beads. J. Biomed. Mater. Res. A **64**(2), 273–281 (2003)
42. Maguire, T., Novik, E., Schloss, R., Yarmush, M.: Alginate-PLL microencapsulation: effect on the differentiation of embryonic stem cells into hepatocytes. Biotechnol. Bioeng. **93**(3), 581–591 (2006)
43. Connelly, J.T., Garcia, A.J., Levenston, M.E.: Inhibition of in vitro chondrogenesis in RGD-modified three-dimensional alginate gels. Biomaterials **28**(6), 1071–1083 (2007)
44. Park, J.S., Woo, D.G., Yang, H.N., Lim, H.J., Park, K.M., Na, K., Park, K.H.: Chondrogenesis of human mesenchymal stem cells encapsulated in a hydrogel construct: neocartilage formation in animal models as both mice and rabbits. J. Biomed. Mater. Res. A **92**(3), 988–996 (2010)
45. Nuttelman, C.R., Tripodi, M.C., Anseth, K.S.: In vitro osteogenic differentiation of human mesenchymal stem cells photoencapsulated in PEG hydrogels. J. Biomed. Mater. Res A **68**(4), 773–782 (2004)
46. Alhadlaq, A., Tang, M., Mao, J.J.: Engineered adipose tissue from human mesenchymal stem cells maintains predefined shape and dimension: implications in soft tissue augmentation and reconstruction. Tissue Eng. **11**(3–4), 556–566 (2005)
47. Hwang, N.S., Varghese, S., Zhang, Z., Elisseeff, J.: Chondrogenic differentiation of human embryonic stem cell-derived cells in arginine-glycine-aspartate-modified hydrogels. Tissue Eng. **12**(9), 2695–2706 (2006)
48. Temenoff, J.S., Park, H., Jabbari, E., Sheffield, T.L., LeBaron, R.G., Ambrose, C.G., Mikos, A.G.: In vitro osteogenic differentiation of marrow stromal cells encapsulated in biodegradable hydrogels. J. Biomed. Mater. Res. A **70**(2), 235–244 (2004)

49. Fok, E.Y., Zandstra, P.W.: Shear-controlled single-step mouse embryonic stem cell expansion and embryoid body-based differentiation. Stem Cells **23**(9), 1333–1342 (2005)
50. Ng, E.S., Davis, R.P., Azzola, L., Stanley, E.G., Elefanty, A.G.: Forced aggregation of defined numbers of human embryonic stem cells into embryoid bodies fosters robust, reproducible hematopoietic differentiation. Blood **106**(5), 1601–1603 (2005)
51. Bauwens, C.L., Peerani, R., Niebruegge, S., Woodhouse, K.A., Kumacheva, E., Husain, M., Zandstra, P.W.: Control of human embryonic stem cell colony and aggregate size heterogeneity influences differentiation trajectories. Stem Cells **26**(9), 2300–2310 (2008)
52. Kramer, J., Hegert, C., Guan, K., Wobus, A.M., Muller, P.K., Rohwedel, J.: Embryonic stem cell-derived chondrogenic differentiation in vitro: activation by BMP-2 and BMP-4. Mech. Dev. **92**(2), 193–205 (2000)
53. Gerecht, S., Brudick, J.A., Ferreira, L., Townsend, S.A., Langer, R., Vunjak-Novakovic, G.: Hyaluronic acid hydrogel for controlled self-renewal and differentiation of human embryonic stem cell. Proc. Natl Acad. Sci. USA **104**(27), 11298–11303 (2007)
54. Siti-Ismail, N., Bishop, A.E., Polak, J.M., Mantalaris, A.: The benefit of human embryonic stem cell encapsulation for prolonged feeder-free maintenance. Biomaterials **29**(29), 3946–3952 (2008)
55. Hwang, N.S., Varghese, S., Theprungsirikul, P., Canver, A., Elisseeff, J.: Enhanced chondrogenic differentiation of murine embryonic stem cells in hydrogels with glucosamine. Biomaterials **27**(36), 6015–6023 (2006)
56. Bauwens, C., Yin, T., Dang, S., Peerani, R., Zandstra, P.W.: Development of a perfusion fed bioreactor for embryonic stem cell-derived cardiomyocyte generation: oxygen-mediated enhancement of cardiomyocyte output. Biotechnol Bioeng. **90**(4), 452–461 (2005)
57. Chayosumrit, M., Tuch, B., Sidhu, K.: Alginate microcapsule for propagation and directed differentiation of hESCs to definitive endoderm. Biomaterials **31**(3), 505–514 (2009)
58. Benoit, D.S., Schwartz, M.P., Durney, A.R., Anseth, K.S.: Small functional groups for controlled differentiation of hydrogel-encapsulated human mesenchymal stem cells. Nat. Mater. **7**(10), 816–823 (2008)
59. Nie, Y., Bergendahl, V., Hei, D.J., Jones, J.M., Palecek, S.P.: Scalable culture and cryopreservation of human embryonic stem cells on microcarriers. Biotechnol. Prog. **25**(1), 20–31 (2009)
60. Fernandes, A.M., Marinho, P.A., Sartore, R.C., Paulsen, B.S., Mariante, R.M., Castilho, L.R., Rehen, S.K.: Successful scale-up of human embryonic stem cell production in a stirred microcarrier culture system. Braz. J. Med. Biol. Res. **42**(6), 515–522 (2009)
61. Lock, L.T., Tzanakakis, E.S.: Expansion and differentiation of human embryonic stem cells to endoderm progeny in a microcarrier stirred-suspension culture. Tissue Eng. Part A **15**(8), 2051–2063 (2009)
62. Fernandes, A.M., Fernandes, T.G., Diogo, M.M., da Silva, C.L., Henrique, D., Cabral, J.M.: Mouse embryonic stem cell expansion in a microcarrier-based stirred culture system. J. Biotechnol. **132**(2), 227–236 (2007)
63. Abranches, E., Bekman, E., Henrique, D., Cabral, J.M.: Expansion of mouse embryonic stem cells on microcarriers. Biotechnol. Bioeng. **96**(6), 1211–1221 (2007)
64. Frauenschuh, S., Reichmann, E., Ibold, Y., Goetz, P.M., Sittinger, M., Ringe, J.: A microcarrier-based cultivation system for expansion of primary mesenchymal stem cells. Biotechnol. Prog. **23**(1), 187–193 (2007)
65. Wang, C., Gong, Y., Zhong, Y., Yao, Y., Su, K., Wang, D.A.: The control of anchorage-dependent cell behavior within a hydrogel/microcarrier system in an osteogenic model. Biomaterials **30**(12), 2259–2269 (2009)
66. Bratt-Leal, A.M., Carpenedo, R.L., McDevitt, T.C.: Engineering the embryoid body microenvironment to direct embryonic stem cell differentiation. Biotechnol. Prog. **25**(1), 43–51 (2009)
67. Alberti, K., Davey, R.E., Onishi, K., George, S., Salchert, K., Seib, F.P., Bornhauser, M., Pompe, T., Nagy, A., Werner, C., Zandstra, P.W.: Functional immobilization of signaling proteins enables control of stem cell fate. Nat. Methods **5**(7), 645–650 (2008)

68. Beckstead, B.L., Santosa, D.M., Giachelli, C.M.: Mimicking cell–cell interactions at the biomaterial–cell interface for control of stem cell differentiation. J. Biomed. Mater. Res. A **79**(1), 94–103 (2006)
69. Jaklenec, A., Wan, E., Murray, M.E., Mathiowitz, E.: Novel scaffolds fabricated from protein-loaded microspheres for tissue engineering. Biomaterials **29**(2), 185–192 (2008)
70. Patel, Z.S., Yamamoto, M., Ueda, H., Tabata, Y., Mikos, A.G.: Biodegradable gelatin microparticles as delivery systems for the controlled release of bone morphogenetic protein-2. Acta Biomater. **4**(5), 1126–1138 (2008)
71. Mahoney, M.J., Saltzman, W.M.: Transplantation of brain cells assembled around a programmable synthetic microenvironment. Nat. Biotechnol. **19**(10), 934–939 (2001)
72. Carpenedo, R.L., Bratt-Leal, A.M., Marklein, R.A., Seaman, S.A., Bowen, N.J., McDonald, J.F., McDevitt, T.C.: Homogeneous and organized differentiation within embryoid bodies induced by microsphere-mediated delivery of small molecules. Biomaterials **30**(13), 2507–2515 (2009)
73. Ferreira, L., Squier, T., Park, H., Choe, H., Kohane, D.S., Langer, R.: Human embryoid bodies containing nano- and microparticulate delivery vehicles. Adv. Mater. **20**(12), 2285–2291 (2008)

Stem Cell Interaction with Topography

Benjamin K. K. Teo, Soneela Ankam and Evelyn K. F. Yim

Abstract The growth and differentiation of stem cells are regulated by biochemical and biophysical cues in the extracellular microenvironment. Increasing evidences have shown that substrate topography, one of the biophysical properties of the microenvironment, can affect stem cell fate, such as the maintenance of embryonic stem cells and the differentiation of adult and embryonic stem cells. The underlying mechanism of how topography influences stem cells remains unknown. Nevertheless, the advancement in technology has enabled the fabrication of synthetic topography with different materials, chemistries, geometries and sizes, allowing systematic studies of the underlying mechanism. Recent studies show that the topography-induced stem cells response can be a result of mechanotransduction via cellular components such as intergrins, focal adhesion and cytoskeleton organization.

1 Introduction

Stem cells are defined as cells with the ability to self-renew and differentiate into specialized cells in response to appropriate signals [1]. Because of its ability to differentiate into different types of functional cells, stem cells possess great value as therapeutics to regenerate and repair damaged tissue. Both the controlled self renewal and directed differentiation are keys to the application of stem cells in

B. K. K. Teo, S. Ankam and E. K. F. Yim (✉)
Division of Bioengineering, National University of Singapore,
9 Engineering Drive 1, Singapore 117576, Singapore
e-mail: eyim@nus.edu.sg

E. K. F. Yim
Department of Surgery, Yong Loo Lin School of Medicine,
National University of Singapore, 10 Medical Drive, Singapore 117597, Singapore

Stud Mechanobiol Tissue Eng Biomater (2010) 2: 61–87
DOI: 10.1007/8415_2010_6

Published online: 25 June 2010

regenerative medicine, which aims to provide a therapeutic platform by creating or controlling the extracellular microenvironment in order to guide tissue growth for functional recovery and/or dictate stem cells differentiation into the appropriate cell types. In their in vivo environment, stem cell fate is controlled by intrinsic factors and microenvironment known as the stem cell niche [2, 3]. The micro-environment is composed of extracellular matrix (ECM), which is a reservoir of biochemical as well as biophysical cues, of which the latter can be presented in the form of nanotopography and matrix stiffness. In the last few years, attention has been drawn to the study of biophysical cues as increasing supporting evidence has shown that cells are regulated not only by chemical signals but also by biophysical properties from the extracellular environment. This chapter will discuss one of the biophysical cues—the topography of the extracellular environment, the micro- and nanofabrication methods to engineer the topography, how stem cells interact with the substrate topography and the current theories and speculations of how topography can direct stem cell fate.

1.1 Extracellular Topography

During natural tissue development, cells are interacting with various nano-scaled topographical and biochemical cues in their microenvironment [4]. Weiss and Garber [5] described the effect of contact guidance, the role of ECM structure in cell orientation and migration, more than 50 years ago. Curtis also proposed the influence of microtopography on cell behavior in 1964 [6], which has also been extensively studied ever since. Recent findings underscore the phenomenon that mammalian cells respond to nanoscale features on synthetic surfaces, and the topic has been described in several recent reviews [7–10].

When stem cells are removed from the in vivo stem cell niche, they can differentiate spontaneously in vitro but this differentiation process is inefficient, uncontrolled and often results in highly heterogeneous cell population [1]. A crucial strategy of regenerative medicine is to understand how to control the microenvironment surrounding the cells to restore the niche equilibrium. A typical strategy is to enrich the biochemical environment in the in vitro culture medium with a combination of soluble growth factors, cytokines and/or serum protein, to induce the stem cells to differentiate preferentially into a particular lineage. However, because the signaling pathways required for stem cells to differentiate into functional adult cell types remains unclear, the culture time, concentration, and the combination of various biochemical cues to improve the differentiation efficiency have yet to be optimized. In addition to enhancing the biochemical signal, gaining a fundamental understanding of the cell–substratum interaction will be required to understand and reconstruct the stem cell niche. While the importance of topographical cues may vary for different cell lineages, its relevance is unquestionable.

1.2 Nanotopography

Stem cells vary in sizes and shapes in the micron range, and different types of stem cells require different biochemical and biophysical cues for differentiation or maintenance; however, it is not surprising to see that various type of stem cells do respond to topography at the nanoscale. Nanotechnology is defined as the manipulation of atoms and molecules in nanometer scale range of <100 nm; nevertheless, as the focus of this chapter is not on the quantum effect of the materials but the cell interaction with substratum features, stem cell's responses to sub-cellular scale topography of >1,000 nm will also be discussed in this chapter to give a more holistic description of the field in discussion.

The fact that the in vivo extracellular matrix or substratum with which cells interact often includes topography at the nanoscale substantiates the importance to investigate cell–substrate interactions at the sub-micron scale. The basement membrane, which is a ubiquitous component of ECM that plays an important role in tissue development and organization, is an example. It possesses a complex mixture of pores, ridges, and fibers, which have sizes in the nanometer range. The ~200 nm thick layer separates tissues such as epithelia, endothelia, muscle fibers, and the nervous system from connective tissue compartments. These nanometer range structures are also seen in the corneal epithelial basement membrane of the Macaque monkey as shown in Fig. 1a [4]. It consists of a porous membrane with a network of cross-linked fibers, with the pores averaging 72 nm and the fibers 77 nm in diameter

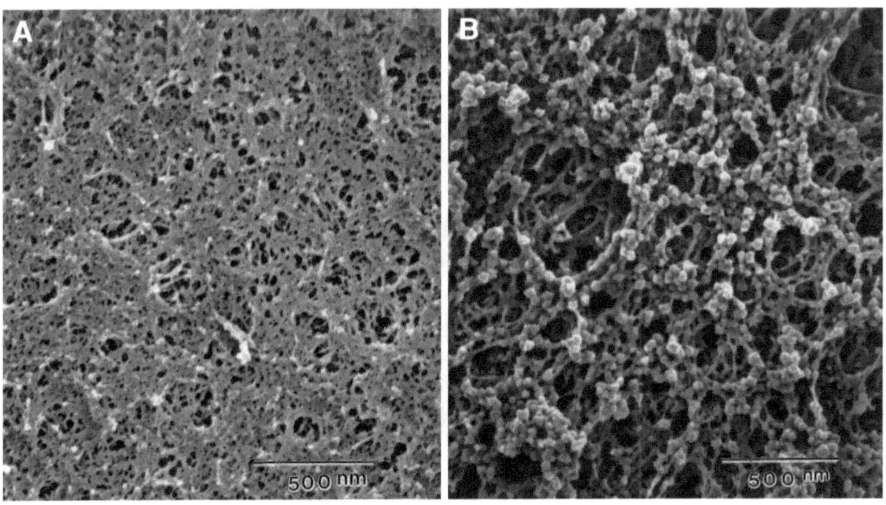

Fig. 1 **a** Scanning electron micrograph of the corneal epithelial basement membrane after incubation in 2.5 mM EDTA and epithelial removal. This shows the overall "felt-like" architecture of the surface, consisting of interwoven fibers and pores. The complex topography of intertwined fibers and pores of varying sizes is evident. **b** Scanning electron micrograph of Matrigel. The nanoscale topography is composed of a meshwork of fibers and pores similar to that found for the "native" anterior corneal basement membrane [4]

respectively. The mean elevation of these features is around 160 nm. The trabecular meshwork of the human cornea is another example [11]. It consists of ECM organized into a network of beams, covered by trabecular endothelial cells.

Similar nanoscale features can be seen in an artificial basement membrane, the commercially available Matrigel^{TM} (BD Biosciences, Fig. 1b) [4], which has been widely used in feeder free expansion of embryonic stem cells (ESCs). Matrigel is a solubilized basement membrane extracted from the Engelbreth–Holm–Swarm (EHS) mouse sarcoma, a tumor rich in ECM proteins. The Matrigel revealed a topography of fibers and pores at the nanoscale when viewed using the SEM, although it still remains debatable whether the protein hydrogel in its swollen state does provide a solid structure to the cells. Nonetheless, these nanofeatures may play an important role in maintaining the pluripotency of ESCs. Another example of nanotopographical influence on stem cell show that nanotopography alone can induce the differentiation of human mesenchymal stem cells into neuronal lineage. The nano-topography induced a more significant upregulation of neuronal markers in hMSCs compared to micro-topography, highlighting the importance of feature size in topography induced differentiation [12].

2 Nanofabrication Techniques

The advancements in micro- and nanofabrication technology enable the studies of cellular response of micro- and nanofeatures with a wide range of chemistry and topography. Most of these technologies were developed initially for the microelectronic industry, but they have since been adapted for cellular studies. Detailed description of the micro- or nanofabrication technology has been covered in several excellent reviews [9, 10, 13–17]. Table 1 summarizes the nanofabrication techniques that are commonly used in studies of cellular response and stem cell niche reconstruction. In general, the techniques can be divided into two categories of fabrication, mainly that of ordered features and random features.

Many fabrication techniques for ordered features are dependent on optical lithography and reactive ion etching. This can be followed by anisotropic etching or UV and glow discharge treatment [15]. Photolithography is limited by diffraction limitations. Its diffraction limitations and thus its resolution is typically that of the wavelength of the light used for the exposure (typically >200 nm). Electron-beam lithography can be used to produce nanoscale patterns, but it is expensive and time-consuming. Some other methods such as laser ablation, X-ray lithography and dip-pen nano-lithography can also be used. Nanoimprint lithography (NIL) is a mechanical lithography process, in which a polymer resist will be embossed with a patterned hard mold, such as a pre-patterned SiO_2 mold. In conventional thermal NIL, a substrate will be spin-coated with a polymer layer before embossing with a hard mold. It is low-cost and can pattern features in large areas with a lateral resolution of <10 nm [19]. In the reversal imprinting, a polymer layer is spin-coated onto the mold only, and then transferred to a bare substrate by imprinting under suitable temperature and pressure [20, 43]. In addition to thermal NIL, UV

Table 1 Fabrication techniques

Fabrication method	Resolution	Features	Advantages	Limitations
(a) Nanofabrication techniques for ordered features				
Electron beam lithography	>3 nm [18]	Pillars, wells Grooves	Precise geometry and pattern No mask needed	Expensive Time consuming Small surface coverage Even lower resolution with negative resist
Nanoimprint lithography	>3 nm (limited by the template) [19]	Pillars, well, grooves 3D [20] Hierarchical structures [21]	Inexpensive Fast Easy to scale up Ability to make complex structure Versatility in choice of material	Expensive equipment Template needed
Soft lithography	>100 nm (commercial PDMS) >50 nm (h-PDMS) [22, 23]	Pillars, well (2D pattern)	Inexpensive Fast Simple	Limitation on material Template/mold required
Dip-pen nanolithography	>100 nm [24]	Line, dots, 2D pattern	Precise control Maskless Pattern and image acquisition simultaneously	Slow Height limitation unless used with particle deposition and/or etching
Direct-write fabrication with multi-photon	>250 nm [25]	3D complex architecture	In situ biocompatible fabrication	Special equipment Limitation in choice of materials
Photolithography	≥250 nm [26]	Pillars, wells, grooves	Precise geometry and pattern Large surface coverage	Expensive equipment Feature size limitation Mask required

(continued)

Table 1 (continued)

Fabrication method	Resolution	Features	Advantages	Limitations
(b) Nanofabrication techniques of random features				
Nanoporous membranes	≥1 nm [27, 28]	Porous membrane	Easy fabrication Inexpensive Precise control over pore size	Insufficient strength for physiological loads
Carbon nanotubes	SWNT: 0.4–2 nm in diameter [29] MWNT: >2 nm in diameter [29]	Woven mesh or vertical arrays Patterns when combined with lithographic methods	Strong, flexible and conductive	Potential toxicity of the carbon nanotubes
Etching: chemical/ reactive ion etching	≥2 nm	Craters- chemical [30] Needles- reactive ion [31] Pores- electrochemical etching [32]	Fast Economical No special equipment required (for chemical etching)	No control in geometry
Electrospining nanofiber	>3 nm [33, 34]	Cylindrical fibers: aligned/ random	Versatility in material choice Composite with nature materials Biological polymer Co-axial fiber enable drug delivery	Only fiber Difficult to control pore size for nanofiber mesh
Self assembly nanofiber	> 10 nm fibers [35, 36]	Hydrated 3D mesh of interwoven nanofibers	Controlled fiber dimension and morphology Self assembly	Required engineering of molecules for self-assembly
Polymer demixing	Vertical ≥13 nm [37–39]	Islands, pits, ribbons	Simple Fast Inexpensive	Limitation in geometry and choice of materials
Colloidal lithography	≥20 nm [40]	Columns	Easy to pattern Large surface coverage area	Limitation in geometry
Carbon nanofibers	>40 nm in diameter [41]	Fiber	Conductive fiber with tunable mechanical and cell adhesion property	Fiber only

(continued)

Table 1 (continued)

Fabrication method	Resolution	Features	Advantages	Limitations
Phase separation	>50 nm fiber [42]	Porous and fibrous network	Simple No special equipment Porosity easily controlled	No organized pattern
Direct-write fabrication with multi-photon	>250 nm [25]	3D complex architecture	In situ biocompatible fabrication	Special equipment Limitation in choice of materials

NIL allows the imprinting of UV-curable materials at room temperature. With the different modes of imprints, NIL provides a wide choice of polymers that can be used for nanofabrication. This can prove to be particularly useful for cellular engineering and the reconstruction of stem cell niches.

Using a template generated by optical or mechanical lithography, the pattern can be replicated rapidly and easily onto elastomers such as polydimethylsiloxane (PDMS) with soft lithography. However, there are limited types of siloxane polymers that can be used in soft lithography.

Although most optical or mechanical lithographies has the capability to generate highly ordered features with high precision, the fabrication process can be expensive and complex. Fabrication techniques of random structures, on the other hand, are relatively simple and more economical. For example, nanoislands of 13–95 nm in height can be fabricated by polymer demixing based on phase separation of polystyrene and poly(4-bromostyrene) spin-coated on silicon wafers [37]. The ability to produce different nanofeatures based on such phase separation phenomenon, however, would be limited.

Another versatile technique of producing nanostructures for cell culture applications is electrospinning [33, 34]. In the electrospinning process, an electrode is placed into a syringe or pipette containing the polymer solution. A high voltage applied discharges an electrically charged jet of polymer solution onto a grounded collector surface. Randomly oriented fibers will be collected, forming a nonwoven mat, while aligned fibers can be collected using a rotating wheel, by matching the rate of rotation with the rate of fiber deposition. Many different types of polymers can be processed into fibers with diameters ranging from several nanometers to tens of microns. Although electrospinning requires relatively simple instruments, many process parameters are needed to be optimized to produce the fiber characteristics of desired size, orientation, and uniformity without bead formation. The processing parameters include viscosity, evaporation rate, surface tension and conductivity of the polymer solution, and processing variables of applied electric voltage and source-to-collector distance.

This section described a generic overview of the nanofabrication techniques that have been used in studying cellular response. Nanotopography can be fabricated using various methods to mimic the stem cell niche with features of different size and geometry. With the constant rapid development of fabrication technologies, more choices are available in selecting substrate materials with different biochemical and biophysical properties, such as rigidity, which allows researchers to optimize the cell–substrate interaction.

3 Stem Cells Reception to Topography

Stem cells can be broadly classified into embryonic stem cells (ESCs) and adult stem cells. ESCs are pluripotent cells derived from the inner cell mass of blastocysts with the potential to maintain an undifferentiated state [44]. The ESCs are hypothetically capable of regenerating all the cell types of the three germ layers—ectoderm, mesoderm and endoderm. Adult stem cells on the other hand, are derived from adult tissues and are usually multipotent, where the lineage is more restricted than ESCs. Recent technologies also allow the re-programming of adult cells into pluripotent stem cells, which are referred to as induced pluripotent stem cells. These induced pluripotent stem cells exhibit properties similar to embryonic stem cells [45]. Stem cells being capable of self-renewal and differentiation have a huge potential for biotechnology and regenerative medicine. In this section, we will restrict our focus on the role of biophysical cues, in particular the role of topography in the differentiation of embryonic stem cells and adult stem cells like neural stem cells/neural progenitor cells (NSCs/NPCs) and mesenchymal stem cells (MSCs).

3.1 Embryonic Stem Cells

ESCs remain as potential candidates for regenerative medicine due to their pluripotent properties. Since undifferentiated ESC generate teratocarcinoma, a germ cell tumor, when implanted in vivo, a directed differentiation of ESCs is required for medical applications. The major challenges of using ESCs for therapeutic applications includes overcoming ESC shortage, obtaining clinical grade cell lines and deriving a pure differentiated population in an efficient and ethical manner.

The propagation of undifferentiated ESCs will be essential to overcome the shortage of ESC supply. Conventionally, ESCs require the support of a feeder layer for their long term propagation in vitro. Many groups have tried different kinds of feeder layers ranging from mouse to human and also transitioned to feeder free cultures like the Matrigel. Nevertheless, the Matrigel is derived from animal sources and may have potential problems in clinical applications. Furthermore, the Matrigel possesses a mixture of topographical cues in addition to a reservoir of biochemical cues. Thus the most opted condition would be a defined feeder free system consisting of appropriate biochemical and topographical cues.

Markert et al. [46] designed topographically distinct micro-structured surface libraries called the BioSurface Structure Arrays. Using murine ESCs, they screened approximately 504 distinct topographical structures with various

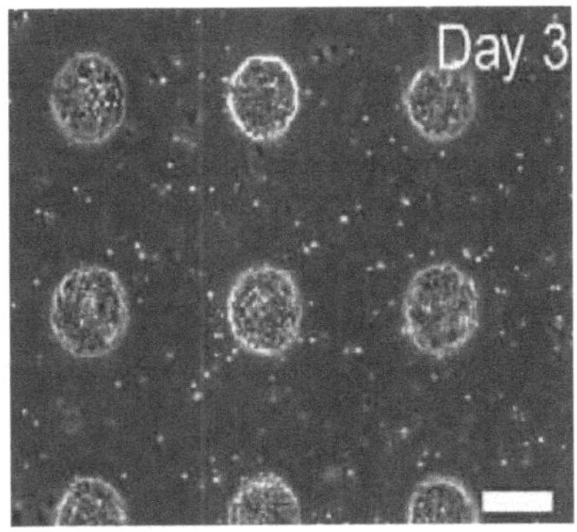

Fig. 2 Micropatterning can potentially serve as a simple and reproducible fabrication method for mass production of size controlled ESCs. Shown here on the *right* is a microscopic image of ESCs cultured on circular micro-patterned substrate, 200 μm in diameter, on day 3. *Scale bars* represent 200 μm [49]

combinations of height, size and gap, etched on a 3 × 3 mm^2 Silicon wafer. They were able to assort the topographies favouring the maintenance of undifferentiated murine ESCs and those promoting differentiation. This study is a breakthrough in realizing the potential of topography to support feeder free ESC culture.

In addition to the ability to propagate sufficient pluripotent ES cells, it is also crucial to control the differentiation of ESCs. In order to achieve this, many approaches have been employed, namely the use of chemically defined media, conditioned media or coating with various types of ECM components. Essentially, researchers have tried to maneuver the in vivo microenvironment to study stem cell differentiation.

To date, most of the researchers have experimented with the chemical factors that aid in directing differentiation of embryonic stem cells into a specific and desired lineage. Murray et al. proposed that mouse ES cell differentiation is regulated by changes in cell–cell interactions observed when the ES cells are cultured as embryoid bodies. On the other hand, in a more recent study by Gerecht et al. [47], it was demonstrated that the nanotopography of the substrate could control human ESC morphology, i.e., elongation and alignment of human ESCs including proliferation via in vitro contact guidance response to the nanoscale grating features. They also showed that nanotopography in the presence of soluble growth factors is able to show more significant effects as compared to substrate signalling alone. These interesting findings have led to the change in the focus of researchers to the eclectic field of micro and nano patterns of various geometries and sizes to design a suitable scaffold and also to observe the response of stem cells to these patterned features [48].

A maskless photolithography system was used by Sasaki et al. [49] to develop adhesive circular micro domains (100–400 μm in diameter) on glass cover slips; 200 μm circular domains were found to be the optimal diameter for the cardiac differentiation of uniform sized murine ESC aggregates (Fig. 2). This is a relatively simple technique to obtain a consistent population of ESC derived cardiac cells for high throughput assays. Smith et al. [50] observed that murine ESCs interacted better with the nanofibrous matrices compared to the films. A marked upregulation of osteogenic markers was also observed in the nanofibers. These two studies highlighted that in addition to chemical factors, the cell–matrix interaction, colony size and shape are also important factors in regulating differentiation. Their work testifies the importance of nanosized topographies in stem cell differentiation.

Here in this section, we put forth selected studies demonstrating the significance of topography in the maintenance of ESCs as well as directing differentiation. These works clearly reiterate the importance of understanding cell–matrix interaction to recreate the stem cell niche using appropriate growth factors and optimal topographies.

3.2 Neural Progenitor Cells/Neural Stem Cells

Neural stem cells are multipotent stem cells and are present in both developing and adult central nervous system. The NSCs are capable of self renewal and differentiation into astrocytes, oligodendrocyte and neurons while the NPCs are lineage restricted and only capable of fewer divisions. The three major types of progenitors found in the adult human brain are ventricular zone neural progenitors, hippocampal neuronal progenitor cells and white matter glial progenitors. Isolation, expansion and in vitro differentiation of these stem cells and progenitors have been studied extensively in the last two decades [51–58]. NSCs and NPCs possess the potential for differentiation into transplantable neurons and so their identification and isolation has given hope for treating debilitating neurodegenerative disorders. Our focus in this section would be on the role of topography in directing differentiation of NSCs and NPCs.

The effect of topography as contact guidance has been studied extensively with a focus on axonal guidance and neuronal regeneration. It was observed that in addition to serving as contact guidance, topography can also affect the differentiation of NSCs. Recently, Christopherson et al. [59] analyzed the oligodendrocyte and neural differentiation of rat hippocampus derived neural stem cells (rNSC) seeded on nanofibers of various sizes. It was observed that in the presence of 1 μM retinoic acid and 1% fetal bovine serum, rNSCs showed a 40% increase in oligodendrocyte differentiation on 283-nm fibers and 20% increase in neuronal differentiation on 749-nm fibers, in comparison to tissue culture polystyrene surface. The smaller fiber diameter also increased proliferation and cell spreading. This study shows the significance of fiber topography and fiber dimension in regulating rNSC differentiation as well as proliferation.

Topography often works synergistically with the appropriate biochemical cues. Recknor et al. [60] studied the effect of topography on neural differentiation of adult rat hippocampal progenitor cells (rNPCs) on micropatterned polystyrene substrates, chemically modified with laminin. When the rNPCs were co-cultured along with the astrocytes on the micropatterns, the percentage of neuronal differentiation and the neurite outgrowth showed a clear increase, suggesting a synergistic effect of nanotopography and the biochemical factors secreted by the astrocytes. The presence of cell–cell contacts can also possibly influence the enhanced differentiation.

In another study performed by Silva et al. [61] NPCs were encapsulated in vitro within a three-dimensional network of nanofibers formed by self-assembly of peptide amphiphile molecules. The self-assembly of the nanofibers was triggered by mixing cell suspensions in media with dilute aqueous solutions of the molecules. The cells not only survived the growth of the nanofibers around them but also rapidly differentiated into neurons; the self-assembled fiber also suppressed the development of astrocytes. This rapid selective differentiation has been attributed to the amplification of bioactive neurite-promoting laminin epitope IKVAV, a biochemical cue presented to the cells by the nanofibers. In a slightly different study, Soen et al. [62] screened the ECM components, morphogens, and other signaling proteins that had a significant effect on the differentiation of neural stem cells. These results taken together, provide a potential way to develop scaffolds with the appropriate combinations of chemical and topographical cues for an efficient differentiation of neural stem cells.

3.3 Mesenchymal Stem Cells

Mesenchymal stem cells (MSCs) are stem cells of mesodermal origin which can be derived from the adult bone marrow. MSCs are multipotent in nature, capable of differentiating into osteoblasts, myoblasts, adipocytes, chondrocyte and skin [63–66].

The most common strategy to differentiate MSCs is to use a combination of growth factors or biochemical cues. Nevertheless, various studies have shown nanotopography alone can induce differentiation of MSC or enhance the MSC differentiation [12, 67–69].

In one of the study, thermosensitive hydroxybutyl chitosan were electrospun into aligned nanofibers, with an average diameter of 437 nm [70], and myogenic induction of human MSCs was observed in response to topographical features without differentiating medium. The maintenance of the cytoskeletal structure and cell–cell contacts in the cell-sheet seems to have a direct relationship in inducing the gene expression of a myogenic lineage. This thermosensitive polymer scaffold could be coupled with the drug encapsulation capacity of electrospinning to serve as a multi-functional platform to encapsulate growth factors apart from providing nanotopographical cues. Thermosensitive polymers have a lower critical solution temperature (LCST) below which they are hydrophilic and water soluble. When the temperature is increased above the LCST, they become hydrophobic and water insoluble. This technique is beneficial in enabling us to differentiate MSCs as well as harvest the differentiated cells as a cell sheet without disrupting the cell–ECM interaction.

Extensive studies have been done on MSCs to understand its response to various kinds of topography due to the ease in obtaining patient specific multipotent MSC. In a comparative study of different patterns by Martino et al., uniform grids and grooved nanopatterns on hydrogenated amorphous carbon film were compared

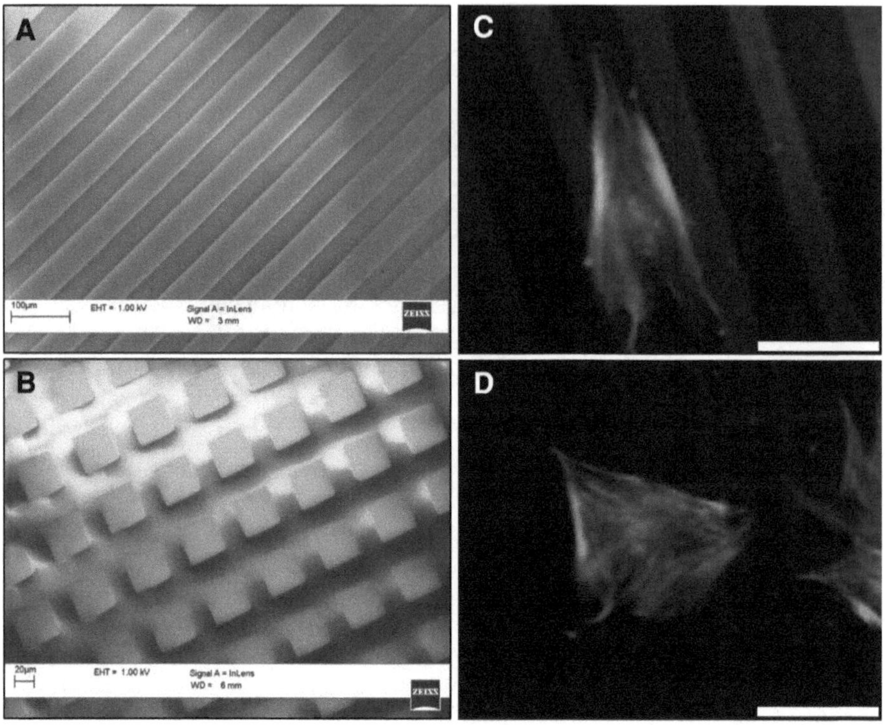

Fig. 3 **a**, **b** The field emission scanning electron microscopy (FESEM) analysis of a hydrogenated amorphous carbon film with groove and grid configurations. **c**, **d** Human bone marrow MSCs grown on the corresponding grids and grooves. Note the differences in the cytoskeletal structure of MSCs on the different patterns. They are immunofluorescently labelled with α-tubulin and F-actin, counter stained for nucleic acid with DAPI [71]

for the differentiation of human bone marrow derived MSCs (huBM–MSCs) into osteogenic and adipogenic lineage (Fig. 3). The grooved nanopatterns exerted a greater effect on differentiation of huBM–MSC into osteogenic lineage due to the alignment and elongation cues presented by the grooves [71].

In yet another study, microstructures imprinted on PMMA in the range of 50–100 μm confirmed the previous results by other researchers that topography indeed plays an important role in directing the differentiation of MSCs into either the osteogenic or adipogenic lineage [10, 72]. Although the exact mechanism remains elusive, this fine tuning is possibly related to the change in cell shape with different topographies, for it was observed that round patterns aided adipogenic lineage differentiation while gratings induced elongation of MSCs in turn promoting osteogenic transition [69].

It was also demonstrated that the heights of topographical features could affect osteogenic differentiation of MSCs. Titanium nanopillar structures with heights of 15, 55 or 100 nm on Ti surfaces were synthesized by anodization through a porous alumina mask. The 15 nm high topography features resulted in the most significant bone matrix nodule formation after 21 days of culture in comparison to the pillars of sizes 55 and 100 nm [73]. In this study, the authors demonstrated that the stem cell behaviour is a function of pillar heights and not of the diameter or gaps. This might serve as another important guiding parameter to consider while designing a scaffold for MSC differentiation. The authors also postulated that nanoscale topography works by presenting adhesive proteins to cells in a preferential/ unpreferential manner, thus controlling the cell response. However, this theory appears less likely as cell alignment can also be observed on round features, with the curvature of the features suggested to be a function of cell alignment [7]. After having understood the significance of pattern geometry and dimensions, in an intriguing research done by Dalby et al. [67], it was shown that random topographies were better than ordered ones in inducing the mesenchymal stem cells to differentiate towards an osteospecific lineage. The same group had also tried to fabricate a controlled disorder nanopit topography by electron beam lithography [74]. It was termed NSQ50 with pits in a square arrangement, 300 nm centre–centre spacing and was used as a scaffold to direct osteoblast differentiation of human bone marrow osteoprogenitor cells. Their results showed that stem cells were able to detect regularity of the nanostructures, and that the symmetry of structures had significant influence on stem cell differentiation.

In addition to the mesenchymal lineage, MSCs have also been shown to acquire atypical neuronal-like phenotypes. In a study conducted by Yim et al. [12] the proliferation, alignment, elongation and transdifferentiation of human mesenchymal stem cell on the nanograting axis was studied. Gratings with line widths of 350 nm, 1 and 10 μm on PDMS were used to study the transdifferentiation of human MSCs into neuronal lineage, with or without retinoic acid induction. They have successfully shown that the effect of nanotopography on the upregulation of neuronal marker is higher compared to the RA induction alone.

The transdifferentiation of bone marrow derived MSCs was also studied on electrospun poly(L-lactic acid)-co-poly-(3-caprolactone)/Collagen (PLCL/Coll) nanofibrous scaffolds, with fiber diameters in the range of 230 ± 31 nm, using neural induction factors like BDNF, EGF, and β-mercaptoethanol [75]. Their results reiterates the need for both topographical cues and chemical cues in order to upregulate the neuronal markers. The mechanism behind transdifferentiation in response to nanotopographical cues is yet to be unraveled. Nonetheless, these studies contribute to materializing the enormous potential for the generation of functional neurons from patient derived mesenchymal stem cells.

In the above section, we have tried to highlight the prominent experiments carried out with topographies of various patterns, height, diameter, gap, materials and order (random and ordered), in combination with biochemical cues to study the differentiation of MSC into osteogenic, adipogenic or neuronal lineage.

We have reviewed selected studies illustrating the effect of various types of topographies on stem cell differentiation and proliferation and are summarized in Table 2. In vivo key topographical features that regulate cellular functions are likely to be tissue specific and possibly required to be investigated on a case by case basis; thus it will be difficult to establish a clear relationship of a cellular

Table 2 Effects of topography on stem cell differentiation

Stem cell	Differentiation	Pattern used	Source
Embryonic stem cells	Differentiated state and undifferentiated state	Biosurface structured arrays [46]	Mouse
	Cardiac cells	Circular microdomains [49]	Mouse
	Osteogenic lineage	Nanofibrous matrices [50]	Mouse
Mesenchymal stem cells	Neuronal lineage	Gratings [12]	Human
	Neuronal lineage	Nanofibrous scaffolds [75]	Human
	Myogenic lineage	Electrospun thermosensitive scaffold [70]	Human
	Osteogenic and adipogenic	Grooves and grids [71]	Human
	Osteogenic and adipogenic	Heat embossed microstructures [10, 72]	Human
	Osteogenic lineage	Random topographies [67, 76]	Human
	Bone regeneration	Titania nanopillar structures [73]	Human
	Osteoblast differentiation	Controlled disorder nanopit [74]	Human
Neural progenitor cells	Neurite outgrowth	Micropatterns [60]	Adult Rat
	Oligodendrocyte and neuronal differentiation	Nanofibers [59]	Rat
	Neuronal differentiation	Nanofibers [61]	Murine

function associated with a topography parameter. However, there is increasing evidence that we can potentially exploit both the topography that is capable of mimicking the in vivo microenvironment and stem cells that are capable of self renewal and multilineage differentiation. Even though researchers have contributed substantially to help us better understand stem cell–matrix interaction, the present scenario necessitates several horizons to be explored. The following section will briefly discuss the current theories and speculation on how topography can direct stem cell fate.

4 Making Sense of Physical Cues in the Extracellular Matrix: Mechanotransduction

4.1 Introduction to the ECM

During development, stem cells in their niche are exposed to diverse yet intricately regulated regular array of structures as previously discussed. The ECM components include proteoglycans that contain large glycosaminoglycan side chains, glycoproteins (e.g. fibronectin, elastin) and collagen. These protein fibers are known to be in the scale of 10–300 nm in diameters [77] and form a meshwork that provides tensile strength to the ECM. Cells in their microenvironment interact closely with the ECM through various adhesions, establishing physical interactions that are often vital for survival. They actively detect and respond to the ECM nanoscale features in their surroundings. One of such important features includes the topography of these scaffolds.

The structural cues from topography changes the orientation of the cells, reorganizes its body, shape and functional state [78–80], which includes stem cell differentiation as shown in the previous sections. Studies employing simplistic 2D topography models to mimic native ECM demonstrated that mechanical cues do play an important role in regulating stem cell differentiation. Intriguingly, the cells must have an intrinsic fundamental mechanism to sense the underlying substrate topography, generating mechanical signals that are translated by intracellular signaling pathways before ultimately regulating the genomic expression and cell function [81]. This is also known as mechanotransduction. It is important however to note that the mechanotransduction involves a complex interplay of different cellular organelles and components (e.g. focal adhesion and cytoskeleton) that by themselves are highly dynamic in vivo, making the process more convoluted [82]. The precise mechanism by which how cells sense and respond to topographical cues remains a mystery, possibly due to both inadequate studies and the complexity of the problem. However, advancement of experimental techniques has helped to provide clues to the mechanism and it appears that cellular components such as the integrins, focal adhesions and cytoskeleton organization collectively play important roles in topography-induced cellular behavior [82–84].

4.2 Mechanotransduction: A Direct Connection?

The extracellular topographical signals have to be transduced from the ECM to the nucleus, where the genome can be regulated as a response to the extracellular signals. Although various other indirect and chemical mechanisms have also been postulated (e.g. G-proteins, ion channels), we have limited our discussion to a direct transmission of forces (via cellular organelles) to the nucleus as a mechanism for gene regulation [85]. Under the effect of nanotopography, the physical substrates do exert differential mechanical stress onto the cells, as indicated by the cytoskeletal reorganization. In the previous section, we have presented an overview of the use of patterned substrates as a form of physical cue to regulate stem cell differentiation into various lineages. In this section, we will discuss about the important components of the mechanotransduction machinery, namely the integrins, focal adhesions and cytoskeleton that translates these topographical cues into secondary pathways, ultimately affecting stem cell differentiation.

4.3 Connecting with the ECM: Cell–Matrix Interactions

Before we try to understand the mechanism of mechanotransduction, it is important to identify the key components involved in the relaying of external topographical cues. Anchorage dependent cells require a favorable substratum for their survival and proliferation and cell shape has been established as a key regulator in the cell physiology [86]. In the studies presented earlier, cells that were cultured on the topographical surfaces adopt different cell morphology as

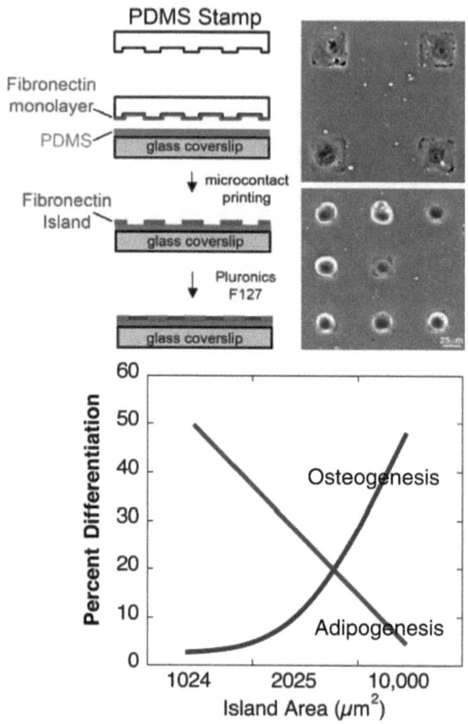

Fig. 4 Cell shape as a regulator of cell fate. An example for the fabrication of ECM (such as fibronectin) islands is shown. The cell shape control can be determined by controlling the size and shape of the islands. *Top right* human MSCs shows different morphology on different ECM patterns. Human MSCs that were allowed to spread on a larger island area preferentially undergo osteogenesis while cells that were restricted to a smaller adhesive island undergoes adipogenesis. Figure is slightly modified from source for B & W print [94]

compared to non-patterned surfaces [8, 10, 12, 72]. In one of our studies, MSCs that were cultured on 350 nm nanogratings pattern demonstrated an elongated morphology as compared to a more spread morphology on the unpatterned surfaces [12]. Importantly, the cell nucleus was also observed to elongate in response to the topography. The change in cell shape thus serves as an early indicator that the cell responds to the underlying nanotopographical cues.

The evident change in cell shape has led to the postulation that it might act as a regulator of stem cell fate. Early studies by Ingber et al. [87] showed that ECM induced changes in the cell shape plays a role in regulating the growth and differentiation of capillary endothelial cells. It is tempting to extend such a mechanism to stem cell differentiation, since the stem cell shape can be physically controlled using patterned substrates as a result of altered adhesions between the cells and the artificial ECM. One such study involved the use of defined "micropatterns" of proteins that was precisely deposited upon a substrate to control the cell attachment area. MSCs exhibited a different morphology on different patterns. While MSCs on a small island exhibited a rounded morphology, cells on larger islands was observed to be flattened [88]. Interestingly the rounded MSCs committed to an adipogenic lineage while the flattened MSCs gave a more osteoblastic phenotype [69], see also Fig. 4. The authors also observed significant changes in the actin cytoskeleton organization and the focal adhesions assemblies, as observed by other groups when inducing the differentiation of similar cell type into other lineages [89, 90].

A large number of studies have investigated the effect of cell shape on differentiation [91–93]. Cell shape is thus one of the earliest observable nanotopographical-induced changes. The changes in cell shape, however, may not be purely topographically induced as interactions with the ECM can be complex [93–96]. Nanotopography induced changes in cell shape can also be a basis for secondary effects from which the genomic expression of the cell is regulated.

4.4 Integrins and Focal Adhesions: Inside Out and Outside In

Anchorage-dependent cells are able to anchor onto the underlying substrate. Various different types of adhesions exist between cells and the ECM, differing in size, shape and biochemical composition. It is therefore no wonder that with a large diversity of adhesions, they perform different and specific functions in cells [97]. These include cell–cell adhesions (e.g. Cadherins) and cell–matrix interactions. Our discussion will focus on the cell–matrix interactions as they are most relevant to stem cells response to nanotopography.

Cells anchor to the ECM through adhesions that are mediated by integrins. Integrins are heterodimeric transmembrane cell adhesion proteins that bind to specific motifs present on the ECM [81, 98]. In relation to nanotopography induced stem cell differentiation, the change in the physical structure of the underlying substrate can influence this clustering of integrins and other adhesion molecules. Arnold et al. [99] was able to use precise nanoscale adhesive islands to establish a relationship between integrin clustering, focal adhesion formation and actin stress fibers which influenced the adhesion and spreading of cells. In fact, the maximal distance where two individual integrin molecules can bind is in the range of 50–70 nm, showing the importance of integrin clustering in the regulation of integrin mediated signal transduction [99].

Using a novel nanoscale ligand spacing gradient, the same group suggested that cells expressed delicate sensitivity to interparticle spacing of about 1 nm, demonstrating the sensitivity of the sensing mechanism [100]. It was also suggested that the sensitivity to minute variations have physiological implications where in vivo, ECM collagen fibers has a 67 nm banding periodicity [101] and fibronectin fibers presents nanoscale epitopes [102, 103]. Upon binding to the ECM ligands, integrins cluster and activates specific signaling pathways that are important for various cellular functions such as migration, proliferation and differentiation. It is therefore likely that the nanotopographical cues modify the extent and activation of integrin clustering as the initial step in subsequent signal transduction in stem cell genomic regulation. However, the role of integrins in gene regulation is complex because these receptors participate in both the sensory and operational functions of the cellular machinery, also commonly known as the outside-in (sensory) and inside-out (operational) signaling activities. The observed dynamics of the integrins on topography can be due to both the response of the cell to the underlying ECM and/or a secondary effect of the actin-cytoskeleton focal adhesion feedback machinery. The complexity in the feedback network connecting the sensory and operational functions is also reflected in the highly intertwined integrin adhesome network [104].

One of the most important integrin-mediated adhesions involved in mechanotransduction is the focal adhesion (FA) [82, 105]. FAs play an important role in linking actin cytoskeleton to the transmembrane integrins [104, 106]. The exact molecular nature of FAs is unclear; however, FAs are composed of a large complex network of adhesion molecules [97, 105]. Some of the important structural proteins include talin, vinculin and focal adhesion kinase (FAK). The formation and maturation of the FA are driven by feedback from the actin cytoskeleton and integrin [82]. Briefly, talin connects integrin dimers with the actin filaments before the recruitment of additional components in the complex [107]. The subsequent maturation of the complex requires a contractile pulling force generated by the actin–myosin machinery [108, 109]. For a more detailed description of the steps and mechanism of the FA assembly, readers can refer to an excellent review by Geiger et al. [82].

Mechanical force thus plays an important role in the promotion of FAs. An important component of FA, vinculin, has been shown to trigger the clustering of activated integrins [110]. The binding of vinculin to talin during the initial stages of FA assembly is shown to be force mediated. Moreover, vinculin–talin interaction in vitro requires a mechanical force mediated unfolding of the flexible talin rod domain [111]. It seems that the contractile stress that actin exerts on the adhesions are essential for the formation of FA. The global forces that are experienced by the cells under nanotopographical cues can alter such forces that the FAs are experiencing, ultimately changing their differentiation lineage. This is one of the many examples of the components that undergo force mediated conformational changes to relay signals. Other components will include p130Cas [112] and fibronectin [103]. All the studies indicate that FAs play an important role in mechanotransduction, possibly also in regulating nanotopography-induced stem cell differentiation.

4.5 Cytoskeleton: Force Transmission

4.5.1 Cell Exerting Forces on the Underlying Substrate

Force generation in the cytoskeleton is required for cell adhesion and migration on ECM. The cytoskeleton consists of actin (microfilaments), microtubules and intermediate filaments. They form a network of filamentous protein that extends throughout the cell cytoplasm in eukaryotic cells. The cytoskeleton has been well studied where it is involved in cellular metabolism and movement ranging from mitosis to migration [113, 114]. The microtubules provide a transport system for vesicles and membrane bound organelles, radiating out from the microtubule organizing centre through the cytoplasm [115]. Intermediate filaments (IF), are another class of cytoskeleton which consists of tough protein fibers. The IFs mainly provides mechanical support for the cell and nucleus [115, 116]. For example, the intermediate filament vimentin binds directly to nuclear lamina proteins, which are regulators of nuclear organization and gene function [117].

The final class of actin microfilaments, on the other hand, provides evidence of cell cytoskeletal response to nanotopography with the observed alignment of actin microfilament to nanogratings and differential organization to the other surface features [12, 83, 118].

The contractile forces present in the actin stress fibers of the cytoskeleton appear to be integral in modulating cell adhesion, shape and even functions like cell differentiation. In a study by Engel et al. [119], the use of matrices with different elasticity regulates the differentiation of MSCs into different lineages. The use of specific non muscle myosin II inhibitor blebbistatin blocked all elasticity-directed lineage specification without strongly affecting cell function and shape significantly, showing the importance of cytoskeletal force generation in ECM sensing. Different groups have tried to investigate the cellular forces that cells exert on the underlying substrate. An early method of cellular force measurement technique used a silicon rubber membrane that deforms and wrinkles upon the exertion of force by adherent cells [120]. Another more recent study employed the use of an array of vertical elastomeric microcantilevers that bends under the exertion of contractile force by cells [121]. These methods of measuring cellular force exertion provided new insight into cytoskeletal forces and their relation to nanotopographical sensing.

The formation of FAs also require external force loading from the ECM [121] or tensile force generation by the actin–myosin contraction of anchoring actin filaments [84, 122]. As the expression of focal contacts and organization of stress fibers in response to topography has effects on cell differentiation, these studies reiterate the importance of actin microfilaments in the cell nanotopographical signal transduction. It appears that the cytoskeletal-mediated nanotopographical sensing mechanism is also present in human ESCs. hESCs aligned and elongated when they were cultured on nanometer scale gratings [47]. More importantly, the topographical cues were able to alter the organization of cytoskeletal components such as F-actin, which in turn affected the ECM–cell force loading profile. It is likely that the stem cells interpret such changes in force signals to regulate cellular functions like differentiation.

4.6 Filopodia: Probing the ECM

The actin cytoskeleton of the cell also aids in the active projections of the cell plasma membrane such as filpodias and lamellipodias. Filopodias are highly motile organelles that have been intimately associated with local topographical cue sensing in cells [123]. A recent study by Satoshi et al. [118] observed that filopodia extensions of MSCs perpendicular to the nanogratings retracted faster than those parallel to the gratings, seemingly demonstrating an ability of filopodia to probe topological differences. It is interesting to note that the perpendicularly directed filopodias also had fragmented and thus weaker vinculin expression as compared to the parallel ones. The role and molecular mechanism of filopodia in nanotopography sensing remains unknown although preliminary studies have been done [124].

4.7 Nucleus: Gene Regulation

The nucleus contains most of the cells genetic material, whereby the regulation of these materials and other associated proteins will be essential for controlling cellular functions. One of the main structures of the nucleus that appears to be important in topography-mediated mechanotransduction is the nuclear lamina. Similar to the cytoskeleton, the nuclear lamina consists of a meshwork of intermediate filaments and lamin proteins [77]. Transmembrane integrins are physically coupled through the focal adhesions to the cytoskeletal filaments that, in turn links to the different components of the nucleus such as the nuclear lamina scaffold, chromatin and DNA. Recently, the LINC, a specialized structure that links both the nucleoskeleton and cytoskeleton has been identified, providing more evidence that mechanical forces that arise due to nanotopography can physically affect the structural organization of the nucleus [125, 126]. Forces that are transmitted to the nuclear scaffolds via LINC complex may regulate critical DNA enzymes or factors. Furthermore, in an earlier study, the disruption of intermediate filaments leads to the mechanical decoupling of the integrins and nuclei [85], demonstrating that a direct physical connection exists between the two. Indeed, the alignment and elongation of MSC nuclei along nanogratings suggests that topography may exert an effect on the structural organization of the nucleus [12].

Although there are evidences of nuclear mechanotransduction, the molecular and biophysical basis for such a mechanism is still left to be understood. It will be interesting to note that such processes are often coupled with chemical signaling mechanisms that make the analysis more complicated. Interested readers can find an excellent review by Wang et al. [127] on nuclear mechanotransduction.

5 Conclusion

The various components that are fundamental to direct mechanotransduction have only been briefly discussed in the context of both nanotopographical cues and generic ECM signal transduction. In this chapter, we have explored the mechanism by looking at studies that have been carried out on individual modules of the cell-topography sensing machinery. Although a substantial amount of work has been

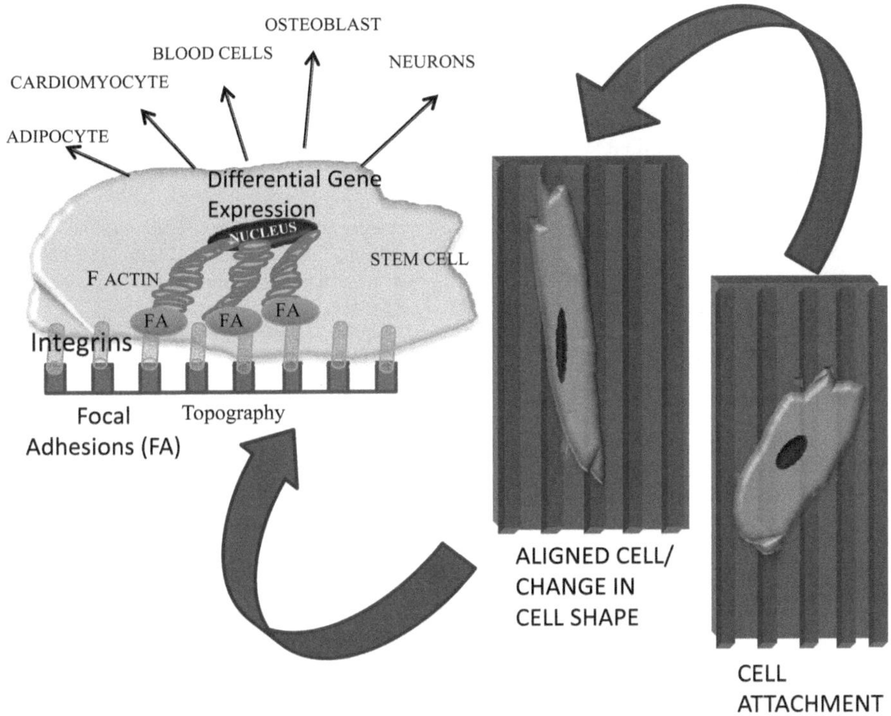

Fig. 5 Brief schematic overview of events in topography-induced gene expression. Although the mechanism is generally unknown, a global morphological change is observed some time after attachment to nanotopographical surface. Adhesions such as transmembrane integrins, focal adhesions and the cytoskeleton appears to be fundamental to the topographical signal transduction

carried out in order to decipher this multifaceted mechanism, the challenge remains that we take heed from the cell's mechanism to put together the various critical mechanosensing modules into a complete model, which is similar to how the cell integrates the different components to orchestrate a concerted response [128–130], see Fig. 5. Many studies have been particularly focused on the individual modules of the mechanism of topography-induced stem cell differentiation, however rigorous work have to be done in integrating all the information together for a complete understanding of stem cell differentiation. It is therefore fundamental that we unravel the intrinsic underlying mechanism for nanotopography-induced stem cell differentiation, to pave a new paradigm in regenerative medicine.

References

1. Ding, S., Schultz, P.G.: A role for chemistry in stem cell biology. Nat. Biotechnol. **22**(7), 833–840 (2004)
2. Fuchs, E., Tumbar, T., Guasch, G.: Socializing with the neighbors: stem cells and their niche. Cell **116**(6), 769–778 (2004)

3. Moore, K.A., Lemischka, I.R.: Stem cells and their niches. Science **311**(5769), 1880–1885 (2006)
4. Abrams, G.A., et al.: Nanoscale topography of the basement membrane underlying the corneal epithelium of the rhesus macaque. Cell Tissue Res. **299**(1), 39–46 (2000)
5. Weiss, P., Garber, B.: Shape and movement of mesenchyme cells as functions of the physical structure of the medium. Proc. Natl Acad. Sci. USA **38**(3), 264–280 (1952)
6. Curtis, A.S., Varde, M.: Control of cell behavior: topological factors. J. Natl. Cancer Inst. **33**, 15–26 (1964)
7. Bettinger, C.J., Langer, R., Borenstein, J.T.: Engineering substrate topography at the micro- and nanoscale to control cell function. Angew. Chem. Int. Ed. **48**(30), 5406–5415 (2009)
8. Yim, E.K.F., Leong, K.W.: Significance of synthetic nanostructures in dictating cellular response. Nanomed. Nanotechnol. Biol. Med. **1**(1), 10–21 (2005)
9. Seidlits, S.K., Lee, J.Y., Schmidt, C.E.: Nanostructured scaffolds for neural applications. Nanomedicine **3**(2), 183–199 (2008)
10. Martínez, E., et al.: Effects of artificial micro- and nano-structured surfaces on cell behaviour. Ann. Anat. Anatomischer Anzeiger **191**(1), 126–135 (2009)
11. Gong, H., et al.: A new view of the human trabecular meshwork using quick-freeze, deep-etch electron microscopy. Exp. Eye Res. **75**(3), 347–358 (2002)
12. Yim, E.K., Pang, S.W., Leong, K.W.: Synthetic nanostructures inducing differentiation of human mesenchymal stem cells into neuronal lineage. Exp. Cell Res. **313**(9), 1820–1829 (2007)
13. Curtis, A., Wilkinson, C.: Nantotechniques and approaches in biotechnology. Trends Biotechnol. **19**(3), 97–101 (2001)
14. Curtis, A., Wilkinson, C.: Topographical control of cells. Biomaterials **18**(24), 1573–1583 (1997)
15. Flemming, R.G., et al.: Effects of synthetic micro- and nano-structured surfaces on cell behavior. Biomaterials **20**(6), 573–588 (1999)
16. Folch, A., Toner, M.: Cellular micropatterns on biocompatible materials. Biotechnol. Prog. **14**(3), 388–392 (1998)
17. Patel, N., et al.: Spatially controlled cell engineering on biodegradable polymer surfaces. FASEB J. **12**(14), 1447–1454 (1998)
18. Vieu, C., et al.: Electron beam lithography: resolution limits and applications. Appl. Surf. Sci. **164**(1–4), 111–117 (2000)
19. Chou, S.Y., Krauss, P.R., Renstrom, P.J.: Imprint lithography with 25-nanometer resolution. Science **272**(5258), 85–87 (1996)
20. Zhang, F.X., Low, H.Y.: Ordered three-dimensional hierarchical nanostructures by nanoimprint lithography. Nanotechnology **17**(8), 1884–1890 (2006)
21. Zhang, F.X., Low, H.Y.: Transfer printing of 3D hierarchical gold structures using a sequentially imprinted polymer stamp. Nanotechnology **19**(41) (2008)
22. Schmid, H., Michel, B.: Siloxane Polymers for High-Resolution, High-Accuracy Soft Lithography. Macromolecules **33**(8), 3042-3049 (2000)
23. Odom, T.W., et al.: Improved Pattern Transfer in Soft Lithography Using Composite Stamps. Langmuir **18**(13), 5314-5320 (2002)
24. Basnar, B., Willner, I.: Dip-pen-nanolithographic patterning of metallic, semiconductor, and metal oxide nanostructures on surfaces. Small **5**(1), 28–44 (2009)
25. Kaehr, B., et al.: Guiding neuronal development with in situ microfabrication. Proc. Natl Acad. Sci. USA **101**(46), 16104–16108 (2004)
26. Norman, J.J., Desai, T.A.: Methods for fabrication of nanoscale topography for tissue engineering scaffolds. Ann. Biomed. Eng. **34**(1), 89–101 (2006)
27. Desai, T.A., et al.: Nanopore technology for biomedical applications. Biomed. Microdev. **2**(1), 11–40 (1999)
28. Moldovan, N.I., et al.: Contribution of monocytes/macrophages to compensatory neovascularization: the drilling of metalloelastase-positive tunnels in ischemic myocardium. Circ. Res. **87**(5), 378–384 (2000)

29. Malarkey, E.B., Parpura, V.: Applications of carbon nanotubes in neurobiology. Neurodegener. Dis. **4**(4), 292–299 (2007)
30. Fan, Y.W., et al.: Culture of neural cells on silicon wafers with nano-scale surface topograph. J. Neurosci. Meth. **120**(1), 17–23 (2002)
31. Turner, S., et al.: Cell attachment on silicon nanostructures. In Papers from the 41st International Conference on Electron, Ion, and Photon Beam Technology and Nanofabrication. AVS, Dana Point (1997)
32. Low, S.P., et al.: Evaluation of mammalian cell adhesion on surface-modified porous silicon. Biomaterials **27**(26), 4538–4546 (2006)
33. Frenot, A., Chronakis, I.S.: Polymer nanofibers assembled by electrospinning. Curr. Opin. Colloid Interface Sci. **8**, 64–75 (2003)
34. Matthews, J.A., et al.: Electrospinning of Collagen Nanofibers. Biomacromolecules **3**(2), 232–238 (2002)
35. Hartgerink, J.D., Beniash, E., Stupp, S.I.: Peptide-amphiphile nanofibers: a versatile scaffold for the preparation of self-assembling materials. Proc. Natl Acad. Sci. USA **99**(8), 5133–5138 (2002)
36. Holmes, T.C., et al.: Extensive neurite outgrowth and active synapse formation on self-assembling peptide scaffolds. Proc. Natl Acad. Sci. USA **97**(12), 6728–6733 (2000)
37. Affrossman, S., et al.: Surface topography and composition of deuterated polystyrene–poly(bromostyrene) blends. Macromolecules **29**(14), 5010–5016 (1996)
38. Affrossman, S., Stamm, M.: Topography and surface composition of thin films of blends of polystyrene with brominated polystyrenes: effects of varying the degree of bromination and annealing. Macromolecules **31**(18), 6280–6288 (1998)
39. Affrossman, S., Stamm, M.: The effect of molecular weight on the topography of thin films of blends of poly(4-bromostyrene) and polystyrene. Colloid Polym. Sci. **278**(9), 888–893 (2000)
40. Dalby, M.J., et al.: Investigating filopodia sensing using arrays of defined nano-pits down to 35 nm diameter in size. Int. J. Biochem. Cell Biol. **36**(10), 2005–2015 (2004)
41. Sen, R., et al.: Preparation of single-walled carbon nanotube reinforced polystyrene and polyurethane nanofibers and membranes by electrospinning. Nano Lett. **4**(3), 459–464 (2004)
42. Smith, L.A., Ma, P.X.: Nano-fibrous scaffolds for tissue engineering. Colloids Surf. B Biointerfaces **39**(3), 125–131 (2004)
43. Huang, X.D., et al.: Reversal imprinting by transferring polymer from mold to substrate. In: Papers from the 46th International Conference on Electron, Ion, and Photon Beam Technology and Nanofabrication. AVS, Anaheim (2002)
44. Stojkovic, M., et al.: Derivation, growth and applications of human embryonic stem cells. Reproduction **128**(3), 259–267 (2004)
45. Takahashi, K., Yamanaka, S.: Induction of pluripotent stem cells from mouse embryonic and adult fibroblast cultures by defined factors. Cell **126**(4), 663–676 (2006)
46. Markert, L.D.A., et al.: Identification of distinct topographical surface microstructures favoring either undifferentiated expansion or differentiation of murine embryonic stem cells. Stem Cells Dev. (2009)
47. Gerecht, S., et al.: The effect of actin disrupting agents on contact guidance of human embryonic stem cells. Biomaterials **28**(28), 4068–4077 (2007)
48. Murray, P., Edgar, D.: The topographical regulation of embryonic stem cell differentiation. Phil. Trans. R. Soc. Lond. Ser. B Biol. Sci. **359**(1446), 1009–1020 (2004)
49. Sasaki, D., et al.: Mass preparation of size-controlled mouse embryonic stem cell aggregates and induction of cardiac differentiation by cell patterning method. Biomaterials **30**(26), 4384–4389 (2009)
50. Smith, L.A., et al.: Enhancing osteogenic differentiation of mouse embryonic stem cells by nanofibers. Tissue Eng. Part A **15**(7), 1855–1864 (2009)
51. Morshead, C.M., et al.: Neural stem cells in the adult mammalian forebrain: a relatively quiescent subpopulation of subependymal cells. Neuron **13**(5), 1071–1082 (1994)
52. Johansson, C.B., et al.: Identification of a neural stem cell in the adult mammalian central nervous system, Cell **96**(1), 25–34 (1999)

53. Goldman, S.A.: Neural progenitor cells of the adult human brain. In: Rao, M.S. (ed.) Neural Development and Stem Cells, Chapter 12, 2nd edn, pp. 267–297. Humana Press, Totowa (2006)
54. Goldman, S.A., et al.: Isolation and induction of adult neural progenitor cells. Clin. Neurosci. Res. **2**(1), 70–79 (2002)
55. Johansson, C.B., et al.: Neural stem cells in the adult human brain. Exp. Cell Res. **253**(2), 733–736 (1999)
56. Svendsen, C.N., Caldwell, M.A., Ostenfeld, T.: Human neural stem cells: isolation, expansion and transplantation. Brain Pathol. **9**(3), 499–513 (1999)
57. Scolding, N., et al.: Oligodendrocyte progenitors are present in the normal adult human CNS and in the lesions of multiple sclerosis. Brain **121**(12), 2221–2228 (1998)
58. Roy, N.S., et al.: Identification, isolation, and promoter-defined separation of mitotic oligodendrocyte progenitor cells from the adult human subcortical white matter. J. Neurosci. **19**(22), 9986–9995 (1999)
59. Christopherson, G.T., Song, H., Mao, H.-Q.: The influence of fiber diameter of electrospun substrates on neural stem cell differentiation and proliferation. Biomaterials **30**(4), 556–564 (2009)
60. Recknor, J.B., Sakaguchi, D.S., Mallapragada, S.K.: Directed growth and selective differentiation of neural progenitor cells on micropatterned polymer substrates. Biomaterials **27**(22), 4098–4108 (2006)
61. Silva, G.A., et al.: Selective differentiation of neural progenitor cells by high-epitope density nanofibers. Science **303**(5662), 1352–1355 (2004)
62. Soen, Y., et al.: Exploring the regulation of human neural precursor cell differentiation using arrays of signaling microenvironments. Mol. Syst. Biol. **2**, 1–14 (2006)
63. Even-Ram, S., Artym, V., Yamada, K.M.: Matrix control of stem cell fate. Cell **126**(4), 645–647 (2006)
64. Ciapetti, G., et al.: Human bone marrow stromal cells: in vitro expansion and differentiation for bone engineering. Biomaterials **27**(36), 6150–6160 (2006)
65. Sanchez-Ramos, J., et al.: Adult bone marrow stromal cells differentiate into neural cells in vitro. Exp. Neurol. **164**(2), 247–256 (2000)
66. Caplan, A.I.: Mesenchymal stem cells. J. Orthop. Res. **9**(5), 641–650 (1991)
67. Dalby, M.J., et al.: The control of human mesenchymal cell differentiation using nanoscale symmetry and disorder. Nat. Mater. **6**(12), 997–1003 (2007)
68. Huang, N.F., Li, S.: Mesenchymal stem cells for vascular regeneration. Regen. Med. **3**(6), 877–892 (2008)
69. McBeath, R., et al.: Cell shape, cytoskeletal tension, and RhoA regulate stem cell lineage commitment. Dev. Cell **6**(4), 483–495 (2004)
70. Dang, J.M., Leong, K.W.: Myogenic induction of aligned mesenchymal stem cell sheets by culture on thermally responsive electrospun nanofibers. Adv. Mater. **19**(19), 2775–2779 (2007)
71. Martino, S., et al.: Hydrogenated amorphous carbon nanopatterned film designs drive human bone marrow mesenchymal stem cell cytoskeleton architecture. Tissue Eng. Part A (2009)
72. Engel, E., et al.: Mesenchymal stem cell differentiation on microstructured poly (methyl methacrylate) substrates. Ann. Anat. Anatomischer Anzeiger **191**(1), 136–144 (2009)
73. Terje, S., et al.: Fabrication of pillar-like titania nanostructures on titanium and their interactions with human skeletal stem cells. Acta Biomater. **5**(5), 1433–1441 (2009)
74. Kantawong, F., et al.: Whole proteome analysis of osteoprogenitor differentiation induced by disordered nanotopography and mediated by ERK signalling. Biomaterials **30**(27), 4723–4731 (2009)
75. Prabhakaran, M.P., Venugopal, J.R., Ramakrishna, S.: Mesenchymal stem cell differentiation to neuronal cells on electrospun nanofibrous substrates for nerve tissue engineering. Biomaterials **30**(28), 4996–5003 (2009)

76. Dalby, M.J., et al.: Osteoprogenitor response to semi-ordered and random nanotopographies. Biomaterials **27**(15), 2980–2987 (2006)
77. Alberts, B., Bray, D., Lewis, J., Raff, M., Watson, J.: Molecular Biology of the Cell. Garland Publishing, New York (1994)
78. Brunette, D.M., Chehroudi, B.: The effects of the surface topography of micromachined titanium substrata on cell behavior in vitro and in vivo. J. Biomech. Eng. **121**(1), 49–57 (1999)
79. Dunn, G.A., Brown, A.F.: Alignment of fibroblasts on grooved surfaces described by a simple geometric transformation. J. Cell Sci. **83**, 313–340 (1986)
80. Bissell, M.J., et al.: Tissue structure, nuclear organization, and gene expression in normal and malignant breast. Cancer Res. **59**(7 suppl), 1757s–1763s; discussion 1763s–1764s (1999)
81. Lutolf, M.P., Hubbell, J.A.: Synthetic biomaterials as instructive extracellular microenvironments for morphogenesis in tissue engineering. Nat. Biotechnol. **23**(1), 47–55 (2005)
82. Geiger, B., Spatz, J.P., Bershadsky, A.D.: Environmental sensing through focal adhesions. Nat. Rev. Mol. Cell Biol. **10**(1), 21–33 (2009)
83. Dalby, M.J., et al.: Fibroblast reaction to island topography: changes in cytoskeleton and morphology with time. Biomaterials **24**(6), 927–935 (2003)
84. Riveline, D., et al.: Focal contacts as mechanosensors: externally applied local mechanical force induces growth of focal contacts by an mDia1-dependent and ROCK-independent mechanism. J. Cell Biol. **153**(6), 1175–1186 (2001)
85. Maniotis, A.J., Chen, C.S., Ingber, D.E.: Demonstration of mechanical connections between integrins, cytoskeletal filaments, and nucleoplasm that stabilize nuclear structure. Proc. Natl Acad. Sci. USA **94**(3), 849–854 (1997)
86. Folkman, J., Moscona, A.: Role of cell shape in growth control. Nature **273**(5661), 345–349 (1978)
87. Ingber, D.E.: Control of capillary growth and differentiation by extracellular matrix. Use of a tensegrity (tensional integrity) mechanism for signal processing. Chest **99**(3 suppl), 34S–40S (1991)
88. Chen, C.S., et al.: Micropatterned surfaces for control of cell shape, position, and function. Biotechnol. Prog. **14**(3), 356–363 (1998)
89. Titushkin, I., Cho, M.: Modulation of cellular mechanics during osteogenic differentiation of human mesenchymal stem cells. Biophys. J. **93**(10), 3693–3702 (2007)
90. Yourek, G., Hussain, M.A., Mao, J.J.: Cytoskeletal changes of mesenchymal stem cells during differentiation. ASAIO J. **53**(2), 219–228 (2007)
91. Hoben, G.M., Koay, E.J., Athanasiou, K.A.: Fibrochondrogenesis in two embryonic stem cell lines: effects of differentiation timelines. Stem Cells **26**(2), 422–430 (2008)
92. Johnstone, B., et al.: In vitro chondrogenesis of bone marrow-derived mesenchymal progenitor cells. Exp. Cell Res. **238**(1), 265–272 (1998)
93. McBride, S.H., Falls, T., Knothe Tate, M.L.: Modulation of stem cell shape and fate B: mechanical modulation of cell shape and gene expression. Tissue Eng. Part A **14**(9), 1573–1580 (2008)
94. Guilak, F., et al.: Control of stem cell fate by physical interactions with the extracellular matrix. Cell Stem Cell **5**(1), 17–26 (2009)
95. Ingber, D.E.: The mechanochemical basis of cell and tissue regulation. Mech. Chem. Biosyst. **1**(1), 53–68 (2004)
96. Lecuit, T., Lenne, P.F.: Cell surface mechanics and the control of cell shape, tissue patterns and morphogenesis. Nat. Rev. Mol. Cell Biol. **8**(8), 633–644 (2007)
97. Zamir, E., Geiger, B.: Molecular complexity and dynamics of cell–matrix adhesions. J. Cell Sci. **114**(Pt 20), 3583–3590 (2001)
98. Ruoslahti, E., Obrink, B.: Common principles in cell adhesion. Exp. Cell Res. **227**(1), 1–11 (1996)

99. Arnold, M., et al.: Activation of integrin function by nanopatterned adhesive interfaces. Chemphyschem **5**(3), 383–388 (2004)
100. Arnold, M., et al.: Induction of cell polarization and migration by a gradient of nanoscale variations in adhesive ligand spacing. Nano Lett. **8**(7), 2063–2069 (2008)
101. Jiang, F., et al.: Assembly of collagen into microribbons: effects of pH and electrolytes. J. Struct. Biol. **148**(3), 268–278 (2004)
102. Little, W.C., et al.: Assay to mechanically tune and optically probe fibrillar fibronectin conformations from fully relaxed to breakage. Matrix Biol. **27**(5), 451–461 (2008)
103. Smith, M.L., et al.: Force-induced unfolding of fibronectin in the extracellular matrix of living cells. PLoS Biol. **5**(10), e268 (2007)
104. Zaidel-Bar, R., et al.: Functional atlas of the integrin adhesome. Nat. Cell Biol. **9**(8), 858–867 (2007)
105. Burridge, K., et al.: Focal adhesions: transmembrane junctions between the extracellular matrix and the cytoskeleton. Annu. Rev. Cell. Biol. **4**, 487–525 (1988)
106. Geiger, B., et al.: Transmembrane crosstalk between the extracellular matrix–cytoskeleton crosstalk. Nat. Rev. Mol. Cell Biol. **2**(11), 793–805 (2001)
107. Gingras, A.R., et al.: The structure of the C-terminal actin-binding domain of talin. EMBO J. **27**(2), 458–469 (2008)
108. Choi, C.K., et al.: Actin and alpha-actinin orchestrate the assembly and maturation of nascent adhesions in a myosin II motor-independent manner. Nat. Cell Biol. **10**(9), 1039–1050 (2008)
109. Even-Ram, S., et al.: Myosin IIA regulates cell motility and actomyosin–microtubule crosstalk. Nat. Cell Biol. **9**(3), 299–309 (2007)
110. Humphries, J.D., et al.: Vinculin controls focal adhesion formation by direct interactions with talin and actin. J. Cell Biol. **179**(5), 1043–1057 (2007)
111. Gingras, A.R., et al.: Structural and dynamic characterization of a vinculin binding site in the talin rod. Biochemistry **45**(6), 1805–1817 (2006)
112. Sawada, Y., et al.: Force sensing by mechanical extension of the Src family kinase substrate p130Cas. Cell **127**(5), 1015–1026 (2006)
113. Berrier, A.L., Yamada, K.M.: Cell–matrix adhesion. J. Cell Physiol. **213**(3), 565–573 (2007)
114. Delon, I., Brown, N.H.: Integrins and the actin cytoskeleton. Curr. Opin. Cell Biol. **19**(1), 43–50 (2007)
115. Lowe, J., van den Ent, F., Amos, L.A.: Molecules of the bacterial cytoskeleton. Annu. Rev. Biophys. Biomol. Struct. **33**, 177–198 (2004)
116. Vale, R.D.: The molecular motor toolbox for intracellular transport. Cell **112**(4), 467–480 (2003)
117. Dechat, T., et al.: Nuclear lamins: major factors in the structural organization and function of the nucleus and chromatin. Genes Dev. **22**(7), 832–853 (2008)
118. Fujita, S., Ohshima, M., Iwata H.: Time-lapse observation of cell alignment on nanogrooved patterns. J. R. Soc. Interface **6**(suppl 3), S269–S277 (2009)
119. Engler, A.J., et al.: Matrix elasticity directs stem cell lineage specification. Cell **126**(4), 677–689 (2006)
120. Harris, A.K., Wild, P., Stopak, D.: Silicone rubber substrata: a new wrinkle in the study of cell locomotion. Science **208**(4440), 177–179 (1980)
121. Tan, J.L., et al.: Cells lying on a bed of microneedles: an approach to isolate mechanical force. Proc. Natl Acad. Sci. USA **100**(4), 1484–1489 (2003)
122. Galbraith, C.G., Yamada, K.M., Sheetz, M.P.: The relationship between force and focal complex development. J. Cell Biol. **159**(4), 695–705 (2002)
123. Galbraith, C.G., Yamada, K.M., Galbraith, J.A.: Polymerizing actin fibers position integrins primed to probe for adhesion sites. Science **315**(5814), 992–995 (2007)
124. Dalby, M.J., et al.: Increasing fibroblast response to materials using nanotopography: morphological and genetic measurements of cell response to 13-nm-high polymer demixed islands. Exp. Cell Res. **276**(1), 1–9 (2002)

125. Crisp, M., et al.: Coupling of the nucleus and cytoplasm: role of the LINC complex. J. Cell Biol. **172**(1), 41–53 (2006)
126. Fey, E.G., Wan, K.M., Penman, S.: Epithelial cytoskeletal framework and nuclear matrix-intermediate filament scaffold: three-dimensional organization and protein composition. J. Cell Biol. **98**(6), 1973–1984 (1984)
127. Wang, N., Tytell, J.D., Ingber, D.E.: Mechanotransduction at a distance: mechanically coupling the extracellular matrix with the nucleus. Nat. Rev. Mol. Cell Biol. **10**(1), 75–82 (2009)
128. Ingber, D.E.: Cellular mechanotransduction: putting all the pieces together again. FASEB J. **20**(7), 811–827 (2006)
129. Bershadsky, A.D., et al.: Assembly and mechanosensory function of focal adhesions: experiments and models. Eur. J. Cell Biol. **85**(3–4), 165–173 (2006)
130. Bershadsky, A., Kozlov, M., Geiger, B.: Adhesion-mediated mechanosensitivity: a time to experiment, and a time to theorize. Curr. Opin. Cell Biol. **18**(5), 472–481 (2006)

The Nanofiber Matrix as an Artificial Stem Cell Niche

Hai-Quan Mao, Shawn H. Lim, Shuming Zhang, Gregory Christopherson, Korey Kam and Stephen Fischer

Abstract Nanofibrous matrices serve as artificial substrates to partially mimic the key physical and structural characteristics of the stem cell niche. Three methods for the preparation of nanofibrous matrices, including electrospinning, solution phase separation and self-assembly are reviewed and contrasted. Their effect on stem cell adhesion, survival, migration, proliferation and differentiation are discussed. In particular the applications of these nanofibrous matrices for the expansion, differentiation and delivery of stem/progenitor cells are highlighted in this chapter. These matrices provide great opportunities to modulate cellular behavior and tissue regeneration at the nanometer scale.

1 The Stem Cell Niche

The term "stem cell niche" was coined in 1978 by Schofield, who proposed the notion that stem cells reside in vivo within specialized compartments, or niches, which maintain these cells in an immature and proliferative phenotype [1]. The

H.-Q. Mao, S. Zhang, G. Christopherson, K. Kam and S. Fischer
Department of Materials Science and Engineering, Whiting School of Engineering, Johns Hopkins University, Baltimore, MD 21218, USA

H.-Q. Mao
Whitaker Biomedical Engineering Institute, Whiting School of Engineering, Johns Hopkins University, Baltimore, MD 21218, USA

S. H. Lim
Department of Biomedical Engineering, School of Medicine, Johns Hopkins University, Baltimore, MD 21205, USA

H.-Q. Mao (✉)
101E Maryland Hall, 3400 North Charles Street, Baltimore, MD 21218, USA
e-mail: hmao@jhu.edu

Stud Mechanobiol Tissue Eng Biomater (2010) 2: 89–118
DOI: 10.1007/8415_2010_2

Published online: 12 June 2010

stem cell niche constitutes the natural microenvironment that surrounds stem cells, and encompasses both cellular and acellular components. The niche is believed to provide a complex array of physical and biochemical signals to stem cells in a temporal and spatial fashion; by integrating both local and systemic cues, the niche actively engages these cells and guides their proliferation, migration, and fate specification [2–5]. Consequently, the niche is both an anatomical and functional entity, and it plays a vital role in regulating how stem cells participate in tissue regeneration and repair [6, 7].

Important lessons can be learned by examining the structure–function relationships of various niche components, and recent studies have provided greater insight into how these niche components act in concert to regulate stem cell behavior [6–9]. For instance, there is mounting evidence demonstrating the importance of cell adhesion molecules in providing selective anchorage to various types of stem cells through both cell–cell and cell–extracellular matrix (ECM) interactions [2, 5, 10, 11]. Such adhesion-based interactions are likely cell-type dependent and are vital to stem cell survival and activity (e.g. phenotype maintenance, migration and proliferation). In addition to providing anchorage, another important function of niche ECM components is the local regulation of "instructional" signaling cues by manipulating the concentration and presentation pattern of signaling molecules (Fig. 1) [12]. In other words, the niche appears to provide the context by which these biochemical cues act on stem cells to orchestrate their biological functions. Thus, these signaling molecules have to be viewed as parts of a multifaceted spatial and temporal network regulated by the ECM and niche cells [13].

In light of the vital role that the stem cell niche appears to play in vivo, the niche concept has been increasingly recognized as a guiding principle for designing systems for the ex vivo manipulation of stem cells and stem cell-based therapies. Engineered artificial matrix systems with a well-controlled presentation of cellular,

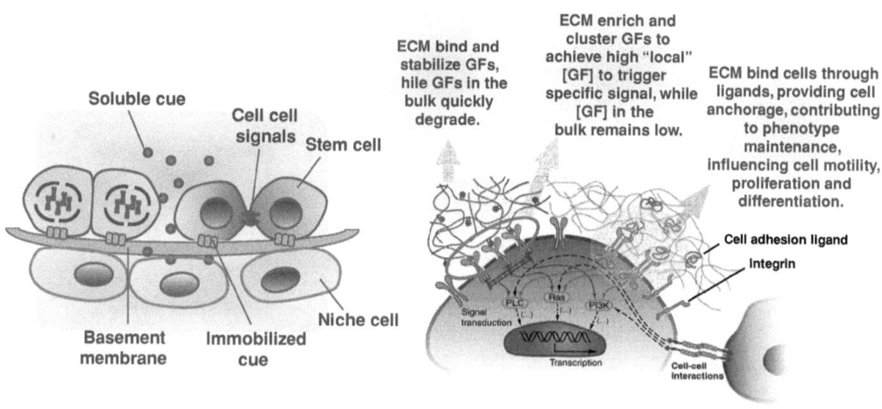

Fig. 1 The concept of the stem cell niche, and role of the extracellular matrix in regulating stem cell survival and signaling. *ECM* extracellular matrix, *GF* growth factor; [*GF*] growth factor concentration. Matrix topography may facilitate these processes. Adapted from [14]

molecular, and physical cues offer a systematic approach to correlate effects of niche components on stem cell behavior. Such systems will not only help to further identify the key niche components and their functions, but also provide feasible matrices for ex vivo stem cell expansion and controlled differentiation. Certainly, the level of complexity of native stem cell niches makes replicating these a great challenge.

2 Nanoscale Topography in the Extracellular Matrix

It has long been recognized that the physical properties of a cell culture substrate, including its topographical, mechanical, and electrical properties, significantly impact cell adhesion, survival, migration, proliferation, and differentiation in vitro. In many cases, the cell type-specific behaviors observed in response to these physical stimuli in vitro can be correlated in some fashion to the native environment in which the cells reside in vivo. For example, recent studies have demonstrated that stem cells can be differentiated towards tissue-specific lineages by matching the mechanical stiffness of a substrate to that of the desired tissue [15, 16]. Likely, the physical properties of stem cell niches play an important role in regulating stem cell behavior in vivo. A major challenge in engineering artificial stem cell niches is to properly identify the relevant physical features of native niches, which are likely different for each stem cell type.

A fairly obvious physical feature of many stem cell niches is the presence of nanoscale topography. For example, ample evidence has shown that the in vivo ECM (e.g. basal lamina) possesses ubiquitous nanoscale fibrous morphology (Fig. 2) [17–21]. Additionally, in-depth microscopic analysis of tissues where stem cells reside has revealed topographical features with both micro- and nanoscale fibrous structure [22–25]. Despite the existence of ECM fibrous features associated with the various niches, the most substantial evidence of the role of topography in regulating stem cell fate has been derived in vitro. A number of studies have shown that various cell types respond to topographical features of the substrate upon which they are cultured [21, 26]. Various model substrates with controlled micro- and nanoscale surface features including islands, pillars, gratings, and fibers have been used to investigate the impact of surface topography on stem cell behavior. For instance, when cultured on parallel grooves, cells elongated and aligned in the direction of the grooves; and this morphological change was facilitated by reorganization of the cytoskeleton structure [27–30]. This reorganization affects cell phenotype and function by triggering cytoskeleton-linked intracellular signaling pathways, for example, the reduction in smooth muscle cell proliferation when cultured on 350 nm-wide gratings [21]. Using the mammalian visual system as a model, Ellis-Behnke et al. have shown that a nanofiber scaffold provided a permissive environment to promote axon regeneration and brain repair, resulting in functional recovery of vision [31]. These results, while highlighting the significance of nano-topographical factors as a general signaling cue, provide impetus for further mechanistic investigation into their influence on stem cells.

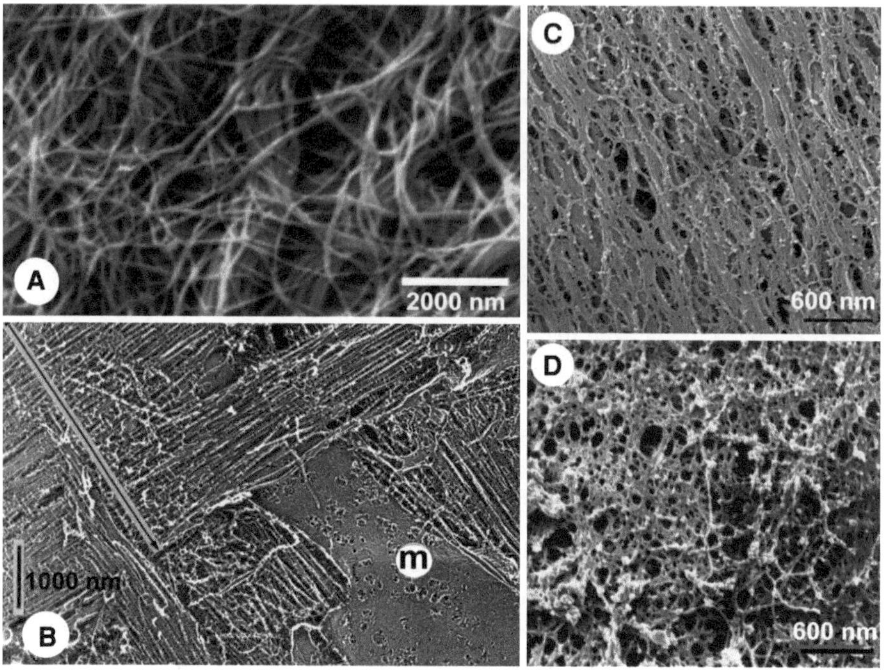

Fig. 2 Examples of extracellular matrices and associated nanotopography. **a** The marginal layer of pars distalis facing the cavum hypophysis in goat; **b** replicas of the developing embryonic corneal stroma at 18 days, and areas with collagen fibrils that are arranged in lamellar structure that criss-cross in a quasi-perpendicular manner. **c** Ultrastructural basement membrane architecture of carotid basement membrane, and **d** inferior vena cava basement membrane. Adapted from **a** [18]; **b** [32]; **c** and **d** [24]

3 Methods to Generate Nanofibrous Matrices

Several methods have been developed to investigate the role of nanoscale topography in regulating stem cell behavior. We will highlight three common techniques, which have been used to generate nanofibrous structures both for ex vivo stem cell culture and in vivo stem cell delivery.

3.1 Electrospinning

Electrospinning is the most widely used method for producing nanofiber matrices, and is prized for its high versatility and capacity to generate nanofibers from a variety of polymer solutions or melts. The diameter of electrospun fibers ranges from tens of nanometers to a few microns [33]. A standard electrospinning setup includes a spinneret (a single or an array of blunt needles or glass pipettes), a

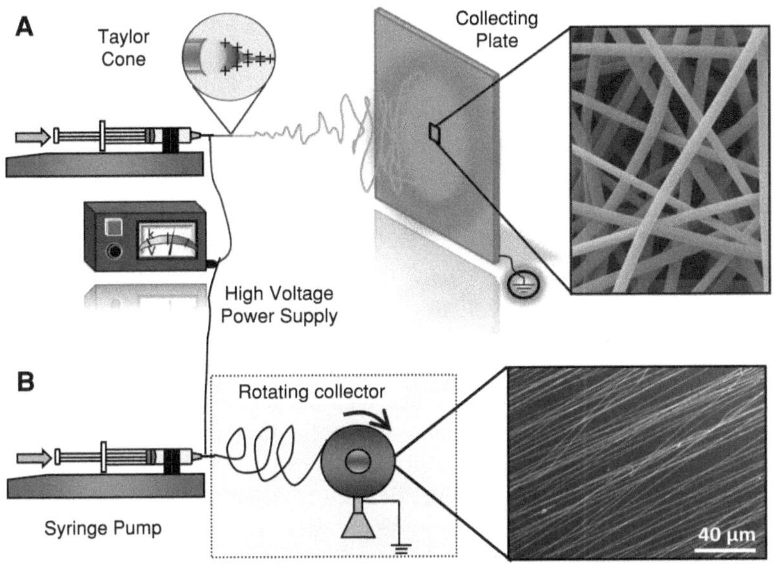

Fig. 3 Electrospinning setup and collection. **a** A typical Electrospinning setup includes a spinneret, a syringe pump, and a collecting plate. Polymer solution or melt is stretched into a thin fiber under applied high voltage, whipping in a random manner before depositing on grounded collecting plate. **b** When a rotating wheel or frame is used for collection, electrospun fibers can be aligned

syringe pump, and a collector plate (Fig. 3a). When the polymer solution or melt is extruded, a high electrostatic potential (usually 5–50 kV) is applied between the spinneret and grounded collector plate, inducing charge accumulation on the surface of the fluid. With increasing electrostatic repulsion, the surface of the fluid at the needle opening forms a conical shape zone (i.e. the Taylor cone) [34]. When the electrostatic repulsion overcomes the intrinsic surface tension of the polymer solution or melt, a charged jet of fluid is ejected from the Taylor cone, whipping in a random orientation before accumulating on the grounded collecting plate. The small diameter of electrospun fibers is mainly achieved by the stretching and acceleration of the viscous jet before fibers land on the collecting plate. Solvent evaporation and polymer melt solidification can occur before or after the fiber collection, depending on the parameters and electrospinning set-up. Due to the bending instability associated with a spinning jet, electrospun fibers usually deposit on the collecting surface as randomly oriented nonwoven mats. However, when these fibers are collected onto rotating wheels or frames, they can form a mesh of aligned nanofibers (Fig. 3b) [35, 36]. Flexibility in controlling the nanofiber spatial orientation is a distinct advantage of electrospinning method.

Although relatively straightforward in design, the electrospinning process itself is rather complicated. Many materials properties (such as solubility, glass-transition temperature, melting point, crystallization velocity, molecular weight, molecular weight distribution, entanglement density, solvent vapor pressure, and pH) and processing parameters (such as polymer concentration, electrical conductivity, surface tension, feed rate, electrode separation and geometry, temperature, and relative humidity) can affect fiber diameter and arrangement of collected nanofibers. Systematic optimization is often required to achieve control over nanofiber morphology, diameter, composition, secondary structures and spatial arrangement. In such a process, the polymer material and solvent combination is generally the first consideration. After selecting an appropriate solvent, parameters such as solution feeding rate, solution concentration, electrical conductivity and inclusion of surfactant can be modulated to control the diameter of produced nanofibers. Generally speaking, lower flow rate and concentration, addition of surfactant [37] and higher electrical conductivity lead to a reduction of fiber diameter [38]. For spatial organization of nanofibers, aligned fibers can usually be obtained by utilizing a rotating collector [36] or creating a quadratic arrangement of four electrodes to force a cross deposition of nanofibers [39]. A shorter distance between spinneret and collection plate (e.g. centimeter or millimeter range) can also achieve a higher degree of alignment [40, 41].

To expand the scope of possible application for electrospun fibers, a variety of improvements have been made to increase the diversity of materials and introduce surface functionality. Polymer blends have been used to prepare electrospun fibers with nanoscale morphologies [42] or a core-sheath structure [43], and natural biomaterials have been electrospun into fibers. The most frequently used natural materials include collagen [44], chitosan [45], gelatin [46] and laminin [47]. Carrier polymers can also be used to facilitate the electrospinning of materials that otherwise cannot be processed alone, and composite or protein nanofibers.

Electrospun fibers have been extensively used as tissue engineering scaffolds and as culture platforms for various types of stem cells. To facilitate cell adhesion, fiber surfaces have been modified using different techniques, including physical adsorption or covalent conjugation of cell adhesion proteins, plasma treatment to introduce surface charge, and grafting polymerization to generate surface functional groups, etc. [48]. For example, electrospun fibers have been coated with positively charged poly-L-ornithine (PLO) and laminin to promote the adhesion of neural stem cells [49]. As one example of chemical functional group grafting, Chua et al. have introduced polyacrylic acid chains to the surface of polyethersulfone nanofibers by UV-initiated surface grafting polymerization, and subsequently converted the carboxylic groups to amino groups to achieve high amine density on fiber surface. These amine-grafted nanofibers have promoted the phenotype maintenance of hematopoietic stem cells during expansion culture [50–52]. Beyond these efforts, electrospun nanofibers have also been covalently conjugated with cell-specific bioactive ligands to enhance cell adhesion, proliferation, and differentiation [53].

3.2 Self-assembly

The formation of functional nanofibers via self-assembly is also an area of growing interest. The term "self-assembly" is used to describe the process in which a disordered system of pre-existing components forms an organized structure or pattern as a consequence of specific, local interactions among the components themselves, without external direction [54, 55]. Using this approach, scientists have developed a family of self-organized nanostructures such as zero-dimensional objects, spherical micelles, cylindrical micelles, ribbons, two-dimensional sheets, and three-dimensional networks [56]. The strategies employed in controlling geometries of self-assembled structures usually require careful design of both molecular shape and molecular interactions [57].

One example of a self-assembled nanofiber matrix is prepared from a de novo-designed water-soluble amphiphilic peptide, consisting of a hydrophobic alkyl tail connected to a short peptide sequence (Fig. 4a) [58–60]. Upon addition of counter ions or pH titration, these molecules can self-assemble into networks of well-defined nanofibers. This process is triggered by charge screening using counter ions and facilitated by the collapse of the alkyl groups from aqueous solution as well as hydrogen bond formation between adjacent peptide residues.

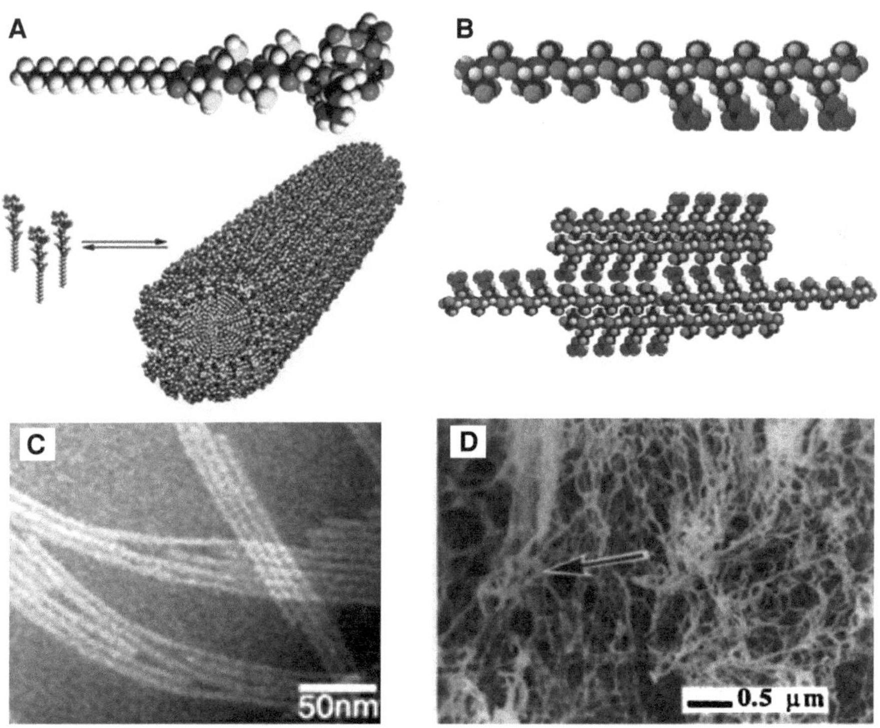

Fig. 4 Self-assembly illustration of **a** a peptide amphiphile [peptide sequence: $C_{15}H_{31}CO$-CCCCGGGS(P)RGD, a transmission electron microscopic image of the assembled fiber is shown in **c**] and **b** a β-sheet forming peptide (DADADADARARARARA, a scanning electron microscopic image of the assembled fibers is shown in **d**). Schematic of the molecules are shown, which further self-assemble together to form a cylindrical micelle. Adapted from [58] and [65]

This nanofiber formation can be initiated at physiological pH and ionic strength, making it possible for cell encapsulation and in vivo applications [61, 62].

Another type of well-studied nanofiber matrix is prepared from peptides that self-assemble into β-sheets in physiological medium (Fig. 4b) [63–65]. The amino acid sequence of these peptides consists of alternating hydrophobic and hydrophilic residues, with the hydrophilic residues carrying alternating positive and negative charges. Upon addition of physiological medium, these molecules self-assemble into networks of well-defined nanofibers, resulting in a hydrogel with about 99% water content [64]. The mechanical moduli of these peptide hydrogels approach that of a hydrogel prepared from type I collagen. Since self-assembly is mediated by intermolecular forces, the hydrogels can be easily deformed and re-assembled under shear, allowing them to be delivered in vivo through injection (with or without cells) to promote tissue regeneration.

In both of these peptide nanofiber systems, ECM-derived short cell adhesion peptides can be incorporated into the sequences of self-assembling peptides to enable cell binding and migration through the matrix. For example, a laminin-derived peptide sequence, isoleucine-lysine-valine-alanine-valine (IKVAV), included in a peptide amphiphile can be decorated onto self-assembling nanofiber surfaces. This IKVAV-conjugated nanofiber matrix enhanced cell adhesion and differentiation of neural progenitor cells [61]. While linking functional short peptide sequences to self-assembling peptides has been successful, the incorporation of larger bioactive proteins is challenging. An alternative approach is to covalently tether biotin to the nanofiber surface, and subsequently immobilize growth factors to biotin moieties using the biotin–streptavidin linkage. Using this approach, Davis et al. have achieved local and sustained release of insulin-like growth factor 1, known to increase cardiac stem cell growth [66].

It is generally difficult to control the alignment of these self-assembled nanofibers due to a lack of long-range order. Recent developments by Zhang et al. have shown that through a thermal induction process, long range alignment of some peptide-based self-assembled nanofibers can be achieved [67]. It remains to be demonstrated whether these aligned nanofibers can be produced in larger scale for stem cell cultures and/or therapeutic delivery.

3.3 Solution Phase Separation

Polymer solutions can be thermally induced to undergo phase separation. Under appropriate conditions, the separated polymer-rich phase can form a nanofibrous structure [68, 69]. This fiber formation occurs through the spinodal liquid–liquid phase separation (Fig. 5) and a consequential crystallization of the polymer-rich phase. When temperature decreases to below the spinodal transition such as that shown in Fig. 5a, the liquid–liquid phase separation spontaneously occurs through a spinodal-decomposition mechanism [70]. A typical procedure of generating nanofiber matrices includes phase separation-gelation, solvent extraction, freezing,

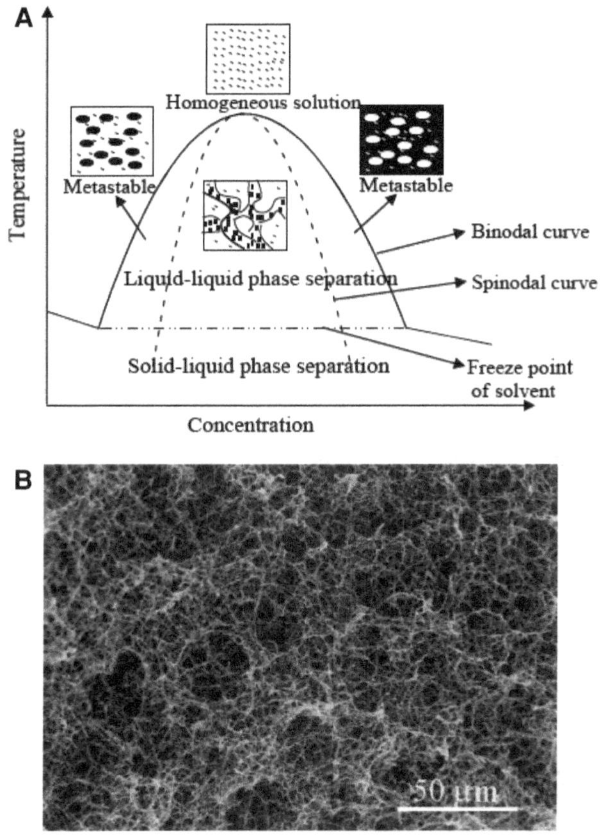

Fig. 5 Phase separation technique used to create nanofibrous matrix. **a** Schematic phase diagram of polymer solution [70] and **b** SEM micrograph of porous nanofibrous matrix made with liquid–liquid phase separation technique [68]. Adapted from [70] and [68]

and lyophilization. The diameters of fibers generated using this method range from tens to hundreds of nanometers. The phase separation-gelation is the most critical step of this process and important for the control over the porosity and fibrous morphology. Low gelation temperature favors the formation of the nanoscale fiber network. Low initial polymer concentration leads to smaller diameter fibers and a more porous network [70]. Based on this principle, many polymer–solvent systems that have similar phase diagrams may be used to generate nanofiber scaffolds with the phase separation technique.

One advantage of phase separation is the ease of incorporating macropores into the nanofiber scaffolds by inducing phase separation in the presence of salt or paraffin particles [71]. These porogen particles are leached-out after lyophilization by soaking the scaffold in appropriate solvents (e.g. water for salt particles and hexane for paraffin particles). This method generates both macroscale (250–425 μm) pores and nanoscale fibers [72]. Recently, this method has been extended to prepare a macroporous and nanofibrous scaffold from gelatin, which mimics both the nanotopography and chemical composition of type I collagen in the ECM. The gelatin nanofiber scaffold has a porosity of higher than 96%, well-connected macropores, and nanofibrous pore wall structures. The nanofiber diameters ranged from 160 to 180 nm [73].

The phase separation method provides a convenient way to generate polymer nanofiber scaffolds without any special equipment; it is easy to scale up; and it is possible to control the porosity and macropore size of the scaffold. Thus far, only preparation of a randomly oriented nanofiber matrix has been reported with this method.

3.4 Comparison of Nanofiber Generation Methods

Electrospinning, self-assembly, and thermally induced phase separation techniques produce nanofibers with slightly different feature sizes and physical properties (Table 1). Electrospinning is the most versatile method and has been used to generate fibers from a wide range of materials. The phase separation method requires more stringent selection of the solvent–polymer pairing that allows spinodal liquid–liquid phase separation to occur. The self-assembly method has only been successful in creating nanofibers from peptide amphiphiles and β-sheet forming peptides. The diameter range of self-assembled nanofibers is typically much smaller than electrospun fibers and nanofibers prepared by solution phase separation. Self-assembled nanofibers are used in hydrogel forms and can be injected, whereas the other two types of fiber matrices are used as solid scaffolds. In addition, surface biofunctionalization techniques have to be specifically tailored for each type of matrix.

Table 1 Comparison of nanofiber matrices prepared by electrospinning, self-assembly and solution phase separation methods

Fiber preparation method	Electrospinning	Self-assembly	Solution phase separation
Materials choice	Wide range of natural and synthetic polymers, ceramic fibers can also be prepared from their precursors	Peptide amphiphiles β-sheet forming peptides	Selected natural and synthetic polymers
Fiber diameter	50 nm to 2 μm	5–20 nm	50–500 nm
Physical form	Solid fiber matrices	Soft hydrogels, injectable	Solid fiber matrices
Incorporation of macropores	Possible	Difficult	Easy to incorporate with porogens
Cell adhesion property	Surface adsorption or covalent conjugation of ECM proteins and other ligands	Direct peptide conjugation or through biotinylation scheme	Surface adsorption or covalent conjugation of ECM proteins and other ligands

4 Nanofibrous Matrices for Stem Cell Expansion

Due to their capacity for long-term self-renewal and ability to differentiate into multiple specialized progeny, stem cells have enormous potential for use in a variety of cellular therapies. In order to achieve therapeutic outcomes in most cases, a much larger number of stem cells is required than can be obtained from a patient, necessitating the use of ex vivo culture systems to expand the initial cell population. In addition to quantity, both the quality (maintaining stem cell phenotype and regeneration potential) and purity (free of contamination from other cell types and animal products) of the expanded stem cells are important. Certainly, robust and efficient methods for producing large populations of high quality stem cells are needed.

In recent years, researchers have engineered a variety of artificial culture systems for the efficient ex vivo expansion of stem cells. Because stem cells are extremely sensitive to their local microenvironment, such systems are designed to provide a well-controlled presentation of cellular, molecular, and physical cues. Here we describe how nanofiber scaffolds have been used to facilitate the undifferentiated expansion of a variety of stem cell types.

4.1 Nanofiber-mediated Expansion of Human Hematopoietic Stem Cells (HSCs)

Human HSCs constantly undergo self-renewal and differentiation within the bone marrow microenvironment, whereas in culture they tend to lose their self-renewal ability. It is believed that HSC self-renewal may be subject to regulation by the biochemical and topographical cues in the bone marrow niche [74–76]. Such biochemical cues include cell-secreted hematopoietic factors (HGFs) and contact with the stroma and ECM molecules [77]; whereas the topographical cues, likely originating from the 3D architecture of the bone marrow, can influence cell–cell and cell–substrate interactions. A successful HSC expansion strategy should ideally mimic these key characteristics of the bone marrow stem cell niche. Current HSC expansion methodologies have incorporated various components of the niche, and can be categorized into either a biological approach or a biomaterials approach.

The biological approach uses a stromal monolayer to provide biological signals to HSCs [76]. In the first few weeks of culture, an adherent layer of stromal cells comprised of fibroblasts, macrophages, adipocytes, endothelial cells and reticular cells is laid down, and serves as a substrate for the subsequent HSC culture. HSC self-renewal can be maintained for months through direct interactions between the HSCs and various elements of the stroma. Although these studies demonstrate that the stromal layer, to a large extent, can promote the self-renewal of human HSCs, it has a number of disadvantages. The stromal layer is fragile thus requires a rigid surface for support, and has a limited lifespan of about 6–8 weeks. More importantly, its use for clinical applications poses logistical problems. First, the need to collect and then grow a layer of the patient's own stromal cells adds significantly to the time, cost, and complexity of the production of autologous HSCs. Second, it could be problematic to obtain healthy stromal cells from patients with other diseases. Third, the composition of stromal layers is poorly defined. It is difficult to harvest and culture the stroma in a reproducible fashion, which renders the expansion outcome less predictable. Allogeneic sources of stroma, although feasible, are unreliable. They pose potential infection risks and contain an insufficient quantity of primary stromal cells. An additional complication is that the stromal layer may produce negative regulators of hematopoiesis such as transforming growth factor TGF-β and chemokines [78, 79].

Due to these concerns, stromal-free culture has gradually emerged as an alternative approach, using various combinations of hematopoietic growth factors to substitute for the regulatory signals provided by the stroma [80, 81]. Stem cell factor (SCF) and Flt-3 ligand (FL) have been shown as the key growth factors for HSC expansion, because c-Kit and Flk-2/Flt-3—the tyrosine kinase receptors for SCF and FL, respectively—have been shown to transduce signals crucial for HSC development. Thrombopoietin (TPO), a ligand for c-Mpl and originally identified as a primary regulator for megakaryopoiesis, has also been shown to act synergistically with SCF and IL-3 to stimulate the expansion of primitive hematopoietic cells [82]. Simplicity is a major advantage of the growth factor-supplemented suspension culture. However, different growth factor and cytokine cocktails have only yielded limited success [74, 83]. This is probably because these suspension cultures lack other components of the unique regulatory microenvironment of bone marrow stroma.

The biomaterials approach aims to partially mimic the three-dimensional (3D) aspects of the bone marrow microenvironment. Several studies have indicated that the 3D culture microenvironment contributes to the regulation of HSC proliferation and differentiation. For example, Bagley et al. have shown that a 3D tantalum porous scaffold supports the maintenance of primitive $CD34^{+}CD38^{-}$ cells for up to 6 weeks and yields a 6.7-fold increase in colony-forming cells (CFCs) without supplementing cytokines in serum-free medium [84]. Li et al. cultured human cord blood-derived HSCs in a nonwoven polyethylene terephthalate (PET) mesh and obtained significantly higher numbers of $CD34^{+}$ cells and colony-forming cells (CFCs) after 7–9 weeks of culture in the scaffold compared with 2D substrate cultures [85].

Independent of physical topography, HSC substrate adhesion through β_1 integrins has been shown to be important for the regulation of HSC survival and homing to the niche [86]. Binding of integrin receptors to surface-coated fibronectin may contribute to improved homing of the expanded HSCs after transplantation, since the engraftment ability is lost if HSCs are cultured in suspension culture. The cell–substrate adhesion through integrin binding is also thought to prevent terminal differentiation in HSCs. Feng et al. prepared FN-conjugated PET fibrous scaffolds and investigated the effect of HSC adhesion in a 3D scaffold on cell expansion [87]. FN-conjugated PET fibrous scaffolds mediated higher cell expansion efficiency than FN-conjugated PET films or tissue culture polystyrene during a 10-day expansion culture in serum-free medium. Human umbilical cord blood $CD34^+$ cells cultured in FN-conjugated scaffolds also gave the highest yield of long-term culture initiating cells compared with all other substrates.

The effect of nanoscale topography on HSC expansion was first demonstrated using a series of electrospun polyethersulfone (PES) fibrous matrices with various types of surface functional groups: carboxylic, hydroxyl, and amino groups in comparison with unmodified fibers [88]. The functionalized PES nanofibers had an average diameter of 529 ± 144 nm with a surface functional group density of 55 nmol/cm^2. Ex vivo culture of human cord blood-derived HSCs showed that surface amino groups in synergy with a nanofiber matrix significantly enhanced the expansion of total $CD34^+$ cells and colony-forming unit (CFU) cells. Moreover, HSC expansion efficiency on the aminated nanofiber matrix was much higher than on similarly aminated PES film surfaces. The CFU assay also confirmed that cells expanded on aminated nanofibers contained a higher frequency of CFU-GEMM cells, which represent more primitive progenitor cells, compared to the aminated PES film.

A striking difference was observed in HSC adhesion to aminated nanofibers versus aminated films. Following a 10-day expansion culture, relatively few cells ($\sim$22%) remained attached to the aminated film and most of the adherent cells were found at the edges of cracks and other film defects. In contrast, cells expanded on an animated nanofiber matrix formed abundant colonies, accounting for more than 45% of the expanded cells. All other conditions (including unmodified, carboxylated and hydroxylated nanofibers and films) yielded negligible levels of adherent cells. More importantly, a much higher fraction (45%) of the adherent cells cultured on aminated PES nanofibers maintained a $CD34^+$ phenotype after 10 days of culture, compared to 17% $CD34^+$ cells among the suspension fraction. The enrichment of $CD34^+$ cell population bound to the animated nanofiber scaffold suggests that substrate topographical cues can improve ex vivo expansion of HSCs through regulated cell adhesion. The clinical application of this technology was further verified by the successful engraftment of the expanded HSCs in NOD/SCID mice [52] and the enhanced neovascularization in NOD/SCID mice in a hind limb vascular injury model [89].

Culture of hHSCs in a collagen nanofiber matrix was performed in a self-assembled type I collagen nanofibrous hydrogel. Human cord blood-derived $CD34^+$ cells were cultured in this collagen type I nanofiber hydrogel in the presence of Flt3-ligand, stem cell factor, and interleukin 3 [90]. 7-day serum-free in vitro culture within the collagen nanofibrous gel produced an increased number of myeloid colony-forming units (CFC-U), although the total expansion factor of $CD34^+$ cells was slightly lower compared to the control suspension cultures, suggesting that collagen I scaffolds may improve the preservation of the multipotency of the $CD34^+$ cells.

It is worth noting that the diameter and elastic modulus of reconstituted collagen nanofibers are different than that of electrospun synthetic fibers. These factors may also contribute to the influence of the fibrous matrix on HSC adhesion strength and the ability to migrate, and, hence, on expansion efficiency.

4.2 Nanofiber-mediated Expansion of Neural Stem Cells (NSCs)

Current protocols for NSC expansion call for either co-culture with stromal cells or neurosphere cultures in growth factor-supplemented media [91–93]. NSCs and neurons provided by these methods are far from satisfactory. The former method poses a serious problem for clinical application due to contamination from xenogenic components; the second method is laborious and inefficient, as the generated neurospheres are heterogeneous and only a small fraction of expanded cells retain their primitiveness. Adult NSCs are associated with basal lamina in the subventricular zone of the lateral ventricle and the dentate gyrus subgranular zone of the hippocampus [94, 95]. Therefore, the nanofibrous topography intrinsic to the basal lamina may benefit the expansion and maintenance of NSCs in culture. Nisbet et al. cultured NSCs isolated from adult rats on electrospun polycaprolactone (PCL) fibers of 750 nm diameter with or without amino-group surface functionalization, and found that the aminated PCL fibers improved NSC expansion over unmodified PCL nanofibers in EGF, FGF-2 and heparin-supplemented medium or under fetal bovine serum-supplemented medium without growth factors [96]. Though amino-functionalization did not appear to impact rat NSC differentiation, the increased hydrophilicity of the PCL nanofibers enhanced cell adhesion, spreading and proliferation.

Laminin is known to promote NSC adhesion in vitro [10] as well as being a critical component of the mammalian basement membrane [97]. To this end, laminin-coated electrospun polyethersulfone (PES) fiber meshes with average fiber diameters of 283, 749 and 1,452 nm were used to investigate the role of nanofiber diameter on rNSC proliferation and differentiation [49]. It was found that rat NSC expansion increased with decreasing fiber diameter. Morphological investigations with scanning electron microscopy (SEM) revealed that rat NSCs on 283 nm fibers proliferated in a colony-like fashion with enhanced cell–cell contact. Conversely, rat NSCs cultured on TCPS were well spread and exhibited far less cell–cell interaction. These observations may point to the role of nanofiber topography in combination with laminin mediated cell–cell and cell–substrate interactions in regulating cell migration and proliferation.

4.3 Nanofiber-mediated Expansion of Embryonic Stem Cells (ESCs)

Until recently, ESCs were cultured exclusively on mouse embryonic fibroblasts, which served as a feeder layer crucial to proper maintenance. A commercially available 3D nanofibrous scaffold of electrospun polyamide nanofibers, known as Ultra-Web, was investigated for its effect on ESC proliferation and self-renewal [98]. ESC colonies were significantly larger in size for cells cultured on Ultra-Web compared to glass coverslips, while still expressing alkaline phosphatase, an undifferentiated stem cell marker. Further mechanistic experiments revealed the critical involvement of Rac (a small GTPase), the PI3K/AKT signaling pathway, and the up-regulation of Nanog and c-Fos in nanofiber-mediated ESC proliferation.

4.4 *Nanofiber-mediated Expansion of Mesenchymal Stem Cells (MSCs)*

Electrospun collagen nanofibers have been evaluated in expansion of MSCs [99]. MSCs were expanded on two sets of collagen fibers with diameters ranging 200–500, and 500–1000 nm, respectively, and on tissue culture polystyrene (TCPS) plates, with the greatest proliferation observed on the 500–1000 nm collagen matrix. Concurrently, there was a decreased amount of vinculin-associated focal adhesion complexes for the large diameter fiber matrix. This suggests a possible involvement of cytoskeletal-linked signaling pathways in hMSC self-renewal, which could be influenced by the topography and mechanical properties of the nanofibrous collagen matrix, in contrast to the rigid 2D TCPS substrates.

The effect of nanofiber topography on MSC expansion was investigated in a similar fashion, examining self-assembled peptide amphiphile nanofiber hydrogels in place of electrospun scaffolds [100]. The peptide nanofibers were functionalized with an arginine-glycine-aspartic acid (RGD) peptide sequence to facilitate cell adhesion. MSCs isolated from the femurs of 3-week-old male Wistar rats were seeded in the peptide nanofiber hydrogels with or without the RGD sequence. The RGD containing PA nanofibers facilitated both significantly higher cell attachment and proliferation than either nanofibers without RGD ligands or TCPS plates. One possible advantage of the self-assembling nanofiber matrix lies with their potential as an in vivo proliferation platform, thereby providing a strategy for possible in vivo expansion before in situ differentiation and tissue regeneration.

5 Nanofiber Matrices for Differentiation of Stem Cells

In addition to facilitating stem cell expansion, nanofiber matrices can be designed to either passively or directly induce differentiation along specific lineages. These matrices are intended to provide provisional support and facilitate stem cell adhesion, differentiation, and tissue organization, either prior to or following implantation in vivo. Conventional approaches towards induction of stem cell differentiation in vitro have been restricted to treatment with a variety of molecular mediators, for example small molecules or growth factors. On the other hand, there is gathering evidence to indicate that stem cell differentiation is sensitive and responsive to the physical signals emanating from their microenvironment. For example, MSCs can be restricted into various geometries by engineering their adhesive contacts to the substrate, resulting in a fate choice between adipogenic (constrained morphology) or osteoblastic (spread out morphology) lineages [101]. It was postulated that these effects were brought about by the associated changes in cytoskeletal arrangement and focal adhesion assembly imposed by the restrictions on cell spreading, and their subsequent impact on Rho/ROCK signaling. Additionally, other studies have emerged to indicate that stem cells can make fate decisions in response to substrates stiffness [15], mechanical stress [102, 103], electrical stimulation [104, 105], and so on. With this in mind, synthetic nanofibrous substrates can be fabricated that are capable of presenting such extracellular signals. In addition to controlling substrate topography, composition and mechanical properties, nanofibrous substrates can be used as a means of presenting surface-immobilized adhesive or signaling ligands, and controlled release of inductive growth factors. The ultimate goal is to be able to sufficiently recapitulate the signaling events involved in the adult stem cell niche or embryonic development, so as to engineer a more functional and well-integrated regenerated tissue.

5.1 Nanofiber-mediated Stem Cell Differentiation into Neuronal Lineages

Development of clinical interventions for the repair of the nervous system is an active area of interest as it is unable to spontaneously recover following traumatic insult or injury. In particular, regeneration in the spinal cord is particularly challenging due to the formation of the glial scar and presence of growth-inhibitory molecules. In contrast to the central nervous system, adult peripheral nerves retain a limited ability to regenerate, as Wallerian degeneration creates a more permissive microenvironment, allowing severed nerves to re-grow over limited distances (typically less than 5 mm). Even in instances of successful nerve re-connection, functional recovery is often less than desired due to the formation of improper axonal contacts and atrophy of target tissues following prolonged denervation [106]. Stem cell transplantation-based strategies for repairing the nervous system have been extensively explored, and shown to promote functional recovery in a variety of disease and injury models, including spinal cord [107] and peripheral nerves.

A number of groups have investigated the use of electrospun nanofibers as niches for the neural differentiation of stem cells. Most of these studies were focused on elucidating the effects of topographical cues presented by nanofiber scaffolds, particularly their influence on cell morphology, attachment, and neurite extension. Mouse ES cells were induced to differentiate via treatment with retinoic acid, and the resultant embryoid bodies were cultured on aligned and random electrospun PCL fiber scaffolds [108]. Neurite fields extending from EBs seeded onto aligned fibers showed directional extension along the axis of fiber alignment, and also had longer average length than neurites observed on random fibers. Interactions of EBs with random fiber topography also seemed to result in a higher proportion of cells that differentiated into astrocytes.

During Wallerian degeneration following peripheral nerve injury, Schwann cells respond to the loss of axonal contacts by dedifferentiation and proliferation within the injury site. The Schwann cells then re-form into bands of Büngner to guide regenerating nerve ends, then finally regain their myelinating phenotypes after axonal reconnection is complete [109, 110]. Chew et al. set out to establish the impact of aligned electrospun fibers on the maturation of these primary Schwann cells by evaluating changes in gene expression [111]. Fiber alignment was found to promote the up-regulation of the pro-myelination genes P0 and MAG as compared with random electrospun fibers and two-dimensional polymeric films, suggesting that fiber alignment was an effective cue in directing cells towards a more differentiated state. The authors hypothesized that this pro-differentiation effect was responsible for the significant functional recovery observed when polymeric conduits presenting an aligned fiber topography were used to guide regenerating axons following rat sciatic nerve injury [112].

In addition to fiber alignment as a potential neurotrophic cue, neural stem cells were demonstrated to be responsive to the dimensions of single electrospun fibers. In one study, the effect of electrospun fiber diameter and alignment on C17.2 mouse neonatal cerebellum stem cells was investigated [113]. Random and aligned mats of fibers with average diameters of 300 nm and 1.5 μm were electrospun from poly-L-lactide (PLLA). The average length of neurites extended from C17.2 cells was significantly longer when cultured on aligned nanofibers than on the other scaffolds, indicating that the fiber dimension in combination with fiber alignment helped to enhance C17.2 differentiation. Further evidence that stem cells are sensitive to electrospun fiber diameter was presented in the work of Christopherson et al. [49]. Rat adult hippocampal-derived NSCs were cultured on laminin-coated PES electrospun fibers and differentiation was induced by retinoic acid treatment. NSCs cultured on fibers with average diameter of 283 nm extended processes that were guided by the underlying nanofiber morphology and showed preferential differentiation into oligodendrocytes. NSCs cultured on larger diameter (average ~749 nm) fibers were restricted in their ability to spread and migrate, and consequently a higher proportion of neuronal progenitors were observed.

Nanofibrous scaffolds resulting from self-assembled peptide amphiphiles can be modified to present signaling epitopes particularly relevant to NSCs, for example epitopes from laminin, which is the predominant ECM protein within the NSC niche. This approach is advantageous over presenting laminin as a component of the scaffold, for example via coating or electrospinning [47], as a supraphysiological density of the signaling epitopes can be achieved. A peptide amphiphile nanofiber network incorporating the laminin-derived pentapeptide epitope IKVAV preferentially enhanced differentiation of embryonic mouse NSCs into neuronal over astrocytic lineage [61]; neuronal differentiation was observed in almost 50% of cells in nanofiber hydrogel, as compared with only 10% of cells cultured on either laminin or poly-D-lysine coated 2D substrates. This effect was attributed to the 1000-fold higher calculated density of IKVAV peptide that could be achieved in the PA gel as compared to a close-packed model of laminin coating.

Electrical stimulation was found to increase regeneration of both motor [114] and sensory [115] peripheral nerves via activation of genes associated with the cellular response to injury [116] as well as secretion of neurotrophic molecules [117]. Sensing of extrinsically applied electrical fields was found to be facilitated by transmembrane ion channels, which appear and become functional even at early stages of embryonic development [118]. Electroactive polymeric materials were investigated as substrates for exposing cells in vitro and in vivo to mild electric potential or current [119, 120], and enhanced neurite outgrowth was observed in PC12 cells seeded on polypyrrole films and subjected to a constant 100 mV potential [121]. Electrically conductive polymers can further be formed into three-dimensional architectures to integrate structural and electrical cues. Lee et al. coated aligned and random electrospun poly-L-lactide-*co*-glycolide (PLGA) fiber matrices with polypyrrole and subjected PC12 cells seeded on the meshes to a constant electrical potential [122], and showed that a higher proportion of stimulated cells extended neurites with a longer median length than cells that did not receive stimulation.

5.2 Nanofiber-mediated Stem Cell Differentiation into Chondrogenic and Osteogenic Lineages

MSCs, most commonly derived from bone marrow, are an attractive cell source for cell replacement therapy due to their relative ease of derivation and expansion, as well as well-established multilineage differentiation potential [123]. For the purposes of musculoskeletal tissue engineering, cell therapy must often be complemented by suitable scaffolding that supports cell migration, repopulation, and matrix deposition, while simultaneously maintaining the desired shape of the final tissue. MSCs have been cultured with a variety of scaffold types including hydrogels, foams, microfiber and nanofiber matrices [124–128]. These studies clearly established the compatibility of various scaffolding materials for MSC culture. However, whether biomimetic scaffolds can be designed that actively instruct cell fates remains an area of active exploration.

Repair of damaged cartilage is a particularly challenging problem because the avascular nature of the tissue offers limited access to both the nutrient supply as well as a cell source for repopulating the injury site. The high porosity inherent within nanofibrous scaffolds offers a microenvironment particularly suited to facilitating biological processes that require intimate cell–cell or cell–matrix contact. Nanofibrous matrices are a logical alternative to the current protocol for chondrogenic induction of MSCs, which calls for high density pellet culture to simulate cartilage development within the embryo [129, 130]. TGF-β1-mediated chondrogenic induction of hMSCs seeded into an electrospun PCL fibrous scaffold proved to be more effective than similar induction of hMSC pellets [125]. Although cartilaginous gene expression as measured by RT–PCR was at a similar extent in both culture systems, hMSCs in fibrous scaffolds consistently produced a greater amount of sulfated glycosaminoglycans, possibly due to more available space within the scaffold for ECM accumulation. Interestingly, chondrocyte-like cells in the fibrous scaffold were also organized into a zonal arrangement that was somewhat reminiscent of native articular cartilage. In an extension of this work, it was also later demonstrated that this platform was amenable to the production of pre-shaped cartilaginous constructs without loss of chondrogenic differentiation potential [131]. Further studies revealed that the mechanical properties of this type of engineered cartilage could also be enhanced when MSC morphology was oriented in response to fiber alignment [132]. MSCs were cultured on aligned PCL fibers in an attempt to replicate the organization of the fibrocartilaginous menisci of the knee, where fibrochondrocytes are surrounded by a tight network of radially aligned collagen I filaments. After 10 weeks in culture, aligned fiber constructs had higher stiffness and modulus compared with random fiber constructs due to the more organized deposition of ECM in the former, even though overall levels of GAG and total collagen production were comparable.

One strategy for designing biomimetic scaffolds for the osteogenic niche is to incorporate ceramics or minerals with similar composition of the inorganic component of the native bone in order to promote osteoinduction and osteointegration between native bone and the scaffold [133, 134]. Bioceramics such as calcium phosphate or hydroxyapatite have demonstrated excellent biocompatibility and bioactivity both in vitro and in vivo [135, 136]. These ceramics can be blended with or used to coat nanofibrous scaffolds in order to ameliorate the intrinsically brittle and fragile nature of the inorganic components. One approach is to incorporate or coat electrospun fibers with calcium phosphate as a means of achieving mineralization in vitro via hydroxyapatite deposition on the fiber surface following immersion in simulated body fluid. Calcium phosphate can be loaded into the polymer solution as aerosolized nanoparticles or crystals prior to electrospinning [137, 138], or simply coated onto mats of nanofibers [139]. These mineralized scaffolds did not have any adverse effects on biocompatibility, as evidenced by their ability to support continued proliferation and osteogenic differentiation of hMSCs. In a study performed by Lee et al., hydroxyapatite encapsulation into PLGA nanofiber matrix resulted in increased alkaline phosphatase activity as well as osteogenic gene expression and further scaffold mineralization through calcium deposition [140].

Osteoblasts and osteoclasts that comprise the cellular component of bone interact with proteins in their niche in a number of ways, such as integrin-mediated binding to osteopontin and bone sialoprotein [141]. Using self-assembling peptides, a nanofiber matrix can be prepared that presents short peptide signaling motifs that underlie these bone matrix interactions [142]. Peptide sequences tested were osteogenic growth peptide ALK [143, 144], the osteopontin cell adhesion motive DGR, and a synthetic sequence containing two RGD cell adhesion motifs. Preosteoblastic MC3T3-E1 cells were cultured on the various hydrogel nanofiber scaffolds and induced to differentiate into bone. The highest degree of osteogenic differentiation was observed in nanofiber matrices presenting the dual-RGD motif, with the highest alkaline phosphatase activity as well as secretion of osteocalcin, although all motif-presenting scaffolds showed greater differentiation than the control nanofiber matrix without any signaling sequence. These results directly implicate the importance of cell–scaffold adhesive interactions in mediating osteogenic differentiation.

5.3 Nanofiber-mediated Stem Cell Differentiation into Myogenic Lineage

Formation of mature skeletal muscle requires the fusion of skeletal myoblasts into bundles of muscle fibers. Although muscle cells are considered to be post-mitotic and terminally differentiated, evidence exists to suggest that injured muscle tissue retains a limited capacity for regeneration. This is believed to occur via activation of a normally quiescent population of progenitor cells known as satellite cells that normally reside below the basal lamina of muscle fibers. Upon activation, satellite cells proliferate, undergo myogenic differentiation to form myoblasts that eventually give rise to new muscle fibers, or fuse with existing damaged fibers [145, 146]. Although the biochemical [147, 148] and biophysical features [149, 150] of the satellite cell niche have been extensively examined elsewhere, less is understood about the effects of the structural features of the niche on myoblast differentiation. Designing nanofiber scaffolds intended to simulate the structure and organization within the basal lamina is one strategy for engineering muscle tissue adopted by several research groups.

The ability of the underlying substrate topography to dictate cell morphology is the driving principle behind the use of aligned electrospun fibers as a scaffold for the regeneration of muscle fibers in vitro. The densely packed, unidirectional fibrous substrate closely mimics the organization of native muscle tissue. A working hypothesis is that satellite cell alignment induced by the substrate can promote myoblast differentiation and facilitate eventual fusion into new muscle fibers. Choi et al. demonstrated that culture on aligned PCL fibers resulted in highly oriented F-actin organization in myotubes [151]. Although the width of myotubes was approximately the same on both random and aligned fibers, myotubes on aligned fibers were on average approximately twice as long as myotubes formed on random fibers [152]. A separate study showed that C2C12 myoblasts cultured on aligned polyesterurethane fibers experienced up-regulation of myogenin and myosin heavy chain (MHC) gene expression, concomitant with the formation of putative multinucleated myotubes in highly parallel arrays, which was not observed on random fibers [153]. The inductive potential of topographical alignment beyond merely facilitating myotube formation and fusion was also suggested in a study by Dang et al., in which hMSCs cultured on aligned hydroxybutyl chitosan fibers showed up-regulation of the myogenic genes collagen IV, desmin, Pax-3, Pax-7 and myogenin compared with hMSCs cultured on films [154].

Electrospun fibers comprised of a blend of poly-L-lactide-*co*-caprolactone (PLCL) and polyaniline (PAN) were used as an electroactive scaffold to facilitate myoblast differentiation into myotubes [155]. PAN is a conducting polymer and was previously shown to support cardiac myoblast proliferation while also retaining its conductive properties under tissue culture conditions [156]. PLCL was blended with PAN to fabricate scaffolds due to the intrinsic lack of mechanical rigidity and stability of PAN alone. C2C12 myoblasts cultured on PAN-containing nanofiber matrix under differentiation conditions showed significantly higher quantities of myotube formation as compared with myoblasts on cultured on matrix without PAN; increasing the proportion of PAN in the matrix from 15 to 30% also resulted in formation of larger myotubes. Furthermore, expression of the myogenic genes myogenin, troponin T and myosin heavy chain were up-regulated in myoblasts cultured on an electroactive nanofiber matrix. Although the cells were not subject to electrical stimulation during the culture period, it was hypothesized that the electrical conductivity of the substrate may facilitate better cell–cell communication. This study implies the importance of exposing skeletal muscle cells to electrical impulses to enhance muscle regeneration.

6 Nanofibrous Matrices for Stem Cell Delivery

Stem cell-based therapy offers tremendous potential for the repair and regeneration of tissues damaged by injury or disease. Using animal models, several groups have shown that they can achieve a significant level of functional recovery simply by injecting the stem cells intravenously [157, 158] or directly into the tissue of interest [159–161]. A basic premise of this strategy is that tissue-specific signals emanating from local microenvironments, both from somatic cells and ECM, direct the implanted stem cells to differentiate towards specific lineages. Despite these promising results, however, bone marrow transplantation remains the only stem cell-based therapy with any real clinical success for humans. Among the reasons indicated for the lack of success with other stem cell therapies are the large-scale cell loss, cell death, and poor engraftment of transplanted stem cells, and the poor control over stem cell fate after implantation [162]. Likely, the local microenvironments present in damaged or diseased tissues do not provide the proper signals for stem cell retention, survival, and differentiation.

To improve the efficacy of stem cell-based therapies, a variety of delivery scaffolds are being developed with the aim of providing microenvironmental support for implanted stem cells [162]. Such scaffolds act as synthetic analogs of the ECM, providing adhesion cues to maintain the viability of implanted cells and retain them at the implant site. Scaffolds may also be designed to encourage stem cell differentiation towards specialized lineages and to provide structural support for the newly developing tissue. They can be seeded with cells prior to implantation (in the case of solid scaffolds) or mixed with the cells and spontaneously assembled in situ (in the case of nanofiber hydrogels).

To date, a limited number of nanofibrous scaffolds have been developed for in vivo stem cell delivery. Electrospun fibers have been used to deliver stem cells for various tissue regeneration applications. Li et al. reported the use of electrospun PCL scaffolds for articular cartilage repair in a swine model [163]. The scaffolds were seeded with either allogenic chondrocytes or xenogenic human MSCs, and maintained in culture for a period of 3 weeks prior to implantation into 7 mm full-thickness cartilage defects created in the distal weight-bearing surface of femoral condyles. At 6 months post-implantation, the MSC-seeded constructs showed the most complete repair of the defects, reproducing a smooth hyaline-like cartilage surface with mechanical properties similar to native hyaline cartilage.

Hashi et al. demonstrated the advantage of using a tubular PLLA nanofiber scaffolds to improve the efficiency of MSC-based vascular tissue engineering in a rat model [164]. After seeding MSCs onto flat, electrospun nanofiber sheets and allowing the cells to attach for 1 day, the authors wrapped the sheets around 0.7 mm mandrels and bound them by sutures to maintain their tubular shape. The cell-seeded constructs were then removed from the mandrels and cultured for a period of 2 days prior to implantation, resulting in tubular scaffolds with embedded and circumferentially aligned MSCs. Similar acellular scaffolds were also prepared as controls. The scaffolds were implanted by suturing them to the common carotid artery using an end-to-end anastomosis procedure. At 2 months post-implantation, large numbers of endothelial cells and smooth muscle cells were present in both the MSC-seeded and acellular scaffolds, and complete endothelialization was observed. However, only the MSC-seeded constructs were non-thrombogenic and they showed excellent long-term patency compared to acellular scaffolds. Surprisingly, the authors attributed the presence of endothelial cells and smooth muscle cells in MSC-seeded constructs solely to recruitment from the surrounding host tissue, even though it is possible that the implanted MSCs may also differentiated into both cell types in situ.

Due to the ease of cell delivery by injectable hydrogels, self-assembled nanofiber hydrogels have also been explored as an artificial matrix to improve the retention, survival, and differentiation of implanted stem cells. Using a β-sheet forming peptide nanofiber hydrogel, Davis et al. showed that a nanofiber hydrogel, when injected into the wall of the left ventricle of adult mice, created a favorable microenvironment within the myocardium for recruitment of host endothelial cells and smooth muscle cells [165]. They also tested the delivery of undifferentiated ESCs to myocardium using this injectable hydrogel, with injected ESCs spontaneously differentiating into cardiac myocytes after 2 weeks, likely due to signals emanating from the myocardial microenvironment. Introducing a local release function of growth factors in the hydrogel may significantly improve the regeneration outcome. This can be achieved by entrapping a sustained release system inside the hydrogel or by immobilization of growth factors to the peptide nanofibers. Davis et al. have achieved local and sustained release of insulin-like growth factor 1 (IGF-1), known to increase cardiac stem cell growth, by using a "biotin-streptavidin sandwich" scheme [66]. Using this method, IGF-1 was loaded without interfering with the self-assembly of peptide nanofibers, and had controlled release for 28 days to the local myocardial tissue. This IGF-1-loaded hydrogel, when injected together with cardiomyoctes, resulted in reduced cardiomyocyte apoptosis and improved restoration of systolic function after experimental myocardial infarction.

The peptide amphiphile nanofiber hydrogel has also been shown as an effective delivery vehicle for stem cells. Cell survival and proliferation in the peptide amphiphile hydrogel can be enhanced by incorporating cell adhesion ligands. Webber et al. constructed a binary nanofiber hydrogel with a mixture of RGDS-conjugated and unconjugated peptide amphiphiles [166]. The luciferase-expressing bone marrow mononuclear cells (BMNCs) were encapsulated in the nanofiber hydrogel and injected subcutaneously in a mouse model. At 4 days post-injection, the cells implanted within the RGDS-containing nanofiber hydrogel showed a significant increase in bioluminescence (315%) when compared to the initial baseline. This increase in bioluminescence was also significantly greater than cells implanted in hydrogel without RGDS ligand (127%) and with saline (147%). This study indicated that cell adhesion cues in nanofiber matrix are crucial to support BMNC survival and proliferation in vivo.

7 Summary

Nanofibrous matrices prepared by electrospinning, solution phase separation, and self-assembly provide a unique set of substrates that can be used to partially recreate stem cell niches for the expansion, differentiation and delivery of stem/progenitor cells. Nanotopography presented by these nanofiber matrices has been shown to significantly influence stem cell adhesion, migration, proliferation and differentiation. These matrices provide great opportunities to modulate cellular behavior and tissue regeneration at the nanometer scale, the same scale where molecular signaling transduction occurs at the cell–cell and cell–substrate interfaces.

Integrating the biochemical cues and topographical cues is an important engineering consideration in designing the nanofiber matrix. A nanofiber matrix without any cell adhesion ligands or other biochemical cues is less likely to exert significant influence to stem cells cultured in the matrix. Specific cell adhesion signals should be an integral component for future design of nanofiber matrix; and the capability to enhance the local stability, concentration and presentation of growth factors relevant to stem cell control will be an important feature of the artificial stem cell niches. In addition, the optimal configuration of topographical cues (diameter, alignment, pore structure, mechanical properties, etc.) and biochemical cues—including matrix bound and supplemented factors—needs to be tailored for each cell type. Similarly, the set of effective cues for expansion and differentiation are likely to be different.

References

1. Schofield, R.: Relationship between spleen colony-forming cell and hematopoietic stem-cell—hypothesis. Blood Cells **4**(1–2), 7–25 (1978)
2. Jones, D.L., Wagers, A.J.: No place like home: anatomy and function of the stem cell niche. Nat. Rev. Mol. Cell Biol. **9**(1), 11–21 (2008)
3. Lathia, J.D., et al.: The microenvironment of the embryonic neural stem cell: lessons from adult niches? Dev. Dyn. **236**(12), 3267–3282 (2007)
4. Lin, H.F.: The stem-cell niche theory: lessons from flies. Nat. Rev. Genet. **3**(12), 931–940 (2002)
5. Orford, K.W., Scadden, D.T.: Deconstructing stem cell self-renewal: genetic insights into cell-cycle regulation. Nat. Rev. Genet. **9**(2), 115–128 (2008)
6. Doetsch, F.: A niche for adult neural stem cells. Curr. Opin. Genet. Dev. **13**(5), 543–550 (2003)
7. Scadden, D.T.: The stem-cell niche as an entity of action. Nature **441**(7097), 1075–1079 (2006)
8. Watts, C., et al.: Anatomical perspectives on adult neural stem cells. J. Anat. **207**(3), 197–208 (2005)
9. Yamashita, Y.M., Fuller, M.T., Jones, D.L.: Signaling in stem cell niches: lessons from the *Drosophila* germline. J. Cell Sci. **118**(Pt 4), 665–672 (2005)
10. Song, H., Stevens, C.F., Gage, F.H.: Astroglia induce neurogenesis from adult neural stem cells. Nature **417**(6884), 39–44 (2002)
11. Campos, L.S.: Beta1 integrins and neural stem cells: making sense of the extracellular environment. Bioessays **27**(7), 698–707 (2005)
12. Hubbell, J.A.: Matrix-bound growth factors in tissue repair. Swiss Med. Wkl. **136**(25–26), 387–391 (2006)
13. Sommer, L.: Growth factors regulating neural crest cell fate decisions. Adv. Exp. Med. Biol. **589**, 197–205 (2006)
14. Lutolf, M.P., Hubbell, J.A.: Synthetic biomaterials as instructive extracellular microenvironments for morphogenesis in tissue engineering. Nat. Biotechnol. **23**(1), 47–55 (2005)
15. Engler, A.J., et al.: Matrix elasticity directs stem cell lineage specification. Cell **126**(4), 677–689 (2006)
16. Saha, K., et al.: Substrate modulus directs neural stem cell behavior. Biophys. J. **95**(9), 4426–4438 (2008)
17. Ojeda, J.L., Piedra, S.: Evidence of a new transitory extracellular structure within the developing rhombencephalic cavity—an ultrastructural and immunoelectron-microscopic study in the chick. Anat. Embryol. **202**(3), 257–264 (2000)
18. Nishimura, S., et al.: Three-dimensional architecture and distribution of collagen components in the goat hypophysis. Anat. Rec.: Part A Discov. Mol. Cell. Evol. Biol. **277A**(2), 275–286 (2004)
19. Trelstad, R., Alberts, B., et al.: Molecular Biology of the Cell. p. 1106. Garland Science, New York (2002)
20. Birk, D.E., Trelstad, R.L.: Extracellular compartments in matrix morphogenesis—collagen fibril, bundle, and lamellar formation by corneal fibroblasts. J. Cell Biol. **99**(6), 2024–2033 (1984)
21. Yim, E.K., Leong, K.W.: Significance of synthetic nanostructures in dictating cellular response. Nanomedicine **1**(1), 10–21 (2005)
22. Kopp, H.G., et al.: The bone marrow vascular niche: home of HSC differentiation and mobilization. Physiology **20**, 349–356 (2005)
23. Pierret, C., et al.: Elements of a neural stem cell niche derived from embryonic stem cells. Stem Cells Dev. **16**(6), 1017–1026 (2007)

24. Liliensiek, S.J., Nealey, P., Murphy, C.J.: Characterization of endothelial basement membrane nanotopography in rhesus macaque as a guide for vessel tissue engineering. Tissue Eng: Part A **15**(9), 2643–2651 (2009)
25. Seri, B., et al.: Cell types, lineage, and architecture of the germinal zone in the adult dentate gyrus (vol 478, p. 359, 2004). J. Comp. Neurol. **480**(4), 427 (2004)
26. Stevens, M.M., George, J.H.: Exploring and engineering the cell surface interface. Science **310**(5751), 1135–1138 (2005)
27. Barbucci, R., et al.: Micro and nano-structured surfaces. J. Mater. Sci. Mater. Med. **14**(8), 721–725 (2003)
28. Curtis, A., Wilkinson, C.: Topographical control of cells. Biomaterials **18**(24), 1573–1583 (1997)
29. Dalby, M.J., et al.: Use of nanotopography to study mechanotransduction in fibroblasts—methods and perspectives. Eur. J. Cell Biol. **83**(4), 159–169 (2004)
30. Flemming, R.G., et al.: Effects of synthetic micro- and nano-structured surfaces on cell behavior. Biomaterials **20**(6), 573–588 (1999)
31. Ellis-Behnke, R.G., et al.: Nano neuro knitting: peptide nanofiber scaffold for brain repair and axon regeneration with functional return of vision. Proc. Natl. Acad. Sci. U.S.A. **103**(13), 5054–5059 (2006)
32. Hirsch, M., et al.: Fine structure of the developing avian corneal stroma as revealed by quick-freeze, deep-etch electron microscopy. Exp. Eye Res. **69**(3), 267–277 (1999)
33. Greiner, A., Wendorff, J.H.: Electrospinning: a fascinating method for the preparation of ultrathin fibres. Angew. Chem. Int. Ed. **46**(30), 5670–5703 (2007)
34. Doshi, J.R., Reneker, D.H.: Electrospinning process and applications of electrospun fibers. J. Electrost. **35**(2–3), 151–160 (1995)
35. Dersch, R., et al.: Electrospun nanofibers: internal structure and intrinsic orientation. J. Polym. Sci.: Part A Polym. Chem. **41**(4), 545–553 (2003)
36. Theron, A., Zussman, E., Yarin, A.L.: Electrostatic field-assisted alignment of electrospun nanofibres. Nanotechnology **12**(3), 384–390 (2001)
37. Lin, K., et al.: Reducing electrospun nanofiber diameter and variability using cationic amphiphiles. Polymer **48**(21), 6384–6394 (2007)
38. Tan, S.H., et al.: Systematic parameter study for ultra-fine fiber fabrication via electrospinning process. Polymer **46**(16), 6128–6134 (2005)
39. Li, D., Wang, Y.L., Xia, Y.N.: Electrospinning nanofibers as uniaxially aligned arrays and layer-by-layer stacked films. Adv. Mater. **16**(4), 361–366 (2004)
40. Sun, D.H., et al.: Near-field electrospinning. Nano Lett. **6**(4), 839–842 (2006)
41. Sundaray, B., et al.: Electrospinning of continuous aligned polymer fibers. Appl. Phys. Lett. **84**(7), 1222–1224 (2004)
42. Bognitzki, M. et al.: Preparation of fibers with nanoscaled morphologies: electrospinning of polymer blends. Polym. Eng. Sci. **41**(6), 982–989 (2001)
43. Sun, Z.C., et al.: Compound core-shell polymer nanofibers by co-electrospinning. Adv. Mater. **15**(22), 1929–1932 (2003)
44. Li, W.J., et al.: Multilineage differentiation of human mesenchymal stem cells in a three-dimensional nanofibrous scaffold. Biomaterials **26**(25), 5158–5166 (2005)
45. Ohkawa, K.: Electrospinning of chitosan. Macromol. Rapid Commun. **25**(18), 1600–1605 (2004)
46. Kidoaki, S., Kwon, I.K., Matsuda, T.: Mesoscopic spatial designs of nano- and microfiber meshes for tissue-engineering matrix and scaffold based on newly devised multilayering and mixing electrospinning techniques. Biomaterials **26**(1), 37–46 (2005)
47. Neal, R.A., et al.: Laminin nanofiber meshes that mimic morphological properties and bioactivity of basement membranes. Tissue Eng.: Part C Methods **15**(1), 11–21 (2009)
48. Yoo, H.S., Kim, T.G., Park, T.G.: Surface-functionalized electrospun nanofibers for tissue engineering and drug delivery. Adv. Drug Deliv. Rev. **61**(12), 1033–1042 (2009)
49. Christopherson, G.T., Song, H., Mao, H.Q.: The influence of fiber diameter of electrospun substrates on neural stem cell differentiation and proliferation. Biomaterials **30**(4), 556–564 (2009)

50. Yao, C., Li, X., Neoh, K.G., Shi, Z., Kang, E.T.: Surface modification and antibacterial activity of electrospun polyurethane fibrous membranes with quaternary ammonium moieties. J. Membr. Sci. **320**(1–2), 259–267 (2008)
51. Chua, K.N. et al.: Stable immobilization of rat hepatocyte spheroids on galactosylated nanofiber scaffold. Biomaterials **26**(15), 2537–2547 (2005)
52. Chua, K.N., et al.: Functional nanofiber scaffolds with different spacers modulate adhesion and expansion of cryopreserved umbilical cord blood hematopoietic stem/progenitor cells. Exp. Hematol. **35**(5), 771–781 (2007)
53. Kim, T.G., Park, T.G.: Biomimicking extracellular matrix: cell adhesive RGD peptide modified electrospun poly(D,L-lactic-co-glycolic acid) nanofiber mesh. Tissue Eng. **12**(2), 221–233 (2006)
54. Stupp, S.I.: Introduction to functional nanostructures. Chem. Rev. **105**(4), 1023–1024 (2005)
55. Whitesides, G.M., Grzybowski, B.: Self-assembly at all scales. Science **295**(5564), 2418–2421 (2002)
56. Lehn, J.M.: Supramolecular chemistry: from molecular information towards self-organization and complex matter. Rep. Prog. Phys. **67**(3), 249–265 (2004)
57. Israelachvili, J.N.: Intermolecular and Surface Forces, 2nd edn. Academic Press, New York (1992)
58. Hartgerink, J.D., Beniash, E., Stupp, S.I.: Self-assembly and mineralization of peptide-amphiphile nanofibers. Science **294**(5547), 1684–1688 (2001)
59. Hartgerink, J.D., Beniash, E., Stupp, S.I.: Peptide-amphiphile nanofibers: a versatile scaffold for the preparation of self-assembling materials. Proc. Natl. Acad. Sci. U.S.A. **99**(8), 5133–5138 (2002)
60. Paramonov, S.E., Jun, H.W., Hartgerink, J.D.: Self-assembly of peptide-amphiphile nanofibers: the roles of hydrogen bonding and amphiphilic packing. J. Am. Chem. Soc. **128**(22), 7291–7298 (2006)
61. Silva, G.A., et al.: Selective differentiation of neural progenitor cells by high-epitope density nanofibers. Science **303**(5662), 1352–1355 (2004)
62. Tysseling-Mattiace, V.M., et al.: Self-assembling nanofibers inhibit glial scar formation and promote axon elongation after spinal cord injury. J. Neurosci.: Off. J. Soc. Neurosci. **28**(14), 3814–3823 (2008)
63. Zhang, S.G., et al.: Spontaneous assembly of a self-complementary oligopeptide to form a stable macroscopic membrane. Proc. Natl. Acad. Sci. U.S.A. **90**(8), 3334–3338 (1993)
64. Zhang, S.G.: Fabrication of novel biomaterials through molecular self-assembly. Nat. Biotechnol. **21**(10), 1171–1178 (2003)
65. Altman, M., et al.: Conformational behavior of ionic self-complementary peptides. Protein Sci. **9**(6), 1095–1105 (2000)
66. Davis, M.E., et al.: Local myocardial insulin-like growth factor 1 (IGF-1) delivery with biotinylated peptide nanofibers improves cell therapy for myocardial infarction. Proc. Natl. Acad. Sci. U.S.A. **103**(21), 8155–8160 (2006)
67. Stupp, S., et al.: Aligned nanofibers and related methods of use. United States Patent Application US 2008/0299657 A1 (2008)
68. Ma, P.X., Zhang, R.Y.: Synthetic nano-scale fibrous extracellular matrix. J. Biomed. Mater. Res. **46**(1), 60–72 (1999)
69. Zhang, R., Ma, P.: Processing of polymer scaffolds: phase separation. In: Atala, A., Lanza, R.P. (eds.) Methods of Tissue Engineering. pp. 715–724. Academic Press, San Diego (2002)
70. Yang, F., et al.: Fabrication of nano-structured porous PLLA scaffold intended for nerve tissue engineering. Biomaterials **25**(10), 1891–1900 (2004)
71. Hu, J., et al.: Chondrogenic and osteogenic differentiations of human bone marrow-derived mesenchymal stem cells on a nanofibrous scaffold with designed pore network. Biomaterials **30**(28), 5061–5067 (2009)
72. Smith, L.A., Ma, P.X.: Nano-fibrous scaffolds for tissue engineering. Colloids Surf B Biointerfaces **39**(3), 125–131 (2004)

73. Liu, X., Ma, P.X.: Phase separation, pore structure, and properties of nanofibrous gelatin scaffolds. Biomaterials **30**(25), 4094–4103 (2009)
74. Hackney, J.A., et al.: A molecular profile of a hematopoietic stem cell niche. Proc. Natl. Acad. Sci. U.S.A. **99**(20), 13061–13066 (2002)
75. Mayani, H., Guilbert, L.J., Janowska-Wieczorek, A.: Biology of the hemopoietic microenvironment. Eur. J. Haematol. **49**(5), 225–233 (1992)
76. Nishikawa, M., et al.: Role of the microenvironment of the embryonic aorta-gonad-mesonephros region in hematopoiesis. Ann. N. Y. Acad. Sci. **938**, 109–116 (2001)
77. Verfaillie, C.M., et al.: The hematopoietic microenvironment: stromal extracellular matrix components as growth regulators for human hematopoietic progenitors. Hematology **4**(4), 321–333 (1999)
78. Sauvageau, G., Iscove, N.N., Humphries, R.K.: In vitro and in vivo expansion of hematopoietic stem cells. Oncogene **23**(43), 7223–7232 (2004)
79. Sorrentino, B.P.: Clinical strategies for expansion of haematopoietic stem cells. Nat. Rev. Immunol. **4**(11), 878–888 (2004)
80. Douay, L.: Experimental culture conditions are critical for ex vivo expansion of hematopoietic cells. J. Hematother. Stem Cell Res. **10**(3), 341–346 (2001)
81. Nakahata, T.: Ex vivo expansion of human hematopoietic stem cells. Int. J. Hematol. **73**(1), 6–13 (2001)
82. Fox, N., et al.: Thrombopoietin expands hematopoietic stem cells after transplantation. J. Clin. Invest. **110**(3), 389–394 (2002)
83. Audet, J., et al.: Common and distinct features of cytokine effects on hematopoietic stem and progenitor cells revealed by dose-response surface analysis. Biotechnol. Bioeng. **80**(4), 393–404 (2002)
84. Bagley, J., et al.: Extended culture of multipotent hematopoietic progenitors without cytokine augmentation in a novel three-dimensional device. Exp. Hematol. **27**(3), 496–504 (1999)
85. Li, Y., et al.: Human cord cell hematopoiesis in three-dimensional nonwoven fibrous matrices: in vitro simulation of the marrow microenvironment. J. Hematother. Stem Cell Res. **10**(3), 355–368 (2001)
86. Dao, M.A., et al.: Adhesion to fibronectin maintains regenerative capacity during ex vivo culture and transduction of human hematopoietic stem and progenitor cells. Blood **92**(12), 4612–4621 (1998)
87. Feng, Q., et al.: Expansion of engrafting human hematopoietic stem/progenitor cells in three-dimensional scaffolds with surface-immobilized fibronectin. J. Biomed. Mater. Res. A. **78**(4), 781–791 (2006)
88. Chua, K.N., et al.: Surface-aminated electrospun nanofibers enhance adhesion and expansion of human umbilical cord blood hematopoietic stem/progenitor cells. Biomaterials **27**(36), 6043–6051 (2006)
89. Das, H., et al.: Ex Vivo nanofiber expansion and genetic modification of human cord blood-derived progenitor/stem cells enhances vasculogenesis. In: 4th International Conference on Cell Therapy for Cardiovascular Diseases. New York, NY (2009)
90. Oswald, J., et al.: Gene-expression profiling of CD34(+) hematopoietic cells expanded in a collagen I matrix. Stem Cells **24**(3), 494–500 (2006)
91. Carpenter, M.K., et al.: In vitro expansion of a multipotent population of human neural progenitor cells. Exp. Neurol. **158**(2), 265–278 (1999)
92. Sen, A., Kallos, M.S., Behie, L.A.: Passaging protocols for mammalian neural stem cells in suspension bioreactors. Biotechnol. Prog. **18**(2), 337–345 (2002)
93. Wachs, F.P., et al.: High efficacy of clonal growth and expansion of adult neural stem cells. Lab. Invest. **83**(7), 949–962 (2003)
94. Correia, A.S., et al.: Stem cell-based therapy for Parkinson's disease. Ann. Med. **37**(7), 487–498 (2005)
95. Roy, N.S., et al.: Functional engraftment of human ES cell-derived dopaminergic neurons enriched by coculture with telomerase-immortalized midbrain astrocytes. Nat. Med. **12**(11), 1259–1268 (2006)

96. Nisbet, D.R., et al.: Characterization of neural stem cells on electrospun poly(epsilon-caprolactone) submicron scaffolds: evaluating their potential in neural tissue engineering. J. Biomateri. Sci. Polym. Ed. **19**(5), 623–634 (2008)
97. Martin, G.R., Timpl, R.: Laminin and other basement-membrane components. Annu. Rev. Cell Biol. **3**, 57–85 (1987)
98. Nur-E-Kamal, A., et al.: Three-dimensional nanofibrillar surfaces promote self-renewal in mouse embryonic stem cells. Stem Cells **24**(2), 426–433 (2006)
99. Shih, Y.R.V., et al.: Growth of mesenchymal stem cells on electrospun type I collagen nanofibers. Stem Cells **24**(11), 2391–2397 (2006)
100. Hosseinkhani, H., Hosseinkhani, M., Kobayashi, H.: Proliferation and differentiation of mesenchymal stem cells using self-assembled peptide amphiphile nanofibers. Biomed. Mater. **1**(1), 8–15 (2006)
101. McBeath, R., et al.: Cell shape, cytoskeletal tension, and RhoA regulate stem cell lineage commitment. Dev. Cell **6**(4), 483–495 (2004)
102. Trepat, X., et al.: Universal physical responses to stretch in the living cell. Nature **447**(7144), 592–595 (2007)
103. Jakkaraju, S., Zhe, X., Schuger, L.: Role of stretch in activation of smooth muscle cell lineage. Trends Cardiovasc. Med. **13**(8), 330–335 (2003)
104. Serena, E., et al.: Electrical stimulation of human embryonic stem cells: cardiac differentiation and the generation of reactive oxygen species. Exp. Cell Res. **315**(20), 3611–3619 (2009)
105. Yamada, M., et al.: Electrical stimulation modulates fate determination of differentiating embryonic stem cells. Stem Cells **25**(3), 562–570 (2007)
106. Fu, S.Y., Gordon, T.: The cellular and molecular basis of peripheral nerve regeneration. Mol. Neurobiol. **14**(1–2), 67–116 (1997)
107. Barnabe-Heider, F., Frisen, J.: Stem cells for spinal cord repair. Cell Stem Cell **3**(1), 16–24 (2008)
108. Xie, J., et al.: The differentiation of embryonic stem cells seeded on electrospun nanofibers into neural lineages. Biomaterials **30**(3), 354–362 (2009)
109. Chen, Z.L., Yu, W.M., Strickland, S.: Peripheral regeneration. Annu. Rev. Neurosci. **30**, 209–233 (2007)
110. Stoll, G., Muller, H.W.: Nerve injury, axonal degeneration and neural regeneration: basic insights. Brain Pathol. **9**(2), 313–325 (1999)
111. Chew, S.Y., et al.: The effect of the alignment of electrospun fibrous scaffolds on Schwann cell maturation. Biomaterials **29**(6), 653–661 (2008)
112. Chew, S.Y., et al.: Aligned protein-polymer composite fibers enhance nerve regeneration: a potential tissue-engineering platform. Adv. Funct. Mater. **17**(8), 1288–1296 (2007)
113. Yang, F., et al.: Electrospinning of nano/micro scale poly(L-lactic acid) aligned fibers and their potential in neural tissue engineering. Biomaterials **26**(15), 2603–2610 (2005)
114. Al-Majed, A.A., et al.: Brief electrical stimulation promotes the speed and accuracy of motor axonal regeneration. J. Neurosci. **20**(7), 2602–2608 (2000)
115. Geremia, N.M., et al.: Electrical stimulation promotes sensory neuron regeneration and growth-associated gene expression. Exp. Neurol. **205**(2), 347–359 (2007)
116. Al-Majed, A.A., Tam, S.L., Gordon, T.: Electrical stimulation accelerates and enhances expression of regeneration-associated genes in regenerating rat femoral motoneurons. Cell Mol. Neurobiol. **24**(3), 379–402 (2004)
117. Karchewski, L.A., et al.: Dynamic patterns of BDNF expression in injured sensory neurons: differential modulation by NGF and NT-3. Eur. J. Neurosci. **16**(8), 1449–1462 (2002)
118. Hotary, K.B., Robinson, K.R.: Endogenous electrical currents and the resultant voltage gradients in the chick embryo. Dev. Biol. **140**(1), 149–160 (1990)
119. Valentini, R.F., et al.: Electrically charged polymeric substrates enhance nerve fibre outgrowth in vitro. Biomaterials **13**(3), 183–190 (1992)
120. Aebischer, P., et al.: Piezoelectric guidance channels enhance regeneration in the mouse sciatic nerve after axotomy. Brain Res. **436**(1), 165–168 (1987)

121. Schmidt, C.E., et al.: Stimulation of neurite outgrowth using an electrically conducting polymer. Proc. Natl. Acad. Sci. U.S.A. **94**(17), 8948–8953 (1997)
122. Lee, J.Y., et al.: Polypyrrole-coated electrospun PLGA nanofibers for neural tissue applications. Biomaterials **30**(26), 4325–4335 (2009)
123. Deans, R.J., Moseley, A.B.: Mesenchymal stem cells: biology and potential clinical uses. Exp. Hematol. **28**(8), 875–884 (2000)
124. Salinas, C.N., Anseth, K.S.: Mesenchymal stem cells for craniofacial tissue regeneration: designing hydrogel delivery vehicles. J. Dent. Res. **88**(8), 681–692 (2009)
125. Li, W.J., et al.: A three-dimensional nanofibrous scaffold for cartilage tissue engineering using human mesenchymal stem cells. Biomaterials **26**(6), 599–609 (2005)
126. Wang, Y., et al.: In vitro cartilage tissue engineering with 3D porous aqueous-derived silk scaffolds and mesenchymal stem cells. Biomaterials **26**(34), 7082–7094 (2005)
127. Yim, E.K., et al.: Proliferation and differentiation of human mesenchymal stem cell encapsulated in polyelectrolyte complexation fibrous scaffold. Biomaterials **27**(36), 6111–6122 (2006)
128. Aung, T., et al.: Chondroinduction of mouse mesenchymal stem cells in three-dimensional highly porous matrix scaffolds. J. Biomed. Mater. Res. **61**(1), 75–82 (2002)
129. Johnstone, B., et al.: In vitro chondrogenesis of bone marrow-derived mesenchymal progenitor cells. Exp. Cell Res. **238**(1), 265–272 (1998)
130. Barry, F., et al.: Chondrogenic differentiation of mesenchymal stem cells from bone marrow: differentiation-dependent gene expression of matrix components. Exp. Cell Res. **268**(2), 189–200 (2001)
131. Janjanin, S., et al.: Mold-shaped, nanofiber scaffold-based cartilage engineering using human mesenchymal stem cells and bioreactor. J. Surg. Res. **149**(1), 47–56 (2008)
132. Baker, B.M., Mauck, R.L.: The effect of nanofiber alignment on the maturation of engineered meniscus constructs. Biomaterials **28**(11), 1967–1977 (2007)
133. LeGeros, R.Z.: Biodegradation and bioresorption of calcium phosphate ceramics. Clin. Mater. **14**(1), 65–88 (1993)
134. Wang, M.: Developing bioactive composite materials for tissue replacement. Biomaterials **24**(13), 2133–2151 (2003)
135. El-Ghannam, A.: Bone reconstruction: from bioceramics to tissue engineering. Expert. Rev. Med. Devices **2**(1), 87–101 (2005)
136. Guda, T., et al.: A cellular perspective to bioceramic scaffolds for bone tissue engineering: the state of the art. Curr. Top. Med. Chem. **8**(4), 290–299 (2008)
137. McCullen, S.D., et al.: Electrospun composite poly(L-lactic acid)/tricalcium phosphate scaffolds induce proliferation and osteogenic differentiation of human adipose-derived stem cells. Biomed. Mater. doi:10.1088/1748-6041/4/3/035002
138. Schneider, O.D., et al.: Cotton wool-like nanocomposite biomaterials prepared by electrospinning: in vitro bioactivity and osteogenic differentiation of human mesenchymal stem cells. J. Biomed. Mater. Res. B Appl. Biomater. **84**(2), 350–362 (2008)
139. Mavis, B., et al.: Synthesis, characterization and osteoblastic activity of polycaprolactone nanofibers coated with biomimetic calcium phosphate. Acta Biomater. **5**(8), 3098–3111 (2009)
140. Lee, J.H., et al.: Control of osteogenic differentiation and mineralization of human mesenchymal stem cells on composite nanofibers containing poly[lactic-co-(glycolic acid)] and hydroxyapatite. Macromol. Biosci. **10**(2), 173–182 (2010)
141. Damsky, C.H.: Extracellular matrix-integrin interactions in osteoblast function and tissue remodeling. Bone **25**(1), 95–96 (1999)
142. Horii, A., et al.: Biological designer self-assembling Peptide nanofiber scaffolds significantly enhance osteoblast proliferation, differentiation and 3-D migration. PLoS One **2**(2), e190 (2007)
143. Bab, I., et al.: Histone H4-related osteogenic growth peptide (OGP): a novel circulating stimulator of osteoblastic activity. EMBO J. **11**(5), 1867–1873 (1992)

144. Fazzi, R., et al.: Bone and bone-marrow interactions: haematological activity of osteoblastic growth peptide (OGP)-derived carboxy-terminal pentapeptide. Mobilizing properties on white blood cells and peripheral blood stem cells in mice. Leuk. Res. **26**(1), 19–27 (2002)
145. Ehrhardt, J., Morgan, J.: Regenerative capacity of skeletal muscle. Curr. Opin. Neurol. **18**(5), 548–553 (2005)
146. Dhawan, J., Rando, T.A.: Stem cells in postnatal myogenesis: molecular mechanisms of satellite cell quiescence, activation and replenishment. Trends Cell. Biol. **15**(12), 666–673 (2005)
147. Collins, C.A., et al.: Stem cell function, self-renewal, and behavioral heterogeneity of cells from the adult muscle satellite cell niche. Cell **122**(2), 289–301 (2005)
148. Kuang, S., Gillespie, M.A., Rudnicki, M.A.: Niche regulation of muscle satellite cell self-renewal and differentiation. Cell Stem Cell **2**(1), 22–31 (2008)
149. Collinsworth, A.M., et al.: Apparent elastic modulus and hysteresis of skeletal muscle cells throughout differentiation. Am. J. Physiol. Cell Physiol. **283**(4), C1219–1227 (2002)
150. Engler, A.J., et al.: Myotubes differentiate optimally on substrates with tissue-like stiffness: pathological implications for soft or stiff microenvironments. J. Cell. Biol. **166**(6), 877–887 (2004)
151. Choi, J.S., et al.: The influence of electrospun aligned poly(epsilon-caprolactone)/collagen nanofiber meshes on the formation of self-aligned skeletal muscle myotubes. Biomaterials **29**(19), 2899–2906 (2008)
152. Huang, N.F., et al.: Myotube assembly on nanofibrous and micropatterned polymers. Nano Lett. **6**(3), 537–542 (2006)
153. Riboldi, S.A., et al.: Skeletal myogenesis on highly orientated microfibrous polyester-urethane scaffolds. J. Biomed. Mater. Res. A **84**(4), 1094–1101 (2008)
154. Dang, J.M., Leong, K.W.: Myogenic induction of aligned mesenchymal stem cell sheets by culture on thermally responsive electrospun nanofibers. Adv. Mater. **19**(19), 2775–2779 (2007)
155. Jun, I., Jeong, S., Shin, H.: The stimulation of myoblast differentiation by electrically conductive sub-micron fibers. Biomaterials **30**(11), 2038–2047 (2009)
156. Bidez, P.R., III, et al.: Polyaniline, an electroactive polymer, supports adhesion and proliferation of cardiac myoblasts. J. Biomater. Sci. Polym. Ed. **17**(1–2), 199–212 (2006)
157. Qu, C., et al.: Treatment of traumatic brain injury in mice with marrow stromal cells. Brain Res. **1208**, 234–239 (2008)
158. Zhao, C.P., et al.: Human mesenchymal stromal cells ameliorate the phenotype of SOD1-G93A ALS mice. Cytotherapy **9**(5), 414–426 (2007)
159. Parr, A.M., et al.: Transplanted adult spinal cord-derived neural stem/progenitor cells promote early functional recovery after rat spinal cord injury. Neuroscience **155**(3), 760–770 (2008)
160. Zhang, S., et al.: Purified human bone marrow multipotent mesenchymal stem cells regenerate infarcted myocardium in experimental rats. Cell Transplant. **14**(10), 787–798 (2005)
161. Daadi, M.M., Maag, A.L., Steinberg, G.K.: Adherent self-renewable human embryonic stem cell-derived neural stem cell line: functional engraftment in experimental stroke model. PLoS ONE **3**(2), e1644 (2008)
162. Mooney, D.J., Vandenburgh, H.: Cell delivery mechanisms for tissue repair. Cell Stem Cell **2**(3), 205–213 (2008)
163. Li, W.J., et al.: Evaluation of articular cartilage repair using biodegradable nanofibrous scaffolds in a swine model: a pilot study. J. Tissue Eng. Regen. Med. **3**(1), 1–10 (2009)
164. Hashi, C.K., et al.: Antithrombogenic property of bone marrow mesenchymal stem cells in nanofibrous vascular grafts. Proc. Natl. Acad. Sci. U.S.A **104**(29), 11915–11920 (2007)
165. Davis, M.E., et al.: Injectable self-assembling peptide nanofibers create intramyocardial microenvironments for endothelial cells. Circulation **111**(4), 442–450 (2005)
166. Webber, M.J., et al.: Development of bioactive peptide amphiphiles for therapeutic cell delivery. Acta Biomater. **6**(1), 3–11 (2010)

Micropatterned Hydrogels for Stem Cell Culture

Sharon K. Hamilton, Hang Lu and Johnna S. Temenoff

Abstract Stem cells represent a new frontier in tissue engineering and regenerative medicine, due, in large part, to their ability to proliferate and differentiate along multiple lineages for use in a myriad of clinical applications. Efforts are currently underway to understand the molecular mechanisms underlying the decision of stem cells to enter mitotic dormancy, undergo self-renewal, or terminally differentiate. Recent advances in the field of micropatterned biomaterials, specifically hydrogels, have generated new approaches that may allow scientists to more easily address the complex questions commonly encountered in stem cell biology. Given the potential power of the combination of hydrogel biomaterials and micropatterning techniques applied to areas of stem cell biology, this chapter begins with a brief overview of important characteristics of stem cells that could be further understood using micropattered hydrogels, as well as a review of basics of hydrogels and micropatterning technologies. The second half of the chapter provides a summary of particular micropatterned gels that have been applied to in vitro cell-based work, with specific attention paid to how these techniques could be applied toward stem cell research.

Abbreviations

ECM	Extracellular matrix
3D	Three-dimensional
ESCs	Embryonic stem cells

S. K. Hamilton and J. S. Temenoff (✉)
Department of Biomedical Engineering, Georgia Institute of Technology and Emory University, 313 Ferst Dr., Atlanta GA, 30332-0535, USA
e-mail: johnna.temenoff@bme.gatech.edu

S. K. Hamilton and H. Lu
School of Chemical and Biomolecular Engineering, Georgia Institute of Technology, 311 Ferst Dr. NW, Atlanta GA, 30332, USA

Stud Mechanobiol Tissue Eng Biomater (2010) 2: 119–152
DOI: 10.1007/8415_2010_12

Published online: 12 June 2010

HSCs	Hematopoietic stem cells
MSCs	Mesenchymal stem cells
NSCs	Neural stem cells
BMPs	Bone morphogenetic proteins
TGF-βs	Transforming growth factor-βs
GDFs	Growth and differentiation factors
HA	Hyaluronic acid
PEG	Poly(ethylene glycol)
PVA	Poly(vinyl alcohol)
RGD	Arg-Gly-Asp
MMP	Matrix metalloproteinase
MEMS	Microelectromechanical systems
LSL	Laser-scanning lithography
SFL	Stop-flow lithography
OFML	Optofluidic maskless lithography
SLMs	Spatial light modulators
PDMS	Polydimethylsiloxane
GPQG	Gly-Pro-Gln-Gly
TPLS	Two-photon laser scanning
LGPA	Leu-Gly-Pro-Ala
HDF	Human dermal fibroblast
PEG-DA	PEG-diacrylate
PEG-mA	PEG-methacrylate
Hep G2	Human hepatoblastoma cells

1 Introduction: Application of Biomaterial Technologies to Stem Cell Research

Stem cells represent a new frontier in tissue engineering and regenerative medicine, due, in large part, to their ability to proliferate and differentiate along multiple lineages for use in a myriad of clinical applications [1–5]. Over the last decade, major advances have been made in the isolation, culture, and differentiation of a variety of stem cell types. Furthermore, stem cells have been identified in every major organ and tissue of the human body [5]. Intense efforts are currently underway to understand the molecular mechanisms underlying the decision of stem cells to enter mitotic dormancy, undergo self-renewal, or terminally differentiate [6, 7]. Understanding these mechanisms will help researchers realize the full therapeutic potential of stem cells. However, much remains to be explored in order to provide biomedical scientists and engineers the knowledge base to reproducibly direct stem cell function or renewal for scale-up to therapeutic applications.

Recent advances in the field of micropatterned biomaterials have generated new approaches that may allow scientists to more easily address the complex questions commonly encountered in stem cell biology [8–10]. For example, it is now becoming possible to study the dynamic responses of stem cells to well-defined model microenvironments [5, 11]. This will allow researchers to elucidate the role of specific niche components and architecture in regulating fundamental behaviors such as mechanisms of cell division, self-renewal, and differentiation [12]. On the other hand, a better understanding of the molecular mechanisms by which specific interactions impact stem cell self-renewal and differentiation is required for the targeted design of biomaterials to promote these functions [13]. Therefore, greater knowledge from today's model systems could lead to the improved design of future stem cell carrier materials, thereby further refining these technologies and decreasing time to translate stem cell therapies to clinical applications.

While many biomaterial carriers have been explored for stem cells, hydrogels are advantageous because they mimic both chemical and mechanical aspects of the native extracellular matrix (ECM) and therefore may be particularly suited for use in model systems that investigate the role of the extracellular environment in stem cell function [6]. In particular, their ability to provide tunable three-dimensional (3D) environments is of particular interest, since a disconnect often exists between traditional two-dimensional cell culture studies and in vivo response to implanted cell-based constructs [11, 14, 15].

As mentioned above, new techniques have recently been developed to pattern biomaterials with different cell or ligand types at the micrometer scale [16–19]. Such microscale control of cellular environments has been used to probe the influence of the spatial effects of cell–ECM, cell–cell, and cell–soluble factor interaction on stem cell fate [8, 12, 20–23]. The recent application of these micropatterning techniques to hydrogel biomaterials is extremely exciting for stem cell research due to the versatility and tailorability of this combination of technologies. For example, cell-laden hydrogels can be arranged, shaped, and layered using microfluidic patterning and micromolding technologies to better replicate the natural organization of cells in tissues [9, 24, 25]. Additionally, many hydrogels are formed through photoinitiation, which enables the sculpting of desired construct shapes prior to gelation and the subsequent setting of these shapes upon exposure of the solution to UV light [26, 27]. Therefore, hydrogels are prime candidate constructs for micropatterning due to the ease of manipulating sample geometry while maintaining the 3D culture environment (Fig. 1) [9, 28–30].

Given the potential power of the combination of hydrogel biomaterials and micropatterning techniques applied to areas of stem cell biology, this chapter begins with a brief overview of important characteristics of stem cells that could be further understood using micropattered hydrogels, as well as a review of basics of hydrogels and micropatterning technologies. The second half of the chapter provides a summary of particular micropatterned gels that have been applied to in vitro cell-based work, with specific attention paid to how these techniques could be applied toward stem cell research.

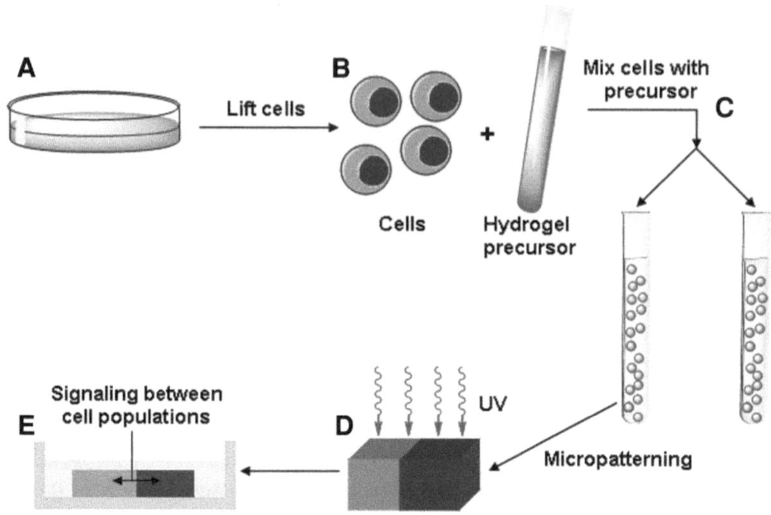

Fig. 1 Combination of stem cells, hydrogels and micropatterning for use in tissue engineering. **a** Stem cells are cultured in plates, then lifted and **b** mixed with a hydrogel precursor. **c** Exogenous factors are added to the cell-laden hydrogel precursor to yield solutions with different compositions (i.e. different ECM molecules, cell types, soluble factors, etc.). **d** Hydrogel precursor solutions are micropatterned to fabricate a heterogeneous hydrogel. **e** Resulting patterned hydrogel is cultured and subsequently assayed for various cell functions

2 Stem Cells

There are many types of stem cells, including embryonic stem cells (ESCs), hematopoietic stem cells (HSCs), mesenchymal stem cells (MSCs), neural stem cells (NSCs), and others [1, 2, 4, 5, 31–37]. Each type of cell may find its own use in specific regenerative medicine applications, but all stem cells share some common characteristics, as discussed in this section. This chapter will focus mainly on MSCs since their encapsulation in micropatterned hydrogels has been explored more than the encapsulation of other stem cells, but many of the characteristics highlighted below are equally applicable to other stem cell types. For more detailed reports of experiments with ESCs [14, 38], HSCs [34, 39], and NSCs [36, 40], please see listed references.

2.1 MSC General Characteristics

Like all other cells, MSCs have several major functions, many of which can be manipulated by micropatterning technologies. These functions are viability, proliferation, migration, cell–cell communication, differentiation, and programmed cell death [41]. Underlying many of these functions is the morphology of the cell as it has been found that cell shape can profoundly influence how the cell responds to cues from its external environment [42–44].

In addition to these basic cell functions, MSCs are of great interest for cell-based therapies because they can be easily isolated and expanded in vitro without phenotypic change before lineage-specific differentiation [1, 2]. Additionally, particularly while resident in local stem cell niche, they can generate identical copies of themselves through mitotic division over extended time periods, and thus are self-renewing [1, 4, 45, 46]. The high proliferation rate of stem cells combined with the ability of these cells to remain in an undifferentiated state can result in a dramatic increase in the expansion of total cells while in culture, making it relatively easy to produce enough cells to fill large tissue defects for clinical applications [1, 2, 4, 44, 47]. For more reading on stem cells and their characteristics please refer to [1, 4, 5, 44, 48].

Further exploration into stem cell functions and how they can be manipulated will not only increase knowledge of stem cell biology, but will help scientists to effectively utilize stem cells in tissue engineering. For example, modifying mesh sizes within 3D hydrogel scaffolds for cell culture could help determine the effects of specific cell morphologies on cellular functions [6, 49, 50]. Also, many scaffolds lend themselves to modifications which could directly influence stem cell proliferation and differentiation; thus carefully modifying scaffolds could shed new light on what induces differentiation into specific cell lines. Many of the hydrogels discussed later can include tethered ECM components, so carefully modifying scaffolds could provide important information on what components of the niche induce cell division and maintenance of pluripotency [51].

2.2 MSC Differentiation and Plasticity

The differentiation potential of MSCs has been studied extensively over the past 31 years [48, 52]. Differentiation of stem cells into phenotypes including bone, cartilage, tendon, muscle, adipose tissue, and hematopoietic-supporting stroma can be affected by both biochemical and mechanical factors [1, 2, 4]. Biochemical influences are usually proteins such as growth factors that can be added exogenously in cell culture media or can be produced by neighboring cells during development [5, 11, 53, 54]. For MSCs, bone and cartilage differentiation have been most extensively studied to date. Soluble factors that are often added to achieve osteogenic (bone) differentiation of MSCs are bone morphogenetic proteins (BMPs), specifically BMP-2 and BMP-6 [55, 56]. Similarly, several signaling molecules including BMPs [56], transforming growth factor-βs (TGF-βs) [57], growth and differentiation factors (GDFs), members of the TGF-β superfamily [58], and Wnt [55] ligands including BMP-4, TGF-β1, and GDF-5 can be supplied to induce chondrogenic (chondrocyte) differentiation. Please see [5, 44, 53, 59] for more detailed discussions on factors that induce stem cell differentiation.

Biomechanical influences can also be generated externally (bioreactors in cell culture/normal activity in development), but can also come in the form of cues internal to the developing tissue, such as matrix stiffness [3]. Similar to application of growth factors, a number of bioreactors have been designed to impart compressive force for osteogenic and chondrogenic differentiation [60–65]. For more information on mechanical influences on MSC differentiation please see [66–70]. Because many of the signaling pathways that induce lineage-specific differentiation, as well as how these pathways are affected by the composition of the extracellular microenvironment, remain unknown, the versatility of patternable 3D hydrogels makes them very useful as platforms to address and control the culture environment for a variety of in vitro stem cell studies.

Initially, it was thought that MSCs could only differentiate into mesenchymal tissues; however, recently, it has been discovered that they are capable of greater plasticity or transdifferentiation than originally recognized [71–73]. Transdifferentiation means that these cells can produce a wide spectrum of cell types regardless of whether the tissues are derived from the same resident tissue [74]. Unfortunately, many of these studies are population based so it is difficult to determine if each individual cell is changing phenotype, or if different subpopulations are responding to different cues [75]. Careful micropatterning of 3D scaffolds, like hydrogels, can provide scientists with ideal systems for the exploration of stem cell plasticity by dynamically tuning the microenvironment of the MSCs to induce differentiation down one lineage and subsequently promote differentiation down an alternate lineage. The results could be carefully analyzed at multiple time points to observe the microenvironment effect on MSCs.

3 Hydrogels

Given the plethora of complex questions that arise in stem cell research, increasingly complicated model systems are required to fully capture the biological events that are occurring as these cells interact with their local extracellular matrix and neighboring cells. The 3D biomimetic microenvironment provided by patterned hydrogels allows for the probing of stem cell response to external stimuli in a well-defined and observable manner and is therefore an excellent candidate for building controlled model systems [5, 15]. Hydrogels are 3D, hydrophilic, polymeric networks that absorb large quantities of water while remaining insoluble in aqueous solutions due to chemical or physical cross-linking of individual polymer chains [76–79]. Specifically, hydrogels are appealing for biological applications due to their cytocompatibility [80–82], modulus similar to many soft tissues [81, 83], and high water content which allows for the formation of thick constructs (up to 1.5 mm) with viable cells embedded throughout the gel [9, 77, 78, 84]. This section provides some basic background information on different types of hydrogels currently used for 3D scaffolds with a focus on those materials employed currently with micropatterning techniques. For a more detailed overview of hydrogels for tissue engineering and drug delivery, please see [29, 77, 78, 80, 83–87].

3.1 Natural Versus Synthetic Polymers

Hydrogels can be prepared from natural or synthetic polymers using various methods (e.g., photo- and thermal-initiated polymerization) that will be discussed later in the chapter [49, 88]. Hydrogels made from natural sources can be derived from polymers such as collagen, hyaluronic acid (HA), fibrin, alginate, agarose, and chitosan [89]. Many natural polymers, such as collagen, HA, and fibrin, have been used in tissue engineering applications because they are either components of or have macromolecular properties similar to the natural extracellular matrix (ECM) [83, 87, 90]. Collagens are composed of three polypeptide strands twisted together to form a triple helix and are the main protein of mammalian ECM [91, 92]. Likewise, hyaluronic acid, an anionic glycosaminoglycan polysaccharide, is also found in nearly all adult animal tissues [92]. Alternatively, alginate, agarose, and chitosan are hydrophilic, linear polysaccharides derived from marine algae sources (alginate and agarose) or crustaceans (chitosan) [93, 94]. Another naturally derived gel, Matrigel™, is derived from soluble basement membrane extract of mouse tumors [95]. Various natural polymers have specific utilities and properties based on their origin and composition [87, 96]. Advantages of natural polymer-based gels include inherent biodegradability, and biologically recognizable moieties that support cellular activities [77, 96]. Disadvantages of some of these hydrogels include mechanical weakness and the possibility of evoking immune/inflammatory responses [77, 87].

Synthetic hydrogels are appealing for tissue engineering due to the amount of control scientists have over structure, such as cross-linking density, and tailored properties, such as biodegradation, mechanical strength, and chemical and biological response to stimuli [87, 96]. Synthetic polymers such as poly(ethylene glycol) (PEG) [97] and other PEG-based polymers [98, 99], or poly(vinyl alcohol) (PVA) [100] can be reproducibly produced with specific molecular weights, block structures, degradable linkages, and cross-linking moieties [101]. These features determine gel formation dynamics, cross-linking density, and mechanical and degradation properties of the material. Hydrogels made from synthetic polymers like PEG, PVA, or their derivatives do not possess the inherent bioactive properties that gels made from natural polymers do. However, they do have well-defined structures and are versatile templates for subsequent modifications that yield tailorable degradability and functionality [77, 87].

3.2 Gelation Mechanisms

Hydrogels undergo a transition from a liquid to semi-solid state (gelation) via a variety of physical and chemical mechanisms. While this is not always the case, for the hydrogel types currently used in micropatterning, those that are naturally derived undergo gelation mainly due to increasing entanglements of the biopolymer chains as a solution of the material is cooled to room or body temperature [83, 93, 102, 103]. On the other hand, synthetic polymers such as PEG require chemical (covalent) cross-links to form a mesh-like structure that is the basis of the gel state [29, 77, 87]. Specific means to induce chemical cross-links that have been utilized in micropatterning processes are highlighted below.

3.2.1 Radical Chain Polymerization

Natural and synthetic biocompatible polymers can be modified to contain two or more carbon–carbon double bonds (C=C) to create multifunctional macromers [27, 87]. In the presence of an initiator, radicals are generated that propagate through multiple carbon–carbon double bonds to form high-molecular-weight kinetic chains that are covalently cross-linked in a network [29]. A variety of cell types have been encapsulated utilizing thermal [104], redox [105, 106], and photoinitiating cross-linking conditions [107, 108]. The cytotoxicity of several redox and photointiating systems has been examined and it was determined that redox-initiating system toxicities are dependent, in part, on the pH of the initiator [109] while photoinitiator system toxicities are based upon initiator chemistry and concentration [110, 111]. Additionally, radical concentration and length of cell exposure to radicals and UV light has significant effects on cell viability [111, 112].

3.2.2 Chemical Cross-linking

Hydrogels can also be fabricated by chemical cross-linking or step-growth polymerization techniques [29]. An attractive feature of this cross-linking mechanism is that it does not require additional components like initiators [113]. Polymers can be modified with hydrated sulfur or thiol (HS–R) side groups that will form sulfur-sulfur or disulfide (S–S) cross-links [114] or they can be modified with esters (R–COOR) that react with sulfur radicals or thiolates (·S–R) to form thioether (R-C–S-R) linkages [113, 115]. Studies utilizing these cross-linking mechanisms verified that the conditions required for chemical cross-linking do not adversely affect cells; however, gelation rates are typically slower compared to radical chain polymerizations [29].

3.3 Functionalization of Hydrogels

3.3.1 Biodegradable Hydrogels

Biodegradable hydrogels are often favored for biomedical applications because they degrade in clinically relevant time-scales under relatively mild conditions, thus eliminating the need for additional surgeries to recover implanted gels [78, 96, 116].

They are advantageous for in vitro applications because they allow the cells to produce extracellular matrix to better mimic native tissue environments [86]. Currently, the fabrication and modeling of hydrolytically degradable hydrogels [117, 118] are well developed and the synthesis and utilization of synthetic gels incorporating biological moieties for enzymatic degradation are under investigation [119, 120]. While hydrogels made from natural polymers are often enzymatically degraded, synthetic hydrogels containing biological moieties often offer more controlled degradation rates due to their tunable physicochemical properties [77].

3.3.2 Biomimetic hydrogels

One of the drawbacks of using synthetic and some natural hydrogels for in vivo applications is that they do not have adhesion or enzyme-responsive sites for supporting cellular activity [80, 121]. However, synthetic polymers can be co-polymerized with other biological molecules to introduce multiple functional moieties into hydrogels representing different aspects of the native 3D tissue [80, 87]. These functionalities often support the suppression or promotion of cell survival and function. For example, the covalent addition of adhesive peptide sequences such as Arg-Gly-Asp (RGD) [26, 122–125] to PEG hydrogels has promoted viability of encapsulated cells [78, 125–127], while the addition of short matrix metalloproteinase (MMP)-sensitive peptide sequences (mentioned above as a means to improve degradation) have allowed cell migration, cell spreading, and tissue formation within the gels [113, 119, 128–131].

4 Micropatterning

As discussed previously, hydrogels have been combined with microfabrication techniques for a variety of tissue engineering applications. Examples include the fabrication of scaffolds with control over features such as shape, size, and architecture [13, 132]. Micropatterning hydrogels allow them to more closely mimic in vivo conditions, particularly by providing a means to culture multiple cell types together in proper spatial relationship [13, 17, 133–137]. These microscale approaches can be used to study cell behavior with tightly regulated control of cell microenvironment interactions such as cell–cell, cell–ECM, and cell-soluble factors in a spatially-regulated manner [2, 138]. Furthermore, patterned hydrogel constructs can be used to define heterogeneous environments in vitro, such as controlling the degree of homotypic and heterotypic cell–cell contact, which is difficult to do in traditional tissue culture [15]. Utilization of these techniques for cell analysis offers many advantages over other devices due to the small device size, high surface-area-to-volume ratio, integration with electronics, high throughput, small sample volumes, the ability for batch processing, geometrical control, and single-cell analysis [139–141]. For example, engineering artificial stem cell niches to determine cell differentiation based on spatially- and temporally-regulated signals may lead to the realization of the proper series of external cues required to efficiently and reproducibly differentiate stem cells into any desired cell type [13, 142–144]. This section provides an overview of microfabrication technologies that have been used to produce patterned, cell-containing constructs. More specific examples of these constructs and the biological knowledge gained from micropatterned culture are found in Sect. 5.

4.1 Microfabrication Technology

Microfabrication for making cell-containing (biomaterial) constructs borrows from some well established processes from integrated circuit fabrication techniques and employs a sequence of steps to produce complex physical structures [139]. The variation of processes and materials used leads to a large range of possible devices and constructs [139, 140]. To fabricate microelectromechanical systems (MEMS) or microfluidic devices, substrates typically composed of silicon, glass, or polymer are utilized to build a device out of the substrate's bulk material in a process known as bulk micromachining, or on the surface of the chosen substrate in surface micromachining processes [145, 146]. These processes can be used individually or in conjunction with one another to build a device [139]. There are four basic methods that are used in microfabrication to create a device: etching, bonding, photolithography, and thin-film growth [139, 147]. Etching is a process to selectively remove materials and thereby create features in the materials that are left, and by bonding two substrates together one could form a closed structure [139]. Photolithography is a method where photon-sensitive materials (such as a class of polymers called photoresists) can be patterned using photomasks [139]. Depending on the tone of the photoresist, photon-exposed areas are either left or removed in a development step; photo-exposure renders positive-tone resist more soluble in solvents [15, 148, 149]. Thin-film growth is an additive technique where materials are grown on substrates by chemical or physical processes [139, 150, 151]. In addition to these above-mentioned techniques, polymer films can be used as thick structural layers and can be deposited in a range of methods including microcontact printing [152, 153] and micromolding [150, 151]. This chapter focuses on adaptations of photolithographic techniques to cell-laden materials, as well as other additive or subtractive techniques and molding specifically developed for fabricating biomaterial constructs.

As described above, traditional photolithography uses masks, photosensitive polymers, photoresists, and UV light to transfer a pattern into a material [139]. When adapted for biomaterials, photolithography works in a similar fashion by cross-linking exposed regions and leaving the unexposed regions largely soluble in the solvent (typically water or salt buffer) [15, 148, 149]. For photopatterning applications, a variety of materials have been developed to be photosensitive, biocompatible, and bioactive. The type of chemistry chosen for each application depends largely on the tolerance of the cells to chemical and photo-toxicity and the biological activities designed in the cellular microenvironment. Other micro-engineering techniques have been utilized to tailor co-cultures and to create microfluidic networks in hydrogels that enable efficient exchange of solute with the bulk of the scaffold. These networks allow quantitative control of the soluble environment experienced by the encapsulated cells in their three-dimensional environment [8–10, 13, 154–161]. There are a multitude of micropatterning techniques that can be employed to engineer these micropatterned constructs (please see [10, 50, 139–141, 157, 162] for in-depth review), but only those currently in use for micropatterning hydrogels will be discussed here.

4.2 Applications in Hydrogel Patterning

4.2.1 Photolithography

In photolithography, a macromer solution is mixed with a photoinitator that catalyzes the cross-linking reaction upon exposure to UV light (Fig. 2) [163, 164]. This technique provides good control over hydrogel size and shape [163]. A photomask containing opaque patterns is often used in photolithography [149]. The mask is placed over the macromer/initiator solution and then exposed to UV light [9, 163]. The hydrogel precursor solution that is exposed to the UV light through the transparent regions of the photomask polymerizes, creating hydrogels in these exposed regions [53, 137]. The shape and size of the resulting hydrogel is based on the features of the mask [9, 149, 164]. This technique of patterning photocross-linkable hydrogels can be used to generate cell-laden microstructures [53, 163]. This micropatterning technique is very versatile as it can be used with a variety of multifunctional macromers [9, 165] and can be adapted for utilization in other lithographic techniques such as laser scanning lithography [157, 158] and stop-flow lithography [10, 159] as well as micromolding techniques [166] (see later sections of this chapter). Photolithography applied in the contact mode where the mask is next to the substrate (e.g. the hydrogel) is usually high throughput

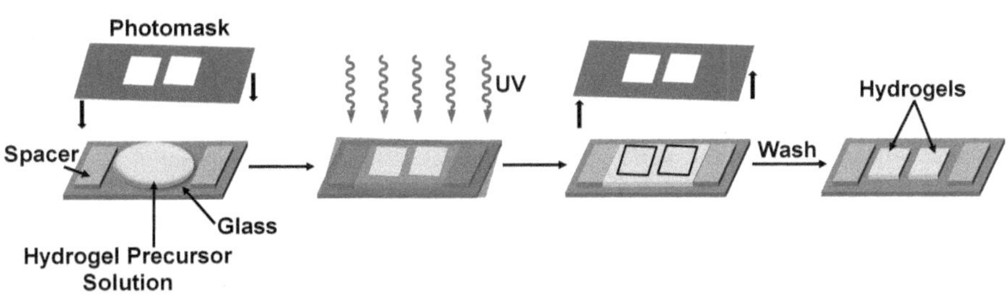

Fig. 2 Formation of hydrogels containing encapsulated cells using photolithography. Cells are encapsulated in hydrogels by exposing the hydrogel precursor solution to UV light through the photomask

because in principle multiple samples can be exposed at the same time. The trade-off is often between the resolution of the pattern and the sample thickness (≤1 mm). This means that to achieve large sample thickness, the resolution must be sacrificed, resulting in a loss of feature fidelity [9, 112, 155]. Contact photolithography is also very inexpensive and accessible compared to some of the specialty techniques as discussed below [50, 139, 140].

4.2.2 Laser-scanning lithography

Laser-scanning lithography (LSL) is a recently developed technology that employs a laser-scanning confocal microscope to pattern photosensitive materials (Fig. 3); it can be used in making actual constructs/devices, or masters for replica molding (as in soft lithography), or cross-linking biomaterials as in the case of photopatternable hydrogels [157, 167]. This technique provides an even exposure across the entire

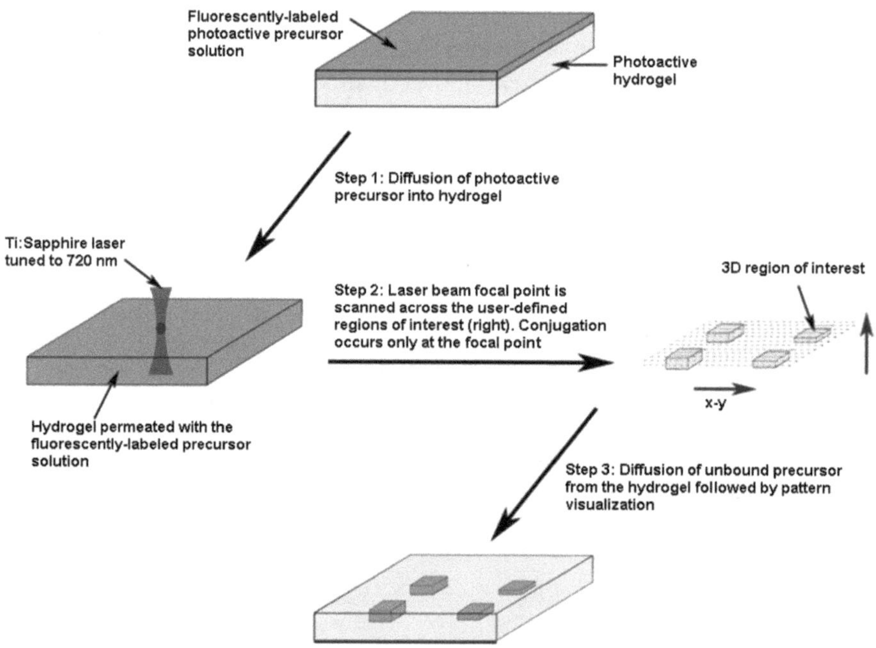

Fig. 3 Formation of hydrogels containing encapsulated cells using laser-scanning lithography. *Step 1* hydrogel precursor solution is layered onto a base hydrogel and allowed to diffuse into the gel network. *Step 2* laser beam is scanned across an *x–y* plane within the hydrogel in the specified regions of interest shown in the ROI template to the right of the gel being patterned. Conjugation of the macromer to the hydrogel substrate only occurs at the laser focal point. Therefore, the microscope stage is incremented axially and patterning is continued within the specific ROIs at this next *x–y* plane and this step repeated until the desired 3D patterns have been formed. *Step 3* unbound precursor is allowed to diffuse out of the hydrogel network. ([167] Copyright Wiley–VCH Verlag GmbH & Co. KGaA. Reproduced with permission)

sample and facilitates accurate alignment of successive exposures to photosensitive materials. Currently, features on the scale of 3 μm have been achieved with a 10× objective (NA 0.45) with high fidelity and greater facility than conventional photomasking [158]. Another advantage of this technology is that various parameters can be controlled by the user to allow maximum flexibility, such as the laser type and power, and the pixel exposure time [158]. Additionally the use of differential interference contrast on the confocal microscope permits immediate verification of each exposure, precise alignment of successive scans and reduced chromatic aberration between imaging and exposure [158, 167]. The trade-off here, however, is that LSL exposes the sample "pixel-by-pixel", and therefore is a serial technique and generally low throughput.

4.2.3 Stop-flow Lithography

Stop-flow lithography (SFL) relies upon the projection of a photomask upon a focal plane with a microfluidic channel that is filled with a photopolymerizable polymer solution (Fig. 4) [10, 159]. Typically, high initiator concentrations are used to ensure rapid cross-linking of the polymers and high throughput. Features of

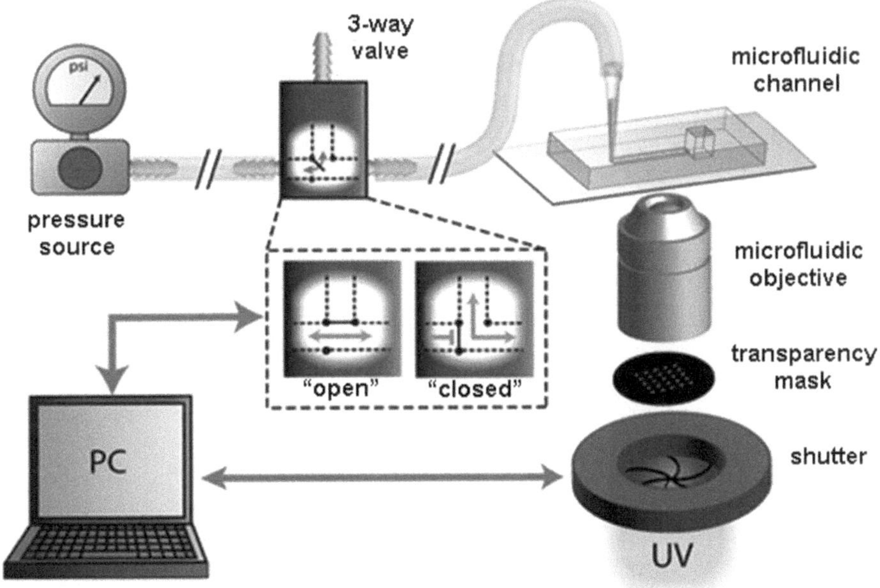

Fig. 4 Formation of hydrogels containing encapsulated cells using stop-flow lithography. Hydrogel precursor solution flow rate in a microfluidic channel is driven by a pressure profile controlled by a computer operated three-way solenoid valve that alternates between closed and opened. The computer also controls the exposure time of the macromer solution to UV light provided by an automated shutter. Polymerization occurs only where the solution is exposed to the UV light. ([159] Reproduced by permission of The Royal Society of Chemistry)

the photocross-linked particles such as size, shape, swelling behavior, and composition can be tailored independently using SFL through mask selection, optical exposure intensity, and polymer composition [10]. Furthermore, SFL allows for the generation of particles with multifunctionalities by flowing multiple streams through a microchannel at the same time and patterning across these streams [10, 159]. Controlling each stream's flow rate and adjusting the choice of photomask and polymerization location can result in the fabrication of a single particle with several orthogonal chemistries with control over the proportion and pattern of each type of chemistry [10, 159, 168, 169]. The throughput of SFL is somewhere between that of contact lithography and LSL, as are the spatial resolution and flexibility of the technique (in terms of the ability to change the shape of the desired particles) [159, 170].

4.2.4 Optofluidic Maskless Lithography

Optofluidic maskless lithography (OFML) is related to SFL. It uniquely combines the concept of maskless and continuous flow lithography techniques in microfluidic channels to provide real-time control of the in situ polymerization process to dynamically synthesize polymeric microstructures with various shapes (Fig. 5)

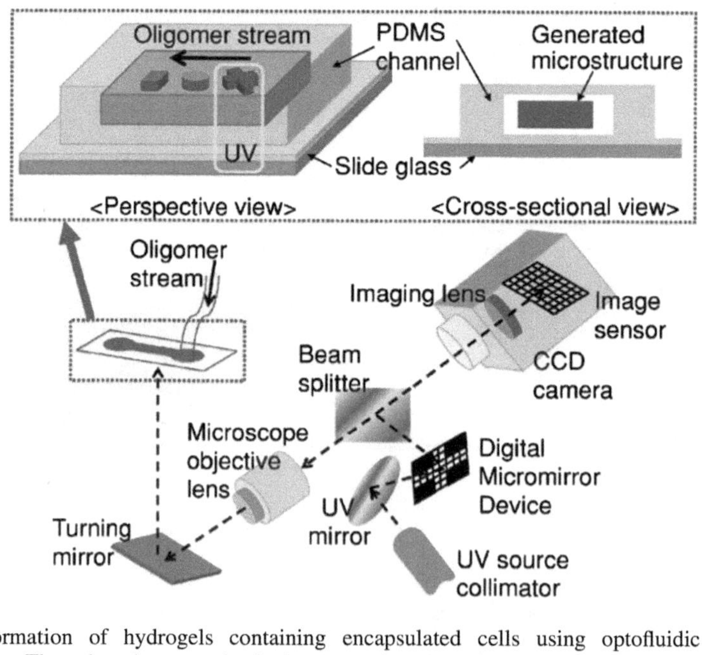

Fig. 5 Formation of hydrogels containing encapsulated cells using optofluidic maskless lithography. The microelectromechanical systems based spatial light modulator (MEMS-based SLM) controls the 2D UV light exposure pattern on the macromer solution with great spatial accuracy to precisely control the polymerization within the field of view. (Reprinted with permission from [162]. Copyright 2007, American Institute of Physics)

[135, 162, 171]. Maskless lithography techniques use programmable exposure patterns with spatial light modulators (SLMs) instead of traditional photomasks and the continuous flow of photosensive polymer in the microfluidic channel supplies the nonpolymerized material into the photopatternable region [162]. The continuous flow also serves to carry the synthesized microstructures for further processing, thus there is no need for addition actuators [162]. This method also takes advantage of microelectromechanical systems (MEMS)-based high-speed SLMs to dynamically control the shape of polymerized microparticles and to improve the throughput, which is especially attractive when generating a large number of constructs [135, 162, 172]. Optofluidic maskless lithography technique is even more flexible than SFL because the shapes of particles can be changed on the fly while in production, but this often comes with the complexity and expensive nature of the setup equipment. Hence, thus far, this technique has only been used for special applications [135].

Also, a method for high-throughput generation of 3D microstructures using a membrane-mounted microfluidic channel and an OFML system has been developed to allow for building more complex and heterogeneous structures. Photopolymerized 3D microstrctures can be fabricated in a layer-by-layer fashion with the thickness of each layer controlled by the deformation of the membrane. The material composition of each layer can be varied using microfluidic control of photocurable polymer to generate composite microstructures with heterogeneity in both lateral and vertical directions. In OFML, the polymer in the UV exposure area polymerizes throughout the cross-section of the channel so the thickness of the particles is determined by the height of the channel [162, 171, 173]. In 3D-OFML, the height of the channel is altered for the generation of multilayered structures. A microfluidic channel is used with a soft polydimethylsiloxane (PDMS) membrane mounted on top, and on the other side of the membrane is a pneumatic chamber. The pneumatic pressure in the chamber causes deformation of the membrane, changing the height of the bottom channel filled with photosensitive polymer. Each layer of the structure is fabricated via projection of the corresponding mask pattern when the membrane is at the desired height. Running the exposure and height control simultaneously for each layer results in 3D structures. A few advantages of this technique are that it is simple, fast ($\sim$0.2 s per exposure), and vertical resolution of gels can be achieved up to $\sim$100 μm using membrane deformation. Furthermore, the photosensitive polymer could be varied for each layer to result in a heterogeneous 3D structure [135].

4.2.5 Photodegradation

While photons are mostly used to cross-link polymers in photo-micropatterning, they can also be used in subtractive processes. Analogous to how positive-tone photoresists work, photodegradation removes materials. Micropatterning hydrogels through photodegradation first requires the preparation of a hydrogel that has a photodegradable macromer unit incorporated into the scaffold [8, 174].

A specific wavelength of light (365 nm) that selectively degrades the photolabile nitrobenzyl ether derivative group in the hydrogel can be utilized to selectively cleave parts of the hydrogel [8, 175]. The degradation of the hydrogel results in a decrease of the local network cross-linking density and often changes hydrogel properties such as stiffness or diffusivity [8]. This technique can be used to create a variety of patterns and different topologies in hydrogel constructs [8]. The resolution of such a technique in creating patterns often depends on the chemistry and how the light is applied. For example, laser scanning would produce higher resolution (at lower throughput) than mask-based methods.

4.2.6 Micromolding

Micromolding is another useful technique for forming micropatterned hydrogels that often utilizes a micropatterned master to mold replicas for repeated fabrication (Fig. 6) [50, 134, 151, 176]. The shape of the master mold determines the shape of the resulting hydrogel structures [134, 151]. Microwells can be formed in an array using micromolding to produce specific patterning of cells [163]. Another micromolding technique, capillary force lithography, can be utilized to create micropatterned hydrogel systems using a master PDMS pattern [141]. The capillary forces and surface tensions and interactions between the master pattern and the macromer solution cause the hydrogel precursor solution to fill the master in the desired shape and structure [160, 177, 178]. Often either a pipette or a syringe pump is employed to fill the PDMS master mold [53, 163, 179]. The advantages of

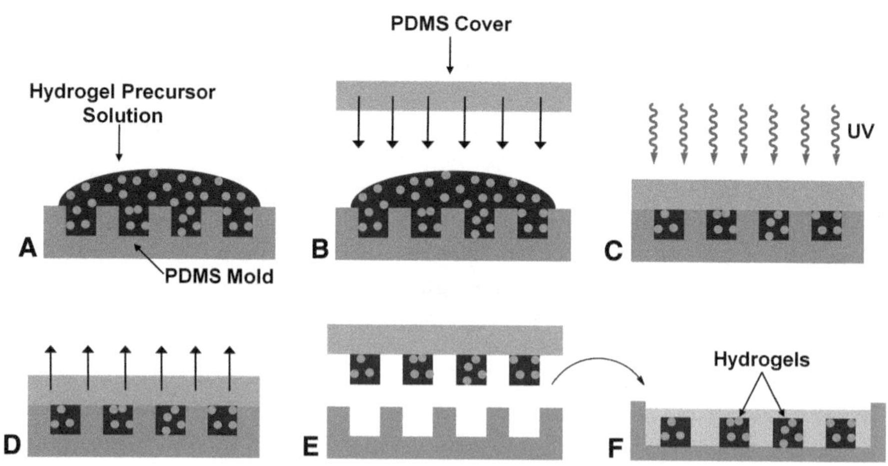

Fig. 6 Formation of hydrogels containing encapsulated cells using micromolding. **a** Hydrogel precursor solution is deposited onto a plasma-cleaned polydimethylsilozane (PDMS) pattern. **b** A PDMS coverslide is placed on top, forming a reversible seal. **c** Hydrogel precursor solution is photopolymerized via exposure to UV light. **d** The PDMS coverslide is removed, **e** removing the gels which can then be **f** harvested and hydrated

the micromolding technique are that it is relatively inexpensive, high-throughput, easy to perform, and the fidelity is well-controlled [134, 151, 180]. The spatial resolution of micromolding technique can be high, but is significantly dependent on the aspect ratio of the structure (width to depth), the cross-linking chemistry of the gel, and the resolution of the technique used to make the master. In general, it is possible to pattern structures with a resolution on the order of a cell.

4.2.7 Two-dimensional Templating

Two-dimensional templating can be performed on 3D constructs like hydrogels using a variety of techniques to create small, spatially-confined regions of deposited material on a hydrogel surface [152, 153, 181–183]. Robotic spotting is a deposition method that allows for the delivery of nanoliter volumes of soluble materials to precise locations on a substrate, including hydrogel surfaces [181]. Soft lithography [140] can be employed to make agarose stamps that can be used to pattern bacteria and cells onto a variety of surfaces including glass and agar [152, 153]. These templating techniques can be utilized to pattern cells or encourage the adhesion of cells though the printing of adhesive ligands onto hydrogel surfaces [152, 181]. These techniques are generally easy to perform and inexpensive, but the resolution is not as good as photolithography-based techniques and therefore are useful for only larger structures ($\leq$1 cm $\times$ 1 cm) [152, 153, 181, 183].

Of the above-mentioned techniques, which to use depends on the specific biology and applications, as well as the chemistry available for the system of interest. Some of the techniques can directly produce the constructs large enough for the final applications (e.g. contact lithography and micromolding), while others produce components that can be assembled post-encapsulation and manipulation. Whether to employ light or other physical constraints to produce patterned cell-laden gels depends on the fidelity and resolution required, as well as the geometry and chemistry requirements. The considerations often used in determining the specifics of the lithographical techniques are light intensity (which translates to macromer choice, photoinitiator concentration, and toxicity considerations), fidelity, and size of the samples.

5 Micropatterning Hydrogels with Embedded Cells

All of the methods described above for micropatterning hydrogels can be directly translated to the exploration of a variety of cell types including stem cells and their cellular functions and interactions [8–10, 179]. Specific chemistry may need to be tailored for stem cells, but the general strategies and trade-offs remain the same. To further clarify the versatility of micropatterned hydrogels as synthetic microenvironments to probe stem cell function and response to various stimuli, particular examples of how these technologies can be applied to stem cell culture are reviewed here. In some cases, stem cells (usually MSCs) have been employed in these experiments. In other studies, differentiated cell types have been used, but it does not preclude these technologies from being applied to stem cell research because the general principles of how they work do not change. These examples are organized by characteristic or function of stem cells (see Sect. 2) they are specially suited to examine.

5.1 Culture of One Cell Type

5.1.1 Cell Viability

To address one of the most basic cell functions, the chemistry and architecture of photopatterned PEG-based hydrogels can be tailored to support specific cell type survival. Specifically, adhesive peptides can be incorporated and the gel architecture modified to minimize barriers to nutrient transport for highly metabolic cells like hepatocytes [9]. Single-layer hydrogels have been fabricated with specific architectures encapsulating cells at high density. The photopatterned features of these gels help reduce diffusive nutrient transport limitations, which effectively increase cell viability in these photopatterned hydrogels as compared to unpatterned gels. Additionally, multilayer constructs with fluidic channels that enable perfusion in a continuous flow bioreactor have been prepared by additive photopatterning of cellular hydrogels. This can be done by patterning the bottom layer of a construct and subsequently patterning other layers on top of it using a specially designed apparatus [9].

The encapsulation of rat hepatocytes in photopatterned gels with a width of 500 μm using additive photopatterning has shown an increase in cell viability as compared to the encapsulation of cells in unpatterned bulk gels. Three-dimensional hepatocyte-laden constructs were photopatterned to reduce nutrient transport limitations and subsequently cultured in a continuous flow bioreactor for 12 days. The photopatterned constructs produced higher levels of albumin and urea, markers of synthetic and metabolic function in the liver, than non-patterned constructs of equal cell number and external scaffold dimensions. Additionally, hepatocytes were photoencapsulated with RGD in single and multilayer hydrogel constructs and hydrogels with photopatterned features of 500 μm showed better cell viability after 24 h as compared to gels without the photopatterned features [9].

While these studies have provided great insight into the field of liver biology, the more universal benefit of these studies is the development of another tool for tissue engineering of organs. A notable advantage of the fabrication of 3D cell-laden gels via additive photopatterning is that it can be generalized to many tissue types, while the use of a multilayer process enables the incorporation of different cell types into each layer which could be utilized in a more complex study of cell–cell interactions, especially towards the controlled differentiation of stem cells (see Sect. 5.2 for further discussion of co-culture technologies) [9].

5.1.2 Cell Migration (and Morphology)

The elegant use of organic materials' chemistries affords synthetic microenvironments that capture critical aspects of extracellular matrices. There are two advantages that spatially tuning material properties provide: the ability (1) to manipulate cell functions, and (2) to directly observe cellular processes including proliferation, morphological changes, and migration. The former includes the ability to allow cells to adhere to the biomaterials and spread/proliferate, providing cues for differentiation, and/or allowing for degradation of the gel locally such that cells can migrate in response to changes in the ECM environment. The latter is important because real-time information concerning encapsulated cell behavior can be tracked optically to gain understanding of how cells are responding to the local environment. The reactive macromer components can be easily exchanged for other functionalities resulting in readily tailorable hydrogels with multiple functionalities for three-dimensional cell studies [165].

Enzymatically-degradable multifunctional PEG-based macromers have been shown to react with one another via a "click" reaction [between an alkyl group (C≡C) and an azide (N=N=N)] to effectively encapsulate cells within a hydrogel equipped with MMP-cleavable peptide sequences (Gly-Pro-Gln-Gly, GPQG) [165, 184]. Furthermore, a photosensitive functionality has been incorporated into the gel that enables patterning of biological functionalities within the gel with micrometer-scale resolution. In this way, the biophysical and biochemical properties of the gel can be independently tailored. Mouse NIH 3T3 fibroblasts have been encapsulated within these degradable gels with a cysteine-containing RGD sequence. It has been found that selectively exposing specific areas of the gels to light affords spatial and temporal control of where the photocoupling reaction of the thiol-containing RGD and the alkene-containing gel occurs in real time. The light exposure can be controlled using conventional photolithography methods including single-photon and multiphoton laser techniques, each affording a high degree of reaction specificity for precision patterning. The coupling reaction is compatible with cells, maintaining a >90% viability at 24 h post encapsulation. It was observed that the presence of the bound RGD within the hydrogels induces localized morphological and migratory changes of the fibroblasts within the patterned regions. This is because the presence of the RGD allowed cells to attach to and locally degrade the surrounding hydrogel network giving rise to a spread cell morphology. When the RGD sequence is absent in the gels, the fibroblasts maintain a rounded morphology [165].

Two-photon laser scanning (TPLS) photolithography (a variation of LSL discussed previously), has also been utilized to affect 3D cell migration through the patterning of acrylate-functionalized biomolecules within degradable, biomimetic hydrogels. This can be done by first encapsulating cells within a PEG-based hydrogel modified with MMP-sensitive peptide sequences (Leu-Gly-Pro-Ala, LGPA). After encapsulation, the hydrogels are incubated in a solution of a desired biomolecule (in this case, selected to promote migration) to allow the molecule to diffuse into the gel. Specific regions of the hydrogel can then be scanned with a two-photon laser to create 3D patterns of the molecule within the gel. The cells eventually are able to migrate through the gel because they secrete MMPs and degrade the gel locally to allow for migration [158, 185].

Specifically, human dermal fibroblast (HDF) clusters in fibrin gels were encapsulated within PEG-diacrylate (PEG-DA) hydrogels containing the collagenase-sensitive peptide sequence, LGPA, and 3D patterns of RGD were created within the gel network using TPLS photolithography. The migration of HDFs into the RGD-patterned gel regions was observed by day 10 of the culture at which point the HDFs had extended long sprouts from the cell clusters into the RGD regions. This technique could also be used to create gradients of biomolecules like RGD inside of hydrogels by varying the exposure settings during the patterning procedure. An advantage of this system is the ability to interchange biomolecules to investigate the different effects various molecules have on a variety of aspects of cell locomotion. Also, it would be feasible to pattern several different biomolecules within a single hydrogel construct to compare the effects of the molecules on encapsulated cells. It is possible that this technology can be easily translated to the field of stem cell research to observe biomolecule effects on stem cell migration and homing [158, 185].

In another study, tracks of biochemical channels were fabricated within modified 3D agarose gels using a focused laser [175]. Originally, agarose was modified with a photolabile cysteine protecting group, which upon cleavage reveals a free thiol functionality on the gel. Using a laser, the free thiol channels can be patterned within a gel and subsequently reacted with thiol-reactive biomolecules that are diffused into the hydrogel after irradiation. Specifically, this technology has been used to engineer patterned 3D RGD channels in an agarose gel [175].

E9 chick dorsal root ganglia were plated on top of the patterned hydrogels containing RGD channels. Due to the poor adhesion properties of the agarose, cells adhered to the gel only on top of the designed channels. It was observed that after 3 days, the cell clusters aggregated on top of the peptide channels extended thick processes into the channels. The migration of cells and the elongation of cell processes were found to turn at the edge of the patterned biochemical domains. The experiment was repeated with a scrambled peptide sequence and cell migration and extension was not observed, which confirms that the cell behavior is sequence-specific [175].

One advantage of this system is that other biological molecules can be easily incorporated into the gels for channel pattering purposes to further explore cellular reactions to specific ligands. These hydrogels containing tracks of biochemical cues could be utilized to elucidate fundamental cell–substrate and cell–ligand adhesion interactions. They could also be employed for the guided regeneration of nerves or other tissues. We believe that this platform could be utilized to explore stem cell interactions with specific substrates and ligands as well as stem cell migration within the designed channels [175].

5.1.3 Cell Differentiation

Recently, the response of MSCs to dynamic presentations of ligands was studied in hydrogels created from photodegradable PEG-based macromers [8]. Postgelation control of the gel properties via selective irradiation has been shown to create arbitrarily-shaped features, introduce temporal changes, and release on-demand pendant functionalities [8, 174]. In these experiments, hMSCs were encapsulated in nondegradable PEG-DA with and without photoreleaseable RGD and cultured in chondrogenic media containing TGF-β. The absence of RGD in culture yielded a statistically significant decrease in cell viability within 7 days in PEG-only gels. At day 10, a portion of the gels with photoreleaseable RGD were irradiated and the release of RGD did not affect cell viability at that time point. However, by day 21, a four-fold statistical increase in glycosaminoglycan production (a marker of chondrogenesis) occurred relative to persistently presented RGD- or PEG-DA-only-based gels, indicating further differentiation of the hMSC cells down the chondrogenic pathway in response to alterations in the extracellular microenvironments. Such photodegradable gels that allow real time manipulation of material properties or chemistry advantageously provide dynamic environments with the scope to answer fundamental questions about material regulation of stem cell differentiation, communication, and proliferation [8, 174].

5.2 Culture of Multiple Cell Types

A great deal of work has emerged over the past several years in the area of co-culture of stem cells and differentiated cells to examine signaling between these cell types and potential implications in cellular differentiation [43, 186–189]. Unfortunately, standard co-culture methods that mix two or more cell types cannot be used to easily control the degree of homotypic and heterotypic cell–cell interactions [166]. Likewise, these co-culture methods are typically carried out in two-dimensional culture plates in a non-separable manner so analysis of individual populations is nearly impossible [188, 190–192]. In addition to allowing growth in a true 3D environment, it is projected that the micropatterning of hydrogels can allow for the selective removal of one population of cells post-culture to further analyze biochemical/gene expression [27, 114, 193, 194]. While the microfabrication techniques described below have not been explored for stem cell co-cultures, they hold great promise as co-culture systems to better understand the role of communication between two or more cell types on various stem cell functions.

5.2.1 Microfluidics

In one example of this type of technology, a device that allows for the patterning of multiple discrete 3D cell-laden hydrogels in sequence has been developed [179].

It is based on a set of parallel microfluidic channels separated by an array of posts that allow selective filling of individual channels, yet permits continuous interfaces between adjacent channels for diffusion of molecules and movement of cells across distinct regions. This microfluidic platform allows for real-time imaging of the interactions between multiple cell types exposed to autocrine and paracrine signaling molecules within the three-dimensional environment [179].

The utility of the system was demonstrated though the investigation of the behaviors of human metastatic breast cancer cells, MDA-MB-231, and tumor-derived RAW 264.1 macrophages in co-cultured well-defined geometries. Over the course of 1 week, the MDA-MB-231 did not invade the neighboring macrophage-containing Matrigel^{TM} gels; however, the macrophages did invade the breast-cancer-containing collagen gels beginning at day 3, increasing in invasiveness as the culture period was extended to day 7. The infiltrating macrophage cells also multiplied in the presence of the MDA-MB-231. Interestingly, it was observed that RAW 264.1 in monoculture were less invasive into the neighboring gels when the breast cancer cells were not present. Therefore, the behavior of the macrophage appears to have been influenced, at least in part, by the signals released by neighboring breast-cancer-derived cells [179].

Compared to currently available models, this system is advantageous because it allows for the patterning of different combinations of hydrogels and culturing of multiple cell types in an easy and straightforward manner. It can also be scaled dimensionally to mimic multicellular in vivo structures and it can be utilized to generate chemically- and mechanically-tailored three-dimensional cellular niches in a reproducible manner. Continued exploration of this system will allow researchers to further investigate paracrine effects on normal and tumor cell functions, as well as stem cell proliferation, self-renewal, and differentiation [179].

A bottom-up approach to generate tissue constructs with tunable architecture and complexity has been developed by directing the assembly of cell-laden microgels [195]. The thermodynamic tendency of multiphase liquid–liquid systems is to minimize their contact surfaces. Therefore, surface tension is the driving force for the induction of microgel assembly. Shape-controlled microgels spontaneously assemble within multiphase reactor systems into predetermined geometric configurations and multicomponent cell-laden constructs can be generated by assembling microgel building blocks and performing a secondary cross-linking reaction [195].

Mouse NIH 3T3 fibroblasts were encapsulated within PEG-methacrylate (PEG-mA) hydrogels and their viability was monitored throughout the preparation steps. The initial cross-linking step did not induce cell death; however, as UV exposure time increased, cell viability decreased. This approach was also shown to generate complex structures by showing that cross- and rod-shaped microgels assembled together, with one cross-shaped gel assembling with one, two, or three rod-shaped gels. This bottom-up approach for the generation of microgels provides a highly scalable approach to form biomimetic 3D tissue constructs. This technique could be used for co-culture involving stem cells by encapsulating two different cell populations within two differently shaped microgels. When the gels assemble together, this would automatically create a patterned co-culture device where cell–cell interactions could be observed [195].

Using the OFML technique described in Sect. 4.1, two different populations of HeLa cells, stained with Fluorescein and Rhodamine B, were encapsulated in PEG-DA and the resulting cell-laden hydrogel generated 3D tissue-like microstructures containing a different population of living cells in each part of the structure [135]. Cell viability was $\geq$90% up to 1 h after the polymerization. Additionally, with a suitable design, this technique should extend to make large scale 3D fabrications on the scale of several millimeters to one centimeter in a few minutes. The ability to generate 3D composite microstructures could be employed for making complex co-culture devices to better understand the cell–cell and cell–scaffold interactions influencing stem cell proliferation and differentiation [135].

5.2.2 Bioreactors

Microfluidic systems containing immobilized hydrogel-encapsulated cells that can be used as cell-based biosensors have been developed [137, 148, 164, 196]. Cell-laden PEG-based hydrogels of various shapes and sizes were photolithographically polymerized in 200 μm-wide channels or in 2 mm × 2 mm chambers of microfluidic devices. Two layer heterogeneous hydrogel microstructures containing different cell types were also generated within the device using the laminar flow properties inside the microchannels in conjunction with a photolithography patterning technique [137, 148, 164, 196].

Cell viability of SV-40 transformed murine peritoneal macrophages within the gels was found to be $\geq$80% after 24 h of culture in the device and ~70% after 7 and 14 days. By taking advantage of the flow characteristics inside microchannels, SV-40-transformed murine peritoneal macrophages and SV-40-transformed murine hepatocytes could be encapsulated side by side within the hydrogels to create heterogeneous hydrogel microstructures. This system offers several advantages over other systems previously discussed since it uses less sample volume and fewer cells. This device offers the ability to explore cell–cell interactions for stem cell biology applications through the separate patterning of hydrogels containing different cell types. Additionally, the number of cell types explored within one of these devices could be increased by patterning several heterogeneous microstructures containing different cells [137, 148, 164, 196].

5.2.3 Micromolding

In another approach, cell aggregates or spheroid microarrays and co-cultures using micromolding technology with chitosan hydrogels have been developed [166]. This method is different from many of the other methods discussed in this chapter in that it utilizes the nonadhesive nature of chitosan to form 3D cell clusters as opposed to encapsulating cells within a gel to form a 3D construct. The chitosan hydrogel surface is altered from cell-repulsive to cell-adhesive during the first few hours of the culture period. While the exact reason for this change in cell adhesion remains unclear, it is hypothesized that proteins included in fetal bovine serum and secreted by the cells adsorbed to the chitosan hydrogel surface and encouraged further cell adhesion to the gel. This time-dependent surface cell attachment occurred in the culture of human hepatoblastoma cells, Hep G2, and mouse NIH 3T3 fibroblasts. It was observed that the Hep G2 began forming aggregates inside the wells 1 day after seeding and the aggregates grew into hemispherical spheroids by day 3 of culture. These aggregates had $\geq$90% viability, and the phenotypic stability of the hepatoblastoma spheroids was verified throughout the 9 days of culture by the increasing production of albumin (one of the markers for liver-specific functions).

For co-culture experiments, Hep G2 were seeded and allowed to aggregate for 24 h. At this point, NIH 3T3 cells were seeded and left to attach to the surrounding chitosan surface of the micromolds. The fibroblasts attached evenly to the chitosan surface surrounding the hepatoblastoma spheroids and eventually proliferated to cover the surface of the entire hydrogel. Some fibroblast cells were observed invading the hepatoblastoma spheroids at day 9 of culture due to the rapid division of fibroblasts; however, both Hep G2 and NIH 3T3 cells were viable throughout the 9 days of culture [166]. This spatially-controlled spheroid co-culture system could provide a useful tool for fabricating biomimetic cellular microenvironments for studies of the effects of heterotypic cell–cell interactions in stem cell biology. It could also be used to perform protein or gene expression analyses in each cell type in culture, thus clarifying the responses of multiple cell types to each other [166].

5.2.4 Stop-flow Lithography

Although less fully utilized for biological applications, stop-flow lithography (SFL) (previously discussed in the Sect. 3) [159, 168], has been employed to synthesize hydrolytically-degradable microscale hydrogel particles from a solution of a macromer containing both PEG and poly(lactic acid). Mouse NIH 3T3 fibroblast cells have been successfully encapsulated in hydrogel microblocks via SFL. It was observed that the majority of encapsulated cells remained viable shortly after microblock fabrication ($\sim$1 h). SFL provides a significant amount of tunability over the fabrication of microgel particles. Although not yet explored for co-culture systems, this versatility could prove to be very beneficial for such culture platforms [169].

6 Future Outlook

Over the last two decades, great strides have been made in stem cell research, especially as it applies to tissue engineering and regenerative medicine. As discussed in this chapter, further progress in basic stem biology may be promoted by using micropatterned hydrogels as a culture milieu to better recreate the 3D architecture and the complex cell–cell and cell–ECM interactions normally occurring with stem cells in native tissue. However, to provide platforms that allow sophisticated biological questions to be explored, additional functionalities may be required in micropatterned hydrogel systems.

From a biomaterials perspective, there is a relatively small cadre of materials that have sufficient cytocompatibility, bioactivity, and patternability to be utilized in these systems. Synthesis of new hydrogels with easily-tailorable degradation rates, mechanical properties, and especially those including orthogonal chemistries for the addition or removal of biological molecules (temporal/spatial control of ligand presentation) during culture is unquestionably needed.

From a micropatterning perspective, since many hydrogels are polymerized in a bulk reaction, amounts of cells and reagents lost during fabrication remain relatively high. Therefore, additional methods (possibly using microfluidic technologies) to reduce reagent usage would be a great advantage for use with stem cells, many of which are difficult to isolate and therefore yield relatively low cell numbers. Furthermore, fabrication methods should be further optimized to allow multiple (more than two) cell types to be cultured in the same construct without cumbersome and complex methods of assembly [13]. In this vein, the development of a separable co-culture system that would promote analysis of each individual cell population after the culture period would provide further insight into the specific effects that various stimuli and interactions have on each cell type.

While there are still areas for improvement, overall, the convergence of micropatterning techniques with novel hydrogel materials that has occurred over the past 5–10 years provides an exciting new technology for use in a variety of biological settings, including further elucidation of means to control stem cell function. The continued merger of engineering, materials science, and biology, particularly at the micrometer scale, will enhance our ability to create in vitro models that can be used for understanding fundamental biological questions or for fabricating replacement tissues [13]. Such technologies and the biological insights they generate have great potential to accelerate the translation of stem-cell based therapies to clinical applications, thereby improving patient care for a wide range of diseases and injuries.

References

1. Kolf, C.M., Cho, E., Tuan, R.S.: Mesenchymal stromal cells. Biology of adult mesenchymal stem cells: regulation of niche, self-renewal and differentiation. Arthritis Res. Ther. **9**, 204–213 (2007)
2. Chen, F.H., Rousche, K.T., Tuan, R.S.: Technology insight: adult stem cells in cartilage regeneration and tissue engineering. Nat. Clin. Pract. Rheumatol. **2**, 373–382 (2006)
3. Discher, D., Dong, C., Fredberg, J.J., Guilak, F., Ingber, D., Janmey, P., Kamm, R.D., Schmid-Schönbein, G.W., Weinbaum, S.: Biomechanics: cell research and applications for the next decade. Ann. Biomed. Eng. **37**, 847–859 (2009)

4. Baksh, D., Song, L., Tuan, R.S.: Adult mesenchymal stem cells: characterization, differentiation, and application in cell and gene therapy. J. Cell. Mol. Med. **8**, 301–316 (2004)
5. Chai, C., Leong, K.W.: Biomaterials approach to expand and direct differentiation of stem cells. Mol. Ther. **15**, 467–480 (2007)
6. Nuttelman, C., Rice, M., Rydholm, A., Salinas, C., Shah, D., Anseth, K.: Macromolecular monomers for the synthesis of hydrogel niches and their application in cell encapsulation and tissue engineering. Prog. Polym. Sci. **33**, 167–179 (2008)
7. Pirone, D.M., Chen, C.S.: Strategies for engineering the adhesive microenvironment. J. Mammary Gland Biol. Neoplasia **9**, 405–417 (2004)
8. Kloxin, A.M., Kasko, A.M., Salinas, C.N., Anseth, K.S.: Photodegradable hydrogels for dynamic tuning of physical and chemical properties. Science **324**, 59–63 (2009)
9. Liu Tsang, V., Chen, A.A., Cho, L.M., Jadin, K.D., Sah, R.L., DeLong, S., West, J.L., Bhatia, S.N.: Fabrication of 3D hepatic tissues by additive photopatterning of cellular hydrogels. Fed. Am. Soc. Exp. Biol. J. **21**, 790–801 (2007)
10. Panda, P., Ali, S., Lo, E., Chung, B.G., Hatton, T.A., Khademhosseini, A., Doyle, P.S.: Stop-flow lithography to generate cell-laden microgel particles. Lab Chip **8**, 1056–1061 (2008)
11. Lund, A.W., Yener, B., Stegemann, J.P., Plopper, G.E.: The natural and engineered 3D microenvironment as a regulatory cue during stem cell fate determination. Tissue Eng. Part B Rev. **15**, 371–380 (2009)
12. Lutolf, M.P., Blau, H.M.: Artificial stem cell niches. Adv. Mater. **21**, 3255–3268 (2009)
13. Khademhosseini, A., Langer, R., Borenstein, J., Vacanti, J.P.: Microscale technologies for tissue engineering and biology. Proc. Natl. Acad. Sci. U.S.A. **103**, 2480–2487 (2006)
14. Reilly, G., Engler, A.: Intrinsic extracellular matrix properties regulate stem cell differentiation. J. Biomech. **43**, 55–62 (2010)
15. Khetani, S.R., Bhatia, S.N.: Engineering tissues for in vitro applications. Curr. Opin. Biotech. **17**, 524–531 (2006)
16. McGuigan, A.P., Bruzewicz, D.A., Glavan, A., Butte, M., Whitesides, G.M.: Cell encapsulation in sub-mm sized gel modules using replica molding. PLoS One **3**, e2258 (2008)
17. Tan, W.H., Takeuchi, S.: Dynamic microarray system with gentle retrieval mechanism for cell-encapsulating hydrogel beads. Lab Chip **8**, 259–266 (2008)
18. Vickerman, V., Blundo, J., Chung, S., Kamm, R.: Design, fabrication and implementation of a novel multi-parameter control microfluidic platform for three-dimensional cell culture and real-time imaging. Lab Chip **8**, 1468–1477 (2008)
19. Chung, S., Sudo, R., Mack, P.J., Wan, C.R., Vickerman, V., Kamm, R.D.: Cell migration into scaffolds under co-culture conditions in a microfluidic platform. Lab Chip **9**, 269–275 (2009)
20. Lii, J., Hsu, W.J., Parsa, H., Das, A., Rouse, R., Sia, S.K.: Real-time microfluidic system for studying mammalian cells in 3D microenvironments. Anal. Chem. **80**, 3640–3647 (2008)
21. Murtuza, B., Nichol, J.W., Khademhosseini, A.: Micro- and nanoscale control of the cardiac stem cell niche for tissue fabrication. Tissue Eng. Part B Rev. **15**, 443–454 (2009)
22. Karp, J.M., Yeh, J., Eng, G., Fukuda, J., Blumling, J., Suh, K.Y., Cheng, J., Mahdavi, A., Borenstein, J., Langer, R., Khademhosseini, A.: Controlling size, shape and homogeneity of embryoid bodies using poly(ethylene glycol) microwells. Lab Chip **7**, 786–794 (2007)
23. Moeller, H.-C., Mian, M.K., Shrivastava, S., Chung, B.G., Khademhosseini, A.: A microwell array system for stem cell culture. Biomaterials **29**, 752–763 (2008)
24. Mapili, G., Lu, Y., Chen, S., Roy, K.: Laser-layered microfabrication of spatially patterned functionalized tissue-engineering scaffolds, J. Biomed. Mater. Res. **75**, 414–424 (2005)
25. Lu, Y., Mapili, G., Suhali, G., Chen, S., Roy, K.: A digital micro-mirror device-based system for the microfabrication of complex, spatially patterned tissue engineering scaffolds. J. Biomed. Mater. Res. **77**, 396–405 (2006)
26. Burdick, J.A., Anseth, K.S.: Photoencapsulation of osteoblasts in injectable RGD-modified PEG hydrogels for bone tissue engineering. Biomaterials **23**, 4315–4323 (2002)

27. Bryant, S.J., Davis-Arehart, K.A., Luo, N., Shoemaker, R.K., Arthur, J.A., Anseth, K.S. Synthesis and characterization of photopolymerized multifunctional hydrogels: water-soluble poly(vinyl alcohol) and chondroitin sulfate macromers for chondrocyte encapsulation. Macromolecules **37**, 6726–6733 (2004)
28. Simms, H.M., Bowman, C.M., Anseth, K.S.: Using living radical polymerization to enable facile incorporation of materials in microfluidic cell culture devices. Biomaterials **29**, 2228–2236 (2008)
29. Nicodemus, G.D., Bryant, S.J.: Cell encapsulation in biodegradable hydrogels for tissue engineering applications. Tissue Eng. Part B Rev. **14**, 149–165 (2008)
30. Gillette, B.M., Jensen, J.A., Tang, B., Yang, G.J., Bazargan-Lari, A., Zhong, M., Sia, S.K.: In situ collagen assembly for integrating microfabricated three-dimensional cell-seeded matrices. Nat. Mater. **7**, 636–640 (2008)
31. Watt, F.M., Hogan, B.L.: Out of eden: stem cells and their niches. Science **287**, 1427–1430 (2000)
32. Ringe, J., Sittinger, M.: Tissue engineering in the rheumatic diseases. Arthritis Res. Ther. **11**, 211 (2009)
33. Gerecht-Nir, S., Cohen, S., Itskovitz-Eldor, J.: Bioreactor cultivation enhances the efficiency of human embryoid body (hEB) formation and differentiation. Biotechnol. Bioeng. **86**, 493–502 (2004)
34. Adams, G.B., Scadden, D.T.: A niche opportunity for stem cell therapeutics. Gene Ther. **15**, 96–99 (2008)
35. Jackson, K.A., Majka, S.M., Wulf, G.G., Goodell, M.A.: Stem cells: a minireview. J. Cell. Biochem. Suppl. **38**, 1–6 (2002)
36. Cordey, M., Limacher, M., Kobel, S., Taylor, V., Lutolf, M.P.: Enhancing the reliability and throughput of neurosphere culture on hydrogel microwell arrays. Stem Cells **26**, 2586–2594 (2008)
37. Ashton, R.S., Banerjee, A., Punyani, S., Schaffer, D.V., Kane, R.S.: Scaffolds based on degradable alginate hydrogels and poly(lactide-co-glycolide) microspheres for stem cell culture. Biomaterials **28**, 5518–5525 (2007)
38. Hwang, N.S., Varghese, S., Elisseeff, J.: Controlled differentiation of stem cells. Adv. Drug Deliv. Rev. **60**, 199–214 (2008)
39. Kobel, S., Limacher, M., Gobaa, S., Laroche, T., Lutolf, M.: Micropatterning of hydrogels by soft embossing. Langmuir **25**, 8774–8779 (2009)
40. Saha, K., Keung, A.J., Irwin, E.F., Li, Y., Little, L., Schaffer, D.V., Healy, K.E.: Substrate modulus directs neural stem cell behavior. Biophys. J. **95**, 4426–4438 (2008)
41. Temenoff, J.S., Mikos, A.G.: Biomaterials: The Intersection of Biology and Materials Science. Pearson Prentice-Hall, Upper Saddle River (2007)
42. McBeath, R., Pirone, D.M., Nelson, C.M., Bhadriraju, K., Chen, C.S.: Cell shape, cytoskeletal tension, and RhoA regulate stem cell lineage commitment. Dev. Cell **6**, 483–495 (2004)
43. Lee, I.-C., Wang, J.-H., Lee, Y.-T., Young, T.-H.: The differentiation of mesenchymal stem cells by mechanical stress or/and co-culture system. Biochem. Biophys. Res. Commun. **352**, 147–152 (2007)
44. Tuan, R.S., Boland, G., Tuli, R.: Adult mesenchymal stem cells and cell-based tissue engineering. Arthritis Res. Ther. **5**, 32–45 (2003)
45. Colter, D.C., Class, R., DiGirolamo, C.M., Prockop, D.J.: Rapid expansion of recycling stem cells in cultures of plastic-adherent cells from human bone marrow. Proc. Natl. Acad. Sci. U.S.A. **97**, 3213–3218 (2000)
46. Bruder, S.P., Jaiswal, N., Haynesworth, S.E.: Growth kinetics, self-renewal, and the osteogenic potential of purified human mesenchymal stem cells during extensive subcultivation and following cryopreservation. J. Cell. Biochem. **64**, 278–294 (1997)
47. Tsutsumi, S., Shimazu, A., Miyazaki, K., Pan, H., Koike, C., Yoshida, E., Takagishi, K., Kato, Y.: Retention of multilineage differentiation potential of mesenchymal cells during proliferation in response to FGF. Biochem. Biophys. Res. Commun. **288**, 413–419 (2001)

48. Pittenger, M.F., Mackay, A.M., Beck, S.C., Jaiswal, R.K., Douglas, R., Mosca, J.D., Moorman, M.A., Simonetti, D.W., Craig, S., Marshak, D.R.: Multilineage potential of adult human mesenchymal stem cells. Science **284**, 143–147 (1999)
49. Drury, J.L., Mooney, D.J.: Hydrogels for tissue engineering: scaffold design variables and applications. Biomaterials **24**, 4337–4351 (2003)
50. Liu Tsang, V., Bhatia, S.N.: Three-dimensional tissue fabrication. Adv. Drug Deliv. Rev. **56**, 1635–1647 (2004)
51. Jia, X., Yeo, Y., Clifton, R.J., Jiao, T., Kohane, D.S., Kobler, J.B., Zeitels, S.M., Langer, R.: Hyaluronic acid-based microgels and microgel networks for vocal fold regeneration. Biomacromolecules **7**, 3336–3344 (2006)
52. Schofield, R.: The relationship between the spleen colony-forming cell and the haemopoietic stem cell. Blood Cells **4**, 7–25 (1978)
53. Morrison, S.J., Spradling, A.C.: Stem cells and niches: mechanisms that promote stem cell maintenance throughout life. Cell **132**, 598–611 (2008)
54. Scadden, D.T.: The stem-cell niche as an entity of action. Nature **441**, 1075–1079 (2006)
55. Hartmann, C.: A Wnt canon orchestrating osteoblastogenesis. Trends Cell Biol. **16**, 151–158 (2006)
56. Gooch, K.J., Blunk, T., Courter, D.L., Sieminski, A.L., Vunjak-Novakovic, G., Freed, L.E.: Bone morphogenetic proteins-2, -12, and -13 modulate in vitro development of engineered cartilage. Tissue Eng. **8**, 591–601 (2002)
57. Massague, J., Blain, S.W., Lo, R.S.: TGFβ signaling in growth control, cancer, and heritable disorders. Cell **103**, 295–309 (2000)
58. Johnstone, B., Hering, T.M., Caplan, A.I., Goldberg, V.M., Yoo, J.U.: In vitro chondrogenesis of bone marrow-derived mesenchymal progenitor cells. Exp. Cell. Res. **238**, 265–272 (1998)
59. Qi, H., Aguiar, D.J., Williams, S.M., La Pean, A., Pan, W., Verfaillie, C.M.: Identification of genes responsible for osteoblast differentiation from human mesodermal progenitor cells. Proc. Natl. Acad. Sci. U.S.A. **100**, 3305–3310 (2003)
60. Ruiz, S.A., Chen, C.S.: Emergence of patterned stem cell differentiation within multicellular structures. Stem Cells **26**, 2921–2927 (2008)
61. Glossop, J.R., Cartmell, S.H.: Effect of fluid flow-induced shear stress on human mesenchymal stem cells: differential gene expression of IL1B and MAP3K8 in MAPK signaling. Gene Expr. Patterns **9**, 381–388 (2009)
62. Kisiday, J.D., Frisbie, D.D., McIlwraith, C.W., Grodzinsky, A.J.: Dynamic compression stimulates proteoglycan synthesis by mesenchymal stem cells in the absence of chondrogenic cytokines. Tissue Eng. Part A **15**, 2817–2824 (2009)
63. Ward, D.F., Jr., Salasznyk, R.M., Klees, R.F., Backiel, J., Agius, P., Bennett, K., Boskey, A., Plopper, G.E.: Mechanical strain enhances extracellular matrix-induced gene focusing and promotes osteogenic differentiation of human mesenchymal stem cells through an extracellular-related kinase-dependent pathway. Stem Cells Dev. **16**, 467–480 (2007)
64. Simmons, C.A., Matlis, S., Thornton, A.J., Chen, S., Wang, C.Y., Mooney, D.J.: Cyclic strain enhances matrix mineralization by adult human mesenchymal stem cells via the extracellular signal-regulated kinase (ERK1/2) signaling pathway. J. Biomech. **36**, 1087–1096 (2003)
65. Liu, J., Zhao, Z., Li, J., Zou, L., Shuler, C., Zou, Y., Huang, X., Li, M., Wang, J.: Hydrostatic pressures promote initial osteodifferentiation with ERK1/2 not p38 MAPK signaling involved. J. Cell. Biochem. **107**, 224–232 (2009)
66. Patwari, P., Lee, R.T.: Mechanical control of tissue morphogenesis. Circ. Res. **103**, 234–243 (2008)
67. Engler, A.J., Sen, S., Sweeney, H.L., Discher, D.E.: Matrix elasticity directs stem cell lineage specification. Cell **126**, 677–689 (2006)
68. Angele, P., Yoo, J.U., Smith, C., Mansour, J., Jepsen, K.J., Nerlich, M., Johnstone, B.: Cyclic hydrostatic pressure enhances the chondrogenic phenotype of human mesenchymal progenitor cells differentiated in vitro. J. Orthop. Res. **21**, 451–457 (2003)

69. Scherer, K., Schunke, M., Sellckau, R., Hassenpflug, J., Kurz, B.: The influence of oxygen and hydrostatic pressure on articular chondrocytes and adherent bone marrow cells in vitro. Biorheology **41**, 323–333 (2004)
70. Altman, G.H., Horan, R.L., Martin, I., Farhadi, J., Stark, P.R.H., Volloch, V., Richmond, J.C., Vunjak-Novakovic, G., Kaplan, D.L.: Cell differentiation by mechanical stress. Fed. Am. Soc. Exp. Biol. J. **16**, 270–272 (2002)
71. De Bari, C., Dell'Accio, F., Tylzanowski, P., Luyten, F.P.: Multipotent mesenchymal stem cells from adult human synovial membrane. Arthritis Rheum. **44**, 1928–1942 (2001)
72. Mezey, E., Chandross, K.J., Harta, G., Maki, R.A., McKercher, S.R.: Turning blood into brain: cells bearing neuronal antigens generated in vivo from bone marrow. Science **290**, 1779–1782 (2000)
73. Wada, M.R., Inagawa-Ogashiwa, M., Shimizu, S., Yasumoto, S., Hashimoto, N.: Generation of different fates from multipotent muscle stem cells. Development **129**, 2987–2995 (2002)
74. Filip, S., English, D., Mokry, J.: Issues in stem cell plasticity. J. Cell Mol. Med. **8**, 572–577 (2004)
75. Weissman, I.L., Anderson, D.J., Gage, F.: Stem and progenitor cells: origins, phenotypes, lineage commitments, and transdifferentiations. Annu. Rev. Cell Dev. Biol. **17**, 387–403 (2001)
76. Cho, E., Kutty, J.K., Datar, K., Lee, J.S., Vyavahare, N.R., Webb, K.: A novel synthetic route for the preparation of hydrolytically degradable synthetic hydrogels. J. Biomed. Mater. Res. Part A **90**, 1073–1082 (2009)
77. Lin, C.-C., Metters, A.T.: Hydrogels in controlled release formulations: network design and mathematical modeling. Adv. Drug Deliv. Rev. **58**, 1379–1408 (2006)
78. Peppas, N.A., Huang, Y., Torres-Lugo, M., Ward, J.H., Zhang, J.: Physicochemical foundations and structural design of hydrogels in medicine and biology. Annu. Rev. Biomed. Eng. **2**, 9–29 (2000)
79. Lowman, A.M., Peppas, N.A.: Hydrogels. Wiley, New York (1999)
80. Tessmar, J.K., Göpferich, A.M.: Customized PEG-derived copolymers for tissue-engineering applications. Macromol. Biosci. **7**, 23–39 (2007)
81. Ratner, B.D., Hoffman, A.S.: Synthetic Hydrogels for Biomedical Applications. American Chemical Society, Washington, DC (1976)
82. Buxton, A.N., Zhu, J., Marchant, R., West, J.L., Yoo, J.U., Johnstone, B.: Design and characterization of poly(ethylene glycol) photopolymerizable semi-interpenetrating networks for chondrogenesis of human mesenchymal stem cells. Tissue Eng. **13**, 2549–2560 (2007)
83. Mano, J.F., Silva, G.A., Azevedo, H.S., Malafaya, P.B., Sousa, R.A., Silva, S.S., Boesel, L.F., Oliveira, J.M., Santos, T.C., Marques, A.P., Neves, N.M., Reis, R.L.: Natural origin biodegradable systems in tissue engineering and regenerative medicine: present status and some moving trends. J. R. Soc. Interface **4**, 999–1030 (2007)
84. Peppas, N.A., Bures, P., Leobandung, W., Ichikawa, H.: Hydrogels in pharmaceutical formulations. Eur. J. Pharm. Biopharm. **50**, 27–46 (2000)
85. Jia, X., Kiick, K.L.: Hybrid multicomponent hydrogels for tissue engineering. Macromol. Biosci. **9**, 140–156 (2009)
86. Tabata, Y.: Tissue regeneration based on tissue engineering technology. Congenit. Anom. **44**, 111–124 (2004)
87. Peppas, N., Hilt, J., Khademhosseini, A., Langer, R.: Hydrogels in biology and medicine: from molecular principles to bionanotechnology. Adv. Mater. **18**, 1345–1360 (2006)
88. Davis, K.A., Anseth, K.S.: Controlled release from crosslinked degradable networks. Crit. Rev. Ther. Drug Carr. Syst. **19**, 385–423 (2002)
89. Lee, K.Y., Mooney, D.J.: Hydrogels for tissue engineering. Chem. Rev. **101**, 1869–1879 (2001)
90. Yang, S., Leong, K.F., Du, Z., Chua, C.K.: The design of scaffolds for use in tissue engineering. Part I. Traditional factors. Tissue Eng. **7**, 679–689 (2001)
91. Lee, C.H., Singla, A., Lee, Y.: Biomedical applications of collagen. Int. J. Pharm. **221**, 1–22 (2001)

92. Alberts, B., Bray, D., Lewis, J., Raff, M., Roberts, K., Watson, J.D.: Molecular Biology of the Cell. 3rd ed., Garland Publishing, Inc., New York, 1994
93. Suh, J.K., Matthew, H.W.: Application of chitosan-based polysaccharide biomaterials in cartilage tissue engineering: a review. Biomaterials **21**, 2589–2598 (2000)
94. Smidsrød, O., Skjåk-Braek, G.: Alginate as immobilization matrix for cells. Trends Biotechnol. **8**, 71–78 (1990)
95. Khademhosseini, A., Ferreira, L., Blumling, III J., Yeh, J., Karp, J.M., Fukuda, J., Langer, R.: Co-culture of human embryonic stem cells with murine embryonic fibroblasts on microwell-patterned substrates. Biomaterials **27**, 5968–5977 (2006)
96. Thomson, R., Wake, M., Yaszemski, M., Mikos, A.: Biodegradable polymer scaffolds to regenerate organs. Adv. Polym. Sci. **122**, 245–274 (1995)
97. Cruise, G.M., Scharp, D.S., Hubbell, J.A.: Characterization of permeability and network structure of interfacially photopolymerized poly(ethylene glycol) diacrylate hydrogels. Biomaterials **19**, 1287–1294 (1998)
98. Hassan, C., Peppas, N.: Structure and morphology of freeze/thawed PVA hydrogels. Macromolecules **33**, 2472–2479 (2000)
99. Brink, K.S., Yang, P.J., Temenoff, J.S.: Degradative properties and cytocompatibility of a mixed-mode hydrogel containing oligo[poly(ethylene glycol) fumarate] and poly(ethylene glycol)dithiol. Acta Biomater. **5**, 570–579 (2009)
100. Nuttelman, C.R., Mortisen, D.J., Henry, S.M., Anseth, K.S.: Attachment of fibronectin to poly(vinyl alcohol) hydrogels promotes NIH3T3 cell adhesion, proliferation, and migration. J. Biomed. Mater. **57**, 217–223 (2001)
101. Ratner, B.D., Hoffman, A.S., Schoen, F.J., Lemons, J.E.: Biomaterials Science: An Introduction to Materials in Medicine. 2nd edn. Elsevier, Amsterdam (2004)
102. Hubbell, J.A.: Materials as morphogenetic guides in tissue engineering. Curr. Opin. Biotech. **14**, 551–558 (2003)
103. Stevens, M.M., Qanadilo, H.F., Langer, R., Prasad Shastri, V.: A rapid-curing alginate gel system: utility in periosteum-derived cartilage tissue engineering. Biomaterials **25**, 887–894 (2004)
104. Temenoff, J.S., Park, H., Jabbari, E., Conway, D.E., Sheffield, T.L., Ambrose, C.G., Mikos, A.G.: Thermally cross-linked oligo(poly(ethylene glycol) fumarate) hydrogels support osteogenic differentiation of encapsulated marrow stromal cells in vitro. Biomacromolecules **5**, 5–10 (2004)
105. Temenoff, J.S., Park, H., Jabbari, E., Sheffield, T.L., LeBaron, R.G., Ambrose, C.G., Mikos, A.G.: In vitro osteogenic differentiation of marrow stromal cells encapsulated in biodegradable hydrogels. J. Biomed. Mater. Res. A **70**, 235–244 (2004)
106. Hong, Y., Song, H., Gong, Y., Mao, Z., Gao, C., Shen, J.: Covalently crosslinked chitosan hydrogel: Properties of in vitro degradation and chondrocyte encapsulation. Acta Biomater. **3**, 23–31 (2007)
107. Bryant, S.J., Anseth, K.S.: Controlling the spatial distribution of ECM components in degradable PEG hydrogels for tissue engineering cartilage. J. Biomed. Mater. Res. Part A **64**, 70–79 (2003)
108. Williams, C.G., Kim, T.K., Taboas, A., Malik, A., Manson, P., Elisseeff, J.: In vitro chondrogenesis of bone marrow-derived mesenchymal stem cells in a photopolymerizing hydrogel. Tissue Eng. **9**, 679–688 (2003)
109. Temenoff, J.S., Shin, H., Conway, D.E., Engel, P.S., Mikos, A.G.: In vitro cytotoxicity of redox radical initiators for cross-linking of oligo(poly(ethylene glycol) fumarate) macromers. Biomacromolecules **4**, 1605–1613 (2003)
110. Bryant, S.J., Nuttelman, C.R., Anseth, K.S.: Cytocompatibility of UV and visible light photoinitiating systems on cultured NIH/3T3 fibroblasts in vitro. J. Biomater. Sci. Polym. Ed. **11**, 439–457 (2000)
111. Williams, C.G., Malik, A.N., Kim, T.K., Manson, P.N., Elisseeff, J.H.: Variable cytocompatibility of six cell lines with photoinitiators used for polymerizing hydrogels and cell encapsulation. Biomaterials **26**, 1211–1218 (2005)

112. Mann, B.K., Gobin, A.S., Tsai, A.T., Schmedlen, R.H., West, J.L.: Smooth muscle cell growth in photopolymerized hydrogels with cell adhesive and proteolytically degradable domains: synthetic ECM analogs for tissue engineering. Biomaterials **22**, 3045–3051 (2001)
113. Park, Y., Lutolf, M.P., Hubbell, J.A., Hunziker, E.B., Wong, M.: Bovine primary chondrocyte culture in synthetic matrix metalloproteinase-sensitive poly(ethylene glycol)-based hydrogels as a scaffold for cartilage repair. Tissue Eng. **10**, 515–522 (2004)
114. Shu, X.Z., Liu, Y., Luo, Y., Roberts, M.C., Prestwich, G.D.: Disulfide cross-linked hyaluronan hydrogels. Biomacromolecules **3**, 1304–1311 (2002)
115. Shu, X.Z., Liu, Y.C., Palumbo, F.S., Lu, Y., Prestwich, G.D.: In situ crosslinkable hyaluronan hydrogels for tissue engineering. Biomaterials **25**, 1339–1348 (2004)
116. Tabata, Y.: Biomaterial technology for tissue engineering applications. J. R. Soc. Interface **6**, S311–S324 (2009)
117. Metters, A.T., Anseth, K.S., Bowman, C.N.: Fundamental studies of a novel, biodegradable. PEG-b-PLA hydrogel. Polymer **41**, 3993–4004 (2000)
118. West, J.L., Hubbell, J.A.: Photopolymerized hydrogel materials for drug-delivery applications. React. Polym. **25**, 139–147 (1995)
119. Lutolf, M.P., Lauer-Fields, J.L., Schmoekel, H.G., Metters, A.T., Weber, F.E., Fields, G.B., Hubbell, J.A.: Synthetic matrix metalloproteinase-sensitive hydrogels for the conduction of tissue regeneration: engineering cell-invasion characteristics. Proc. Natl. Acad. Sci. U.S.A. **100**, 5413–5418 (2003)
120. Hu, B.H., Messersmith, P.B.: Rational design of transglutaminase substrate peptides for rapid enzymatic formation of hydrogels. J. Am. Chem. Soc. **125**, 14298–14299 (2003)
121. Shin, H., Jo, S., Mikos, A.: Biomimetic materials for tissue engineering. Biomaterials **24**, 4353–4364 (2003)
122. Hern, D.L., Hubbell, J.A.: Incorporation of adhesion peptides into nonadhesive hydrogels useful for tissue resurfacing. J. Biomed. Mater. Res. Part A **39**, 266–276 (1998)
123. Yang, F., Williams, C.G., Wang, D.-A., Lee, H., Manson, P.N., Elisseeff, J.: The effect of incorporating RGD adhesive peptide in polyethylene glycol diacrylate hydrogel on osteogenesis of bone marrow stromal cells. Biomaterials **26**, 5991–5998 (2005)
124. DeLong, S.A., Gobin, A.S., West, J.L.: Covalent immobilization of RGDS on hydrogel surfaces to direct cell alignment and migration. J. Control. Release **109**, 139–148 (2005)
125. Benoit, D.S.W., Tripodi, M.C., Blanchette, J.O., Langer, S.J., Leinwand, L.A., Anseth, K.S.: Integrin-linked kinase production prevents anoikis in human mesenchymal stem cells. J. Biomed. Mater. Res. Part A **81**, 259–268 (2007)
126. Lin, C.-C., Anseth, K.S.: PEG hydrogels for the controlled release of biomolecules in regenerative medicine. Pharm. Res. **26**, 631–643 (2009)
127. Nuttelman, C.R., Tripodi, M.C., Anseth, K.S.: In vitro osteogenic differentiation of human mesenchymal stem cells photoencapsulated in PEG hydrogels. J. Biomed. Mater. Res. **68**, 773–782 (2004)
128. Gobin, A.S., West, J.L.: Effects of epidermal growth factor on fibroblast migration through biomimetic hydrogels. Biotechnol. Prog. **19**, 1781–1785 (2003)
129. Raeber, G.P., Lutolf, M.P., Hubbell, J.A.: Molecularly engineered PEG hydrogels: a novel model system for proteolytically mediated cell migration. Biophys. J. **89**, 1374–1388 (2005)
130. Lee, S.-H., Miller, J.S., Moon, J.J., West, J.L.: Proteolytically degradable hydrogels with a fluorogenic substrate for studies of cellular proteolytic activity and migration. Biotechnol. Prog. **21**, 1736–1741 (2005)
131. Seliktar, D., Zisch, A.H., Lutolf, M.P., Wrana, J.L., Hubbell, J.A.: MMP-2 sensitive, VEGF-bearing bioactive hydrogels for promotion of vascular healing. J. Biomed. Mater. Res. A **68**, 704–716 (2004)
132. Yu, Q., Bauer, J.M., Moore, J.S., Beebe, D.J.: Responsive biomimetic hydrogel valve for microfluidics. Appl. Phys. Lett. **78**, 2589–2591 (2001)
133. Bhatia, S.N., Balis, U.J., Yarmush, M.L., Toner, M.: Probing heterotypic cell interactions: hepatocyte function in microfabricated co-cultures. J. Biomater. Sci, Polym. Ed. **9**, 1137–1160 (1998)

134. Yeh, J., Ling, Y., Karp, J.M., Gantz, J., Chandawarkar, A., Eng, G., Blumling, J., Langer, R., Khademhosseini, A.: Micromolding of shape-controlled, harvestable cell-laden hydrogels. Biomaterials **27**, 5391–5398 (2006)
135. Lee, S.A., Chung, S.E., Park, W., Lee, S.H., Kwon, S.: Three-dimensional fabrication of heterogeneous microstructures using soft membrane deformation and optofluidic maskless lithography. Lab Chip **9**, 1670–1675 (2009)
136. Bruzewicz, D.A., McGuigan, A.P., Whitesides, G.M.: Fabrication of a modular tissue construct in a microfluidic chip. Lab Chip **8**, 663–671 (2008)
137. Koh, W.G., Pishko, M.V.: Fabrication of cell-containing hydrogel microstructures inside microfluidic devices that can be used as cell-based biosensors. Anal. Bioanal. Chem. **385**, 1389–1397 (2006)
138. Spradling, A., Drummond-Barbosa, D., Kai, T.: Stem cells find their niche. Nature **414**, 98–104 (2001)
139. Voldman, J., Gray, M.L., Schmidt, M.A.: Microfabrication in biology and medicine. Annu. Rev. Biomed. Eng. **1**, 401–425 (1999)
140. Whitesides, G., Ostuni, E., Takayama, S., Jiang, X., Ingber, D.: Soft lithography in biology and biochemistry. Annu. Rev. Biomed. Eng. **3**, 335–373 (2001)
141. Kim, S.M., Lee, S.H., Suh, K.Y.: Cell research with physically modified microfluidic channels: a review. Lab Chip **8**, 1015–1023 (2008)
142. Semino, C.: Can we build artificial stem cell compartments? J. Biomed. Biotechnol. **2003**, 164–169 (2003)
143. Czyz, J., Wobus, A.: Embryonic stem cell differentiation: the role of extracellular factors. Differentiation **68**, 167–174 (2001)
144. Zandstra, P.W., Nagy, A.: Stem cell bioengineering. Annu. Rev. Biomed. Eng. **3**, 275–305 (2001)
145. Kovacs, G.T.A., Maluf, N.I., Petersen, K.E.: Bulk micromachining of silicon. Proc IEEE **86**, 1536–1551 (1998)
146. Bustillo, J.M., Howe, R.T., Muller, R.S.: Surface micromachining for microelectromechanical systems. Proc IEEE **86**, 1552–1574 (1998)
147. Park, T.H., Shuler, M.L.: Integration of cell culture and microfabrication technology. Biotechnol. Prog. **19**, 243–253 (2003)
148. Revzin, A., Russell, R.J., Yadavalli, V.K., Koh, W.G., Deister, C., Hile, D.D., Mellott, M.B., Pishko, M.V.: Fabrication of poly(ethylene glycol) hydrogel microstructures using photolithography. Langmuir **17**, 5440–5447 (2001)
149. Hahn, M.S., Taite, L.J., Moon, J.J., Rowland, M.C., Ruffino, K.A., West, J.L.: Photolithographic patterning of polyethylene glycol hydrogels. Biomaterials **27**, 2519–2524 (2006)
150. Kim, E., Xia, Y., Whitesides, G.M.: Polymer microstructures formed by molding in capillaries. Nature **376**, 581–584 (1995)
151. Xia, Y., Whitesides, G.M.: Soft lithography. Annu. Rev. Mater. Sci. **28**, 153–184 (1998)
152. Stevens, M.M., Mayer, M., Anderson, D.G., Weibel, D.B., Whitesides, G.M., Langer, R.: Direct patterning of mammalian cells onto porous tissue engineering substrates using agarose stamps. Biomaterials **26**, 7636–7641 (2005)
153. Weibel, D.B., Lee, A., Mayer, M., Brady, S.F., Bruzewicz, D., Yang, J., DiLuzio, W.R., Clardy, J., Whitesides, G.M.: Bacterial printing press that regenerates its ink: contact-printing bacteria using hydrogel stamps. Langmuir **21**, 6436–6442 (2005)
154. Tan, W., Desai, T.A.: Microfluidic patterning of cells in extracellular matrix biopolymers: effects of channel size, cell type, and matrix composition on pattern integrity. Tissue Eng. **9**, 255–267 (2003)
155. Liu, V., Bhatia, S.: Three-dimensional photopatterning of hydrogels containing living cells. Biomed. Microdev. **4**, 257–266 (2002)
156. Yu, T., Ober, C.K.: Methods for the topographical patterning and patterned surface modification of hydrogels based on hydroxyethyl methacrylate. Biomacromolecules **4**, 1126–1131 (2003)

157. Hahn, M., Miller, J., West, J.: Laser scanning lithography for surface micropatterning on hydrogels. Adv. Mater. **17**, 2939–2942 (2005)
158. Miller, J.S., Béthencourt, M.I., Hahn, M., Lee, T.R., West, J.L.: Laser-scanning lithography (LSL) for the soft lithographic patterning of cell-adhesive self-assembled monolayers. Biotechnol. Bioeng. **93**, 1060–1068 (2006)
159. Dendukuri, D., Gu, S.S., Pregibon, D.C., Hatton, T.A., Doyle, P.S.: Stop-flow lithography in a microfluidic device. Lab Chip **7**, 818–828 (2007)
160. Khademhosseini, A., Eng, G., Yeh, J., Fukuda, J., Blumling, J., Langer, R., Burdick, J.A.: Micromolding of photocrosslinkable hyaluronic acid for cell encapsulation and entrapment. J. Biomed. Mater. Res. Part A **79**, 522–532 (2006)
161. Chung, K.-Y., Mishra, N.C., Wang, C.-C., Lin, F.-H., Lin, K.-H.: Fabricating scaffolds by microfluidics. Biomicrofluidics **3**, 022403/1–022403/8 (2009)
162. Chung, S,E,, Park, W., Park, H., Yu, K., Park, N., Kwon. S.: Optofluidic maskless lithography system for real-time synthesis of photopolymerized microstructures in microfluidic channels. Appl. Phys. Lett. **91**, 041106/1-041106/3 (2007)
163. Rivest, C., Morrison, D., Ni, B., Rubin, J.: Microscale hydrogels for medicine and biology: synthesis, characteristics and applications. J. Mech. Mater. Struct. **2**, 1103–1119 (2007)
164. Koh, W.-G., Itle, L.J., Pishko, M.V.: Molding of hydrogel microstructures to create multiphenotype cell microarrays. Anal. Chem. **75**, 5783–5789 (2003)
165. DeForest, C.A., Polizzotti, B.D., Anseth, K.S.: Sequential click reactions for synthesizing and patterning three-dimensional cell microenvironments. Nat. Mater. **8**, 659–664 (2009)
166. Fukuda, J., Khademhosseini, A., Yeo, Y., Yang, X., Yeh, J., Eng, G., Blumling, J., Wang, C.- F., Kohane, D.S., Langer, R.: Micromolding of photocrosslinkable chitosan hydrogel for spheroid microarray and co-cultures. Biomaterials **27**, 5259–5267 (2006)
167. Hahn, M.S., Miller, J.S., West, J.L.: Three-dimensional biochemical and biomechanical patterning of hydrogels for guiding cell behavior. Adv. Mater. **18**, 2679–2684 (2006)
168. Jang, J.-H., Dendukuri, D., Hatton, T.A., Thomas, E.L., Doyle, P.S.: A route to three-dimensional structures in a microfluidic device: stop-flow interference lithography. Angew. Chem. Int. Ed. **46**, 9027–9031 (2007)
169. Hwang, D.K., Oakey, J., Toner, M., Arthur, J.A., Anseth, K.S., Lee, S., Zeiger, A., Van Vliet, K.J., Doyle, P.S.: Stop-flow lithography for the production of shape-evolving degradable microgel particles. J. Am. Chem. Soc. **131**, 4499–4504 (2009)
170. Dendukuri, D., Hatton, T.A., Doyle, P.S.: Synthesis and self-assembly of amphiphilic polymeric microparticles. Langmuir **23**, 4669–4674 (2007)
171. Chung, S.E., Park, W., Shin, S., Lee, S.A., Kwon, S.: Guided and fluidic self-assembly of microstructures using railed microfluidic channels. Nat. Mater. **7**, 581–587 (2008)
172. Chung, K., Cho, J.K., Park, E.S., Breedveld, V., Lu, H.: Three-dimensional in situ temperature measurement in microsystems using Brownian motion of nanoparticles. Anal. Chem. **81**, 991–999 (2009)
173. Chung, S.E., Lee, S.A., Kim, J., Kwon, S.: Optofluidic encapsulation and manipulation of silicon microchips using image processing based optofluidic maskless lithography and railed microfluidics. Lab Chip **9**, 2845–2850 (2009)
174. Kloxin, A., Tibbitt, M., Kasko, A.: Tunable hydrogels for external manipulation of cellular microenvironments through controlled photodegradation. Adv. Mater. **22**, 61–66 (2010)
175. Luo, Y., Shoichet, M.S.: A photolabile hydrogel for guided three-dimensional cell growth and migration. Nat. Mater. **3**, 249–253 (2004)
176. Ling, Y., Rubin, J., Deng, Y., Huang, C., Demirci, U., Karp, J.M., Khademhosseini, A.: A cell-laden microfluidic hydrogel. Lab Chip **7**, 756–762 (2007)
177. Khademhosseini, A., Yeh, J., Jon, S., Eng, G., Suh, K.Y., Burdick, J.A., Langer, R.: Molded polyethylene glycol microstructures for capturing cells within microfluidic channels. Lab Chip **4**, 425–430 (2004)
178. Tang, M.D., Golden, A.P., Tien, J.: Molding of three-dimensional microstructures of gels. J. Am. Chem. Soc. **125**, 12988–12989 (2003)

179. Huang, C.P., Lu, J., Seon, H., Lee, A.P., Flanagan, L.A., Kim, H.-Y., Putnam, A.J., Jeon, N.L.: Engineering microscale cellular niches for three-dimensional multicellular co-cultures. Lab Chip **9**, 1740–1748 (2009)
180. Xia, Y., McClelland, J.J., Gupta, R., Qin, D., Zhao, X.M., Sohn, L.L., Celotta, R.J., Whitesides, G.M.: Replica molding using polymeric materials: a practical step toward nanomanufacturing. Adv. Mater. **9**, 147–149 (1997)
181. Flaim, C.J., Chien, S., Bhatia, S.N.: An extracellular matrix microarray for probing cellular differentiation. Nat. Methods **2**, 119–125 (2005)
182. Di Benedetto, F., Biasco, A., Pisignano, D., Cingolani, R.: Patterning polyacrylamide hydrogels by soft lithography. Nanotechnology **16**, S165–S170 (2005)
183. Ilkhanizadeh, S., Teixeira, A.I., Hermanson, O.: Inkjet printing of macromolecules on hydrogels to steer neural stem cell differentiation. Biomaterials **28**, 3936–3943 (2007)
184. Nagase, H., Fields, G.B.: Human matrix metalloproteinase specificity studies using collagen sequence-based synthetic peptides. Biopolymer **40**, 399–416 (1996)
185. Lee, S.H., Moon, J.J., West, J.L.: Three-dimensional micropatterning of bioactive hydrogels via two-photon laser scanning photolithography for guided 3D cell migration. Biomaterials **29**, 2962–2968 (2008)
186. Tapp, H., Deepe, R., Ingram, J.A., Kuremsky, M., Hanley, E.N., Gruber, H.E.: Adipose-derived mesenchymal stem cells from the sand rat: transforming growth factor beta and 3D co-culture with human disc cells stimulate proteoglycan and collagen type I rich extracellular matrix. Arthritis Res. Ther. **10**, R89 (2008)
187. Song, I.-H., Caplan, A.I., Dennis, J.E.: Dexamethasone inhibition of confluence-induced apoptosis in human mesenchymal stem cells. J. Orthop. Res. **27**, 216–221 (2009)
188. Mo, X.- t., Guo, S.-c., Xie, H.-q., Deng, L., Zhi, W., Xiang, Z., Li, X.-q., Yang, Z.-m.: Variations in the ratios of co-cultured mesenchymal stem cells and chondrocytes regulate the expression of cartilaginous and osseous phenotype in alginate constructs. Bone **45**, 42–51 (2009)
189. Hwang, N.S., Varghese, S., Elisseeff, J.: Derivation of chondrogenically-committed cells from human embryonic cells for cartilage tissue regeneration. PLoS One **3**, e2498 (2008)
190. Bandara, G., Georgescu, H.I., Lin, C.W., Evans, C.H.: Synovial activation of chondrocytes: evidence for complex cytokine interactions. Agents Actions **34**, 285–288 (1991)
191. Gleghorn, J.P., Jones, A.R.C., Flannery, C.R., Bonassar, L.J.: Boundary mode frictional properties of engineered cartilaginous tissues. Eur. Cell Mater. **14**, 20–29 (2007)
192. Lima, E.G., Tan, A.R., Tai, T., Bian, L., Stoker, A.M., Ateshian, G.A., Cook, J.L., Hung, C.T.: Differences in interleukin-1 response between engineered and native cartilage. Tissue Eng. Part A **14**, 1721–1730 (2008)
193. Rydholm, A.E., Bowman, C.N., Anseth, K.S.: Degradable thiol-acrylate photopolymers: polymerization and degradation behavior of an in situ forming. Biomaterials **26**, 4495–4506 (2005)
194. Tsai, M.F., Tsai, H.Y., Peng, Y.S., Wang, L.F., Chen, J.S., Lu, S.C.: Characterization of hydrogels prepared from copolymerization of the different degrees of methacrylate-grafted chondroitin sulfate macromers and acrylic acid. J. Biomed. Mater. Res. A **84**, 727–739 (2008)
195. Du, Y., Lo, E., Ali, S., Khademhosseini, A.: Directed assembly of cell-laden microgels for fabrication of 3D tissue constructs. Proc. Natl. Acad. Sci. U.S.A. **105**, 9522–9527 (2008)
196. Koh, W.G., Revzin, A., Pishko, M.V.: Poly(ethylene glycol) hydrogel microstructures encapsulating living cells. Langmuir **18**, 2459–2462 (2002)

Microengineering Approach for Directing Embryonic Stem Cell Differentiation

Hojae Bae, Jason W. Nichol, Amir Foudeh, Behnam Zamanian, Cheong Hoon Kwon and Ali Khademhosseini

Abstract The microenvironment plays an integral role in directing the differentiation of stem cells. The ability to control and manipulate systems on the microscale can be used to control the cellular microenvironment to direct stem cell behavior. For stem cells, this control greatly improves our ability to study cell–microenvironment interactions in a rapid and precise manner to regulate stem cell behaviors such as differentiation and proliferation. Combining microscale technologies with high throughput techniques could also greatly increase the possibility for probing the multivariable complexity of biological systems. In this chapter, microengineering approaches to control the cellular microenvironment and to influence embryonic stem cell (ESC) self-renewal and differentiation are introduced and specific examples of the use of microfabrication technologies for directing ESC fate decisions are discussed.

1 Introduction

The establishment of embryonic stem cell (ESC) lines derived from both mouse and human cells has generated the possibility of cell therapies based on an unlimited and renewable source of cells [1, 2]. However, the use of human

H. Bae, J. W. Nichol, A. Foudeh, B. Zamanian, C. H. Kwon and A. Khademhosseini
Department of Medicine, Center for Biomedical Engineering, Brigham and Women's Hospital, Harvard Medical School, Cambridge, MA 02139, USA

H. Bae, J. W. Nichol, A. Foudeh, B. Zamanian, C. H. Kwon and A. Khademhosseini (✉)
Harvard-MIT Division of Health Sciences and Technology, Massachusetts Institute of Technology, Cambridge, MA 02139, USA
E-mail: alik@mit.edu

Stud Mechanobiol Tissue Eng Biomater (2010) 2: 153–171
DOI: 10.1007/8415_2010_5

Published online: 16 June 2010

embryos to generate ESC lines is somewhat controversial [3]. A breakthrough to this problem was established when Yamanaka and co-workers demonstrated in vitro reprogramming of murine fibroblasts into induced pluripotent stem (iPS) cells [4]. These iPS cells were proven to be functionally and molecularly similar to ESCs [5, 6] offering new opportunities in regenerative cell therapy [7–9]. However, there are still many requirements to fulfill before these cells can be used clinically. One such challenge is to controllably direct stem cell differentiation [10].

Stem cells are sensitive to a variety of microenvironmental stimuli that regulate both self-renewal and differentiation [11]. Therefore stem cells can be characterized by their capacity to differentiate into specific cell lineages in response to temporally and spatially regulated extrinsic and intrinsic signals. Microengineering enables the regulation of cell–microenvironment interactions, such as cell–cell, cell–extracellular matrix (ECM), cell–soluble factors, and cell–mechanical stimuli interactions [12]. Thus, by utilizing microscale engineering, cell differentiation can be guided through the controlled interplay between regulatory factors down to the level of individual cells. Also, microengineering can be used to generate three dimensional (3D) microenvironments for the development of physiologically relevant tissue models for tissue engineering [13–15]. This can be achieved through the merger of microfabrication techniques with biomaterials to create the desired tissue structures [16]. Advanced biomaterials are being used to study stem cells, the interactions between cells and biomaterials, and as tissue engineering scaffolds. The field of tissue engineering is driven by the need to provide functional equivalents of native tissues that can be used for the in vitro study of tissue physiology and pathology, as well as for implantation where no native or artificial transplantable materials are currently available in sufficient quantities to repair damaged tissues [17]. This interdisciplinary field is at the intersection of engineering, biology, and medicine and aims to develop biological substitutes that restore, maintain, or improve tissue function.

Recent research has confirmed the existence of stem cells residing in niches, unique to the tissues and organs in which they reside that contain highly ordered microarchitectures, cellular compartmentalization, and arrangement [18]. The merger of microfabrication and advanced biomaterials are useful to recreate many of these complex features, to create in vitro microenvironments with the ability to effectively direct ESC behavior. These approaches are becoming powerful tools for the study of stem cells. The following chapter will highlight the current techniques for using microscale biomaterials both for investigating and directing stem cell behavior and for creating engineered tissues using 3D cell-laden scaffolds. These two applications of microscale engineering may be of great benefit to the future of regenerative medicine. The clinical success of regenerative medicine is highly dependent on successes in both tissue engineering and stem cell differentiation, suggesting that the development of technologies to achieve the desired outcome will hopefully shorten the time needed to bring these techniques to the clinic.

2 Control of the Cellular Microenvironment

The ability to control the cellular microenvironment is key to controlling cell viability, growth, migration, and differentiation. Microenvironmental control can be performed at the level of interactions with surrounding cells, substrate mechanics, biomaterial chemistry, applied physical forces, and the degradation of surrounding materials [19–21]. For ESCs, self-renewal and differentiation pathways appear to be controlled by several interconnected pathways based on cell–cell, cell–ECM, and cell–soluble factors interactions (Fig. 1). We will discuss these interactions and subsequently discuss how microengineering techniques could control these interactions.

2.1 Cell–cell Contacts

Cell–cell contacts are a major class of regulatory signals that control cell behavior [22]. For instance, the differentiation of skeletal muscle cells has been reported to depend on cell–cell contacts that induce cell cycle arrest and subsequent gene expression [23]. One representative factor that participates in and controls cell–cell adhesions are cadherins. Cadherins are a class of type I transmembrane protein with extracellular Ca^{2+}-binding domains. There are several classes of cadherin molecules, and different classes are found in different cell types and tissues. For instance, E-cadherins are generally found in epithelial tissues, and preferentially bind to other E-cadherin molecules. This homophilic binding has been attributed to the NH_2-terminal EC1 domain [24]. However, several observations have suggested

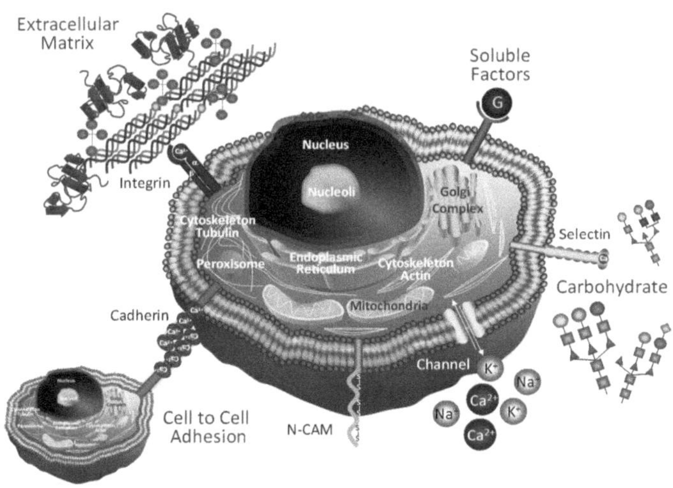

Fig. 1 Modifications of various factors (cell–cell contacts, mechanics, ligand and materials chemistry, etc.) can be used to control the stem cell microenvironment

that cadherins can also interact in a heterophilic fashion [25, 26]. In addition to E-cadherins, other cell types and tissues express other classes of cadherin: P-cadherin in the placenta, N-cadherin in neural cells, R-cadherin in renal tissue, VE-cadherin in vascular endothelial cells, M-cadherin in myotubules, OB-cadherin in osteoblasts, and K-cadherin in the kidney [27, 28]. Cadherin binding leads to specific signaling cascades that affect cellular behavior and function. Upon binding of the extracellular domain of cadherins, the intracellular domain binds p120-catenin and beta-catenin containing a highly phosphorylated region. Activated catenin also binds to alpha-catenin, resulting in regulation of actin-containing cytoskeletal filaments, ultimately leading to specific control of gene transcription [27, 29].

The loss or decrease in E-cadherin expression and function has been reported to be associated with high cancer cell progression and metastasis in which the decreased strength of cellular adhesion within a tissue results in an increase in cellular motility and may allow cancer cells to cross the basement membrane and invade the surrounding tissues [30–32]. For example, the addition of anti-E-cadherin antibodies restored cell migratory behavior by blocking cadherin-dependent adhesion [33]. Another study showed that transfecting fibroblasts with E-cadherin similarly suppressed cell infiltration of collagen gels in an E-cadherin-dependent manner [34]. Recently, it has been reported that E-cadherin regulates cell motility by both adhesion-dependent and adhesion-independent mechanisms in which E-cadherin-dependent reduction in epithelial cell motility depends both on the E-cadherin expression level and on the E-cadherin density on the migratory substratum [35]. These mechanisms are important in wound healing as cell–cell interactions, cell migration, and cell proliferation as it plays key roles in the tissue's ability to repair a defect [36]. In addition to its modulation of wound healing mechanisms, similar cadherin modulated cell–cell interactions also play a key role in ESC differentiation and growth. For example, E-cadherin has been shown to induce cell to cell adhesion for the formation of embryoid bodies (EBs), which are 3D spherical aggregates containing differentiated cells of all three germ layers. Although E-cadherin is constitutively expressed in early stage embryos, it is down-regulated as the cells differentiate [37, 38]. Other factors that can similarly modulate stem cell attributes such as aggregation, migration, proliferation, or differentiation can have a profound impact on the control of stem cell behavior.

2.2 Cell–soluble Factor Interactions

Soluble factors, such as growth factors, play an important role in influencing cell growth and differentiation. Growth factors are proteins that bind to receptors on the cell surface, and the binding activates a specific series of cellular mechanisms leading to events such as cellular proliferation and/or differentiation. These regulatory molecules are naturally derived, and can either be added to the culture medium or more specifically delivered to the targeted cells. Many growth factors are quite versatile, stimulating cellular responses in numerous cell types; while others act specifically on a particular cell-type. Based on their ability to control and direct cell behavior, the use of growth factors has been investigated for directing the lineage specific differentiation of ESCs. For example, Activin A was shown to mediate dorsoanterior mesoderm differentiation, while bone morphogenetic protein 4 (BMP-4) was shown to mediate the formation of hematopoietic cells [39–41]. Also, it has been demonstrated that binding of IGF-II, a member of the insulin-like growth factor (IGF) family, to its signaling receptor, IGF1R, at the surface of mesoderm precursor cells increased mesoderm formation [42]. In addition, EGF, BMP-4, and FGF have been shown to induce ectodermal and mesodermal differentiation [43].

2.3 Cell–extracellular Matrix Interactions

The ECM consists of the structural proteins that surround cells in mammalian tissues. This serves to provide structural support and anchorage to the cells in addition to compartmentalizing cells and tissues, regulating intercellular communications, and sequestering growth factors [44, 45]. Cell–ECM interactions play a critical role in cell function and physiology through cell binding to components of the ECM. Cell–ECM adhesion is regulated by specific molecules of the surface of cells called cellular adhesion molecules (CAM), such as integrins. ECM provides a template for cell adhesion, proliferation, migration, differentiation, and tissue formation. The phenotype of originated cells has also been observed to be sensitive to tissue elasticity and stiffness. For example, to direct the differentiation of mesenchymal stem cells (MSCs) soft matrices that mimic the brain are found to be neurogenic, while stiffer matrices that mimic muscle are myogenic, and comparatively rigid matrices that mimic collagenous bone were found to be osteogenic [46].

Recently, researchers have designed various biomaterials to provide more complex and biomimetic environments for ESC expansion and differentiation. For instance, Lammers and colleagues tested the ability of various biomaterials for ESC proliferation and differentiation, and reported that scaffolds made of insoluble collagenous bone matrix in combination with β-tricalciumphosphate were more effective in directing ESC differentiation into osteogenic lineage [47]. In addition, Langer and colleagues have demonstrated that rigid polymeric scaffolds supported ESC differentiation. When ESCs were surrounded by a soft biomaterial such as Matrigel, tissue was not formed, however, as Matrigel was combined with a rigid polymeric scaffold, tissue-like structures were formed [48]. These data suggested that the mechanical properties of polymeric scaffolds can be used to direct ESC differentiation and induce tissue formation. In addition, it was also found that fibronectin (FN) stimulated ESCs to differentiate into endothelial cells whereas laminin induced ESCs to differentiate into cardiomyocytes [49]. Thus, showing that the ECM chemistry works alongside mechanics in directing ESC fates.

3 Microengineering the Environment

Microengineering techniques can be used to regulate the cell's interactions with its microenvironment [50]. For example, by restricting the cellular geometry through patterned substrates, controlled microenvironments were created to investigate specific cell behaviors [51–53]. Microengineering approaches have also been used to control the cellular microenvironment by using microscale channels [54], microwells [55], and cell-laden hydrogels [50, 56, 57] (Fig. 2). In addition, biomaterials have been engineered that are capable of controlling cell behavior [50, 58, 59]. These biomaterials have tunable property such as a controlled degradation rate, hydrophilicity, and mechanical properties. The ability to use microengineering to manipulate biomaterials is emerging as a powerful tool in regulating ESC behavior. These manipulations can involve methods of fabricating microscale units of biomaterials or spatial control of material properties. Furthermore, microarray techniques can be used to identify biomaterials that can direct ESC fate responses by enabling massively parallel synthesis and characterization of cell–materials interactions. These studies of cell behavior in response to various natural or synthetic stimuli have applications in a wide variety of fields ranging from basic biological

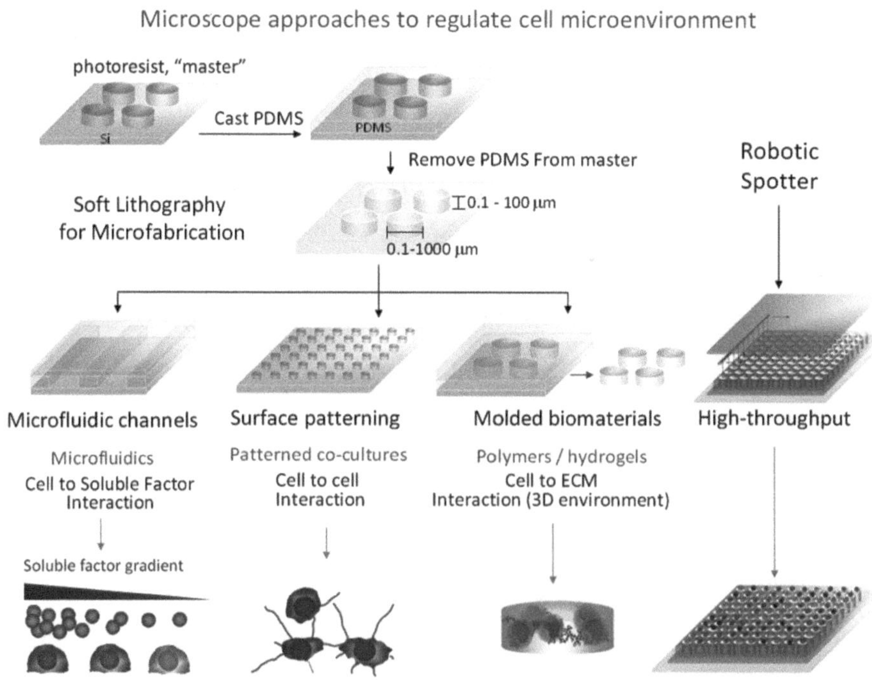

Fig. 2 Microengineering techniques (i.e. microfluidics, surface patterning, and high-throughput microarray) can be used to control cell–microenvironment interactions, such as cell–cell, cell–ECM, and cell–soluble factor interactions

studies to drug discovery. Here, we will review these techniques and where appropriate provide examples of their use in directing ESC differentiation.

3.1 *Microfluidic Platforms for Controlling Cell–soluble Factor Interactions*

The field of microfluidics involves the manipulation and processing of micro- and nanoliter volumes of fluids within small channels that are typically in the range of 10–100 μm [60]. Fluid behavior on the micron scale is different from that occurring on the macroscale [61]. For example, viscosity and molecular diffusion are of high importance at the microscale, while having less effect at the macroscale. This is because the small dimensions of microfluidic systems cause the Reynolds number to be low, making the flow in the laminar range. The Reynolds number is determined as $Re = \rho * v * d / \mu$, where ρ is the fluid density, v is the fluid velocity, d is the hydraulic diameter of the channel, and μ is the fluid viscosity. Therefore small changes to viscosity is important at the microscale because of their substantial impact on the Reynolds number and whether the flow remains laminar. In addition to laminarly flowing fluids, microchannels offer other advantages compared to conventional systems. These include reduced analysis times, the need for lower sample/reagent volumes as well as the ability to run multiple assays on a single device [62, 63].

With the introduction of microfluidic platforms, stem cell research has entered a new era in which the cell–soluble factor interactions can be controlled in a much more controlled manner. Microfluidic technologies, which often use polydimethylsiloxane (PDMS) [62], have been used in a number of different applications for studying cells such as fluidic based cell patterning, subcellular localization of media components, high throughput drug screening, analysis of fluid shear stress on cells, and creation of soluble factor gradients [64–67].

One of the most valuable applications of microfluidic systems is the ability to expose cells to laminar flows to control fluid mixing and shear stress. Kim et al. showed that microfluidic devices can produce a logarithmic scale of flow rates and logarithmic concentration gradients. It has been shown that cell morphology and proliferation for ESCs varies based on these gradients [68]. This ability to control the flow rate and shear stress has been utilized to study ESC self-renewal and proliferation, to show that high flow rates result in increased proliferation [69]. Since microfluidic systems use low sample/reagent volumes they are useful tools for the screening of soluble components such as serum components (growth factors), conditioned media, or different chemical formulations. In addition, microfluidic systems have been used for making gradients growth factors or oxygen concentrations to test their effects on cells in an efficient manner [70].

Microfluidic systems have also been used to mechanically stimulate cells. For example, Park et al. showed that compressive cyclic loading enhanced the osteogenic differentiation of human MSCs [71]. In another study, shear stress gradients were studied on hESC-derived endothelial cells to show that these cells were capable of responding to shear stress changes through varying gene expression [72]. The combination of microfluidic systems with controlled flow rates or concentration gradients have been shown to be important for regulating cell differentiation [64, 73]. For example, Chung et al. showed a microfluidics system for neural stem cell differentiation and proliferation, demonstrating that different growth factor combinations consisting of epidermal growth factor (EGF), fibroblast growth factor 2 (FGF2) and platelet-derived growth factor (PDGF) could be used for directing differentiation towards astrocyte lineage. Microfluidic systems have also been used for analyzing single cell gene expression profiles with increased sensitivity compared to bulk reactions [74]. Moreover, microfluidic devices can be used to deliver extremely precise concentrations of signaling molecules to cells [11]. With these systems, there is an improved ability to generate complex microenvironments with strict control over spatial and temporal resolution of soluble factors. This high degree of control on the microscale makes these systems ideal for studying the response of ESCs. In addition the volumes are much smaller than in traditional culture systems reducing the use of expensive soluble factors to enable testing in a high throughput manner.

3.2 Controlled Microbioreactors

In a developing organism, tissues emerge from coordinated sequences of cell renewal, differentiation, and assembly within a dynamic environment characterized by spatial and temporal gradients of multiple factors. Thus, to direct cells to differentiate at the right time, in the right place, and into the right phenotype, it is important to recreate the appropriate microenvironment.

One way to accomplish this complicated feat is to characterize the native tissue environments and attempt to mimic these features in vitro. In tissues, cells are surrounded by other cells and embedded in an ECM that defines the architecture, signaling, and biomechanics of the microenvironment [75]. Cells respond to their immediate microenvironment, via phenotypic or heterotypic interactions with neighboring cells [76]. This can be difficult to recreate completely in vitro since much of the complex interplay of mechanical and molecular factors present in vivo are absent [77]. It has been argued that a new generation of 3D culture systems are needed that would be "something between a Petri dish and a mouse" to correctly present a cell's environment in a living organism and be more predictive of in vivo systems [78].

For stem cells in particular, to unlock their full potential and obtain biologically sound and relevant data in vitro, at least some aspects of the dynamic 3D environments that are associated with their renewal, differentiation, and assembly into tissues need to be reconstructed. A fundamental approach to the generation of engineered tissues is to direct the 3D organization of cells (via biomaterial scaffolds) and to establish the conditions necessary for the cells to reconstruct a functional tissue structure (via bioreactors). This approach is based on a premise that cells' responses to environmental factors are predicable, and that the cell function in vitro can be modulated by the same complex factors known to play a role during tissue development and remodeling.

ESCs can greatly benefit from being cultured in bioreactors that controllably recreate cell and tissue specific environmental conditions. Bioreactor systems may better facilitate the transition from laboratory to clinical scale production due to their scalability, while providing a means to quantitatively study cell behavior in response to various stimuli. Some of the important considerations during bioreactor design are: (1) rapid and controllable expansion of cells, (2) enhanced cell seeding of 3D scaffolds at a desired cell density, yield, kinetic rate, and spatial uniformity, (3) efficient exchange of oxygen, nutrients, and metabolites, and (4) provision of physiological stimuli [11]. Current designs incorporate cascades of biological and physical stimuli to exert greater influence over cellular differentiation and development into functional tissue constructs [79]. Bioreactors are typically custom engineered to account for specific mechanisms of nutrient transfer and specific physical factors inherent in the desired tissue. For instance, simple control of the transport rate of oxygen could monitor the pH of the environment in which stem cells are cultured [80, 81]. These bioreactors are often designed to be modular, mini-scaled, and multi-parametric, to economize cells and reagents, and to increase the number of samples that can be analyzed.

3.3 Surface Micropatterning for Controlling Cell–cell Contacts

Microscale surface patterning techniques enable the control of cell–cell contacts on geometrically defined 2D surfaces [12, 82]. A variety of microscale technologies including microtopography [55, 56], microfabricated stencils [83], micro-contact printing [84], and layer-by-layer deposition [85] have been developed to pattern the ESCs on 2D substrates. In one approach, hydrogel microwell arrays that generate low shear stress regions enabled the docking and positioning of ESCs. Such microwells could be seeded with feeder cells to maintain human ESCs in an undifferentiated state [86]. Microwell arrays that were fabricated from polyethyleneglycol (PEG) hydrogels have been demonstrated as a useful research tool for uniform cell seeding and EB formation [87] (Fig. 3). Similar microwell arrays were used to generate spatially and temporally synchronized beating human EBs [88]. Microwell arrays can also be used to screen and characterize ESCs, while homogeneous EBs obtained from this microfabricated microwell can direct EB-mediated differentiation [84]. In another example, a Bio Flip Chip (BFC) containing thousands of microwells with sizes optimized to trap single murine

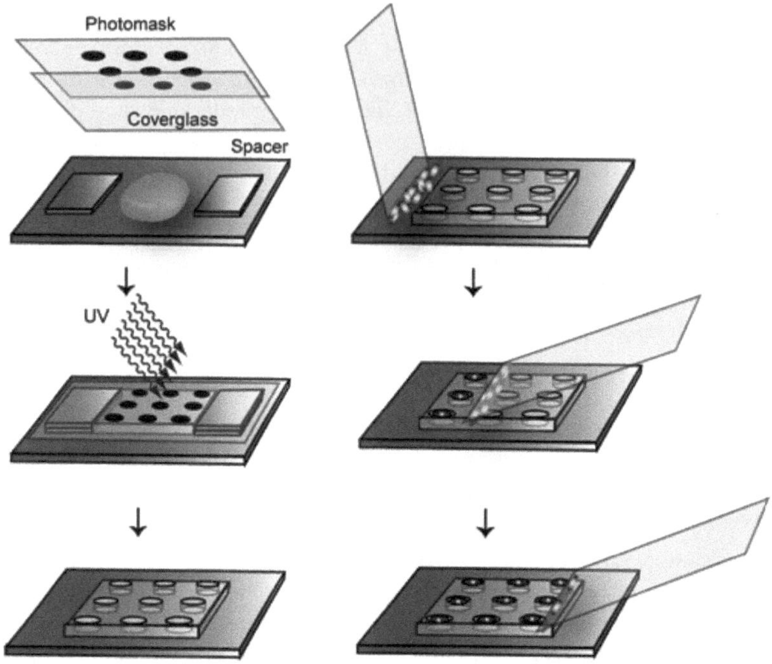

Fig. 3 (*Left*) PEG microwell fabrication through UV cross-linking. (*Right*) Localizing cells within arrays of microwells using a wiping technique. This method produces cell seeding densities that vary consistently with microwell geometry and cell concentration. Reprinted with permission from the *Journal of Biomedical Materials Research* [87]

ESCs were fabricated to provide both incremental and independent control of contact-mediated signaling [89].

Cell–cell communications in aggregates play a significant role in controlling ESC proliferation and differentiation. A microfabricated PDMS stencil can regulate the initial size of ESC aggregates, which is a critical factor for controlling ESC differentiation. By using this technique, it was demonstrated that larger aggregates resulted in more mesoderm and endoderm differentiation than small aggregates. Microfabricated parylene-C stencils were also used to micropattern co-cultures in both static and dynamic conditions by sequentially using several stencils with differently modified surface properties [90]. In addition to microfabricated parylene-C stencils, microcontact printing was used to study the differentiation of micropatterned human ESCs [84]. The gene and protein expression analysis of micropatterned human ESCs demonstrated that the endodermal and neuronal expression increased inversely with colony size. Moreover, greater mesoderm and cardiac induction was observed for larger EBs. This analysis demonstrated that heterogeneity in human ESC colony and aggregate size generated subsets of appropriate conditions for differentiation.

3.4 High-throughput Microarrays for Screening Microenvironments

ESC differentiation can be directed through different combinations of factors that influence the microenvironment. To study the combination of various stem cell differentiation factors is an enormous combinatorial problem that would be extremely difficult to perform without the facilitation of high-throughput analysis, screening, and imaging. This goal can be achieved more readily through the effective use of microarray systems for analysis and screening of multiple factors individually and in combination. Because of the significant number of possible combinations, high-throughput approaches can be used to rapidly test and discover important combinations of parameters in ESC differentiation. Parameters such as ECM proteins, soluble molecules (e.g. cytokines, growth factors), and biomaterial interactions can influence the pathway and efficiency of ESC differentiation. To improve our understanding of ESC differentiation, studying the combined influence of parameters that may regulate stem cell expansion and specialization in an efficient and cost effective manner using high-throughput analysis, screening, and imaging may prove beneficial.

In addition to growth factors and cell-secreted morphogenetic factors, synthetic small molecules can direct ESC differentiation. Small permeable, naturally occurring molecules such as vitamin C, sodium pyruvate and retinoic acid have been used to regulate stem cell fate [91]. Similarly, new synthetic, heterocyclic small molecules that can alter stem cell fate have recently been studied to control stem cell differentiation. Ding et al. reported ESC differentiation into various cell types employing small molecules [92]. One manner by which synthetic molecules can be used to selectively control and regulate stem cell differentiation and proliferation is through adjusting the activities of proteins. Such processes have successfully been employed to cause controllable neurogenetic and cardiomyogenetic induction in murine ESCs, osteogenesis induction in MSCs, and skeletal muscle cell differentiation [91, 93]. Further investigation into small molecule discovery could play a substantial role in furthering the ability to reliably differentiate stem cells down specific pathways.

The interactions between ESCs and the surrounding matrix environment can profoundly influence stem cell behavior and direct ESC differentiation. A library of 576 different combinations of acrylate-based polymers was used to determine the effect on stem cell behavior in a biomaterial array format [94]. Combinations of the different polymers were mixed in 384-well plates and were robotically printed on coated glass slides. After printing, the slides were exposed to long wave UV to initiate polymerization, dried, sterilized with UV, and washed with PBS and cell culture media. ESCs were then seeded onto the slides and the influence of the materials on ESC differentiation was analyzed. Based on these observations, it was concluded that ESC differentiation may be induced toward epithelial cell lineage through interactions with specific combinations of materials. This study demonstrated the potential for creating a library of synthetic materials to direct ESC fate decisions.

3.5 Three Dimensional Scaffolds for Culturing ESCs

Biocompatible and biodegradable polymer scaffolds can be used to control growth and differentiation of ESCs [95]. Scaffolds provide structural support and physical cues for cell attachment, orientation, alignment, and spreading. A fundamental approach to the generation of engineered tissues is to direct the 3D organization of cells (via biodegradable scaffolds) and to establish the conditions necessary for the cells to reconstruct a functional tissue structure (via bioreactors). This approach is based on the premise that the cellular response to environmental factors are predictable, and that the cell function, in vitro can be modulated by the same complex factors known to play a role during development and remodeling in vivo.

Various scaffold features, such as physical cues can have an impact on the differentiation of ESCs. ESCs cultured within 3D scaffolds can be differentiated into specific tissue lineages based on the scaffold characteristics. For example, ESC differentiation towards an osteogenic lineage was achieved through seeding cells onto appropriately designed biodegradable scaffolds as demonstrated by increased expression of osteo-specific markers and bone nodules [96]. The effect of the 3D environment on ESC growth and differentiation was further studied by culturing EBs in different polymer networks [49]. Various compositions and structures of polymer matrices were evaluated, demonstrating that both material structure and stiffness of the 3D scaffolds influenced ESC fate. Using analogous systems, ESC differentiation down a neuronal differentiation pathway, was also studied [97]. It was shown that the number and maturity of neural-like structures identified by neuronal marker (i.e. βIII-tubulin) were increased within these 3D tissue constructs. Therefore, specifically designed 3D scaffolds have the potential for creating 3D tissue constructs using stem cells differentiated into the desired tissue type.

3.6 Tissue Engineering Using Assembly of Microengineered Building Blocks

Similar to their use in controlling the behavior of stem cells, microscale technologies can be used to control and direct engineered tissue development by creating microgels with specific microarchitectural features containing stem cells. Tissue engineering, using a bottom-up approach, creates macroscale tissues with controlled microarchitectural features through assembly of microscale cell-laden building blocks. Through control of the microenvironment, the goal is to create macroscale tissues with specific microarchitectural features, amenable to directing tissue behavior.

One such technique employed a layer-by-layer approach, where arrays of cells and ECM/polymer were additively photocrosslinked [98–100]. In one example, rat hepatocytes were mixed with RGD functionalized PEG prepolymer and the resulting mixture was polymerized through exposure to light through one of three photomasks, the first of which resulted in a three pointed star alignment. Subsequent layers were created using complementary photomasks, creating a honeycomb enclosure around two-layer cell-laden tissue units, mimicking native hepatic tissue. Tissue from this technique demonstrated many benchmarks of hepatic tissue function, such as urea production. The next step would be to recreate this technique using stem cells to determine whether recreation of the native architecture using these protocols would lead to functional hepatic tissues with a more clinically relevant cell source.

A major challenge to overcome in engineered tissues is the creation of functional, integrated microvasculature. Previously, random packing technique was used to create tissues with perfusable capillary networks. To create these building blocks, HepG2 cells were mixed with collagen, allowed to gel in specific shapes and then mixed with HUVEC cells which formed an endothelial monolayer around the surfaces of the hydrogel blocks [101]. The HepG2-HUVEC modules were packed into unidirectional perfused tubing with a porous plug to contain and allow for module aggregation and compaction. The engineered tissues were successfully perfused with blood to demonstrate cell viability as well as endothelial function, as the tissues perfused without clotting. One major advantage of this technique was the ability to create perfusable tissues with functional endothelialized channels, with the potential to contain any number of cell types in the hydrogel module cores. Employing stem cells in this system could yield functional tissues designed primarily for filtration, such as the liver or kidney, while more tissue types could be possible if the culture conditions are optimized to allow for removal from the perfused tubing.

Recent work in our laboratory has demonstrated the use of directed assembly to create tissue structures using cell-laden microgels. This technique harnesses the surface tension properties of hydrophilic hydrogels to assemble cell-laden microgels into tissue structures [102]. Cell-laden hydrogels of varying aspect ratios were placed into hydrophobic mineral oil to induce aggregation of hydrophilic microgels as they attempt to minimize surface energy. These assemblies were subsequently crosslinked into macroscale engineered tissues. The tissue shape and dimensions were controllable based on the building blocks' aspect ratios, while even greater control, and the potential for making more complex structures, was demonstrated by using lock-and-key type geometries. These results introduce a useful approach to create larger and more complex tissues with controlled co-culture conditions, using a number of cell types.

These are just a few of the techniques that could be used to create engineered tissues with microscale features containing stem cells (Fig. 4). With the ability to not only create specific microarchitectural features to control and direct stem cell behavior, but also to assemble these microgels into macroscale structures while retaining control over the microenvironment, the future for creation of stem cell based engineered tissues for use in vivo appears promising.

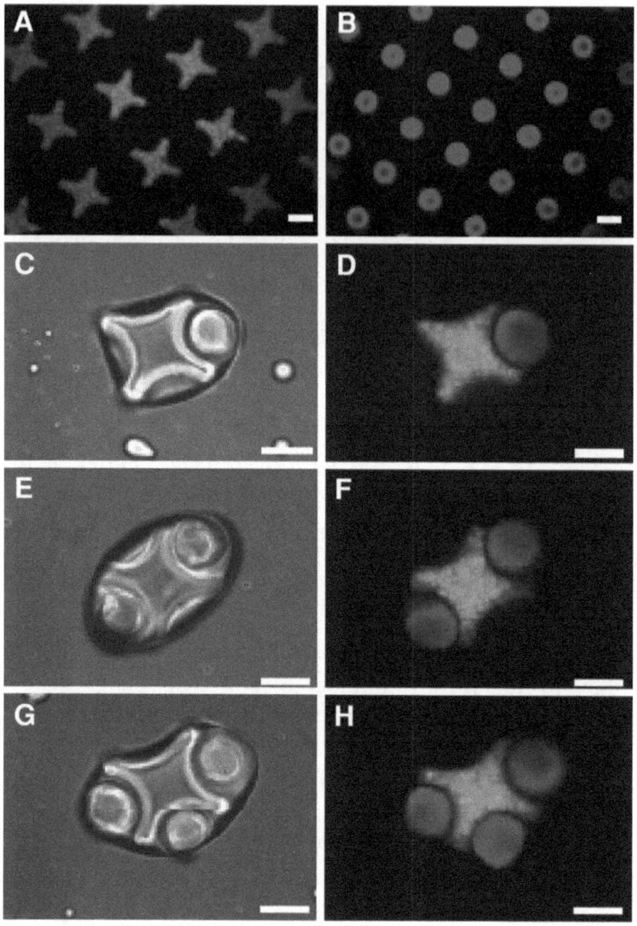

Fig. 4 Creating engineered tissues using directed assembly. Lock (**a**) and key (**b**) shaped hydrogel modules were created through UV exposure through a photomask, then aggregated and assembled in mineral oil, into single (**c, d**), double (**e, f**), triple (**g, h**) arrangements demonstrating control of co-cultured structures. *Scale*: 200 μm. Reprinted with permission from the National Academy of Science [102]

4 Conclusions

Although ESCs have been considered as a potential cell source for tissue engineering, some major challenges still remain in directing stem cell differentiation to specific cell types to create tissues with specific functions. It is believed that the merger of microengineering with biomaterials will lead to more precise understanding of cell biology and greatly contribute to the therapeutic potential of stem cells for tissue engineering. The combination of these fields contains great potential for the advancement of the field of regenerative medicine. The improvements in the fields mentioned above may aid in realizing the dream of regenerating diseased or defective tissues to cure millions of current and future patients.

References

1. Martin, G.R.: Isolation of a pluripotent cell line from early mouse embryos cultured in medium conditioned by teratocarcinoma stem cells. Proc. Natl. Acad. Sci. USA **78**(12), 7634–7638 (1981)
2. Thomson, J.A. et al.: Embryonic stem cell lines derived from human blastocysts. Science **282**(5391), 1145–1147 (1998)
3. Rolletschek, A., Wobus, A.M.: Induced human pluripotent stem cells: promises and open questions. Biol. Chem. **390**(9), 845–849 (2009)
4. Takahashi, K., Yamanaka, S.: Induction of pluripotent stem cells from mouse embryonic and adult fibroblast cultures by defined factors. Cell **126**(4), 663–676 (2006)
5. Amabile, G., Meissner, A.: Induced pluripotent stem cells: current progress and potential for regenerative medicine. Trends Mol. Med. **15**(2), 59–68 (2009)
6. Hochedlinger, K., Plath, K.: Epigenetic reprogramming and induced pluripotency. Development **136**(4), 509–523 (2009)
7. Ebert, A.D., et al.: Induced pluripotent stem cells from a spinal muscular atrophy patient. Nature **457**(7227), 277–280 (2009)
8. Tateishi, K., et al.: Generation of insulin-secreting islet-like clusters from human skin fibroblasts. J. Biol. Chem. **283**(46), 31601–31607 (2008)
9. Park, I.H., et al.: Disease-specific induced pluripotent stem cells. Cell **134**(5), 877–886 (2008)
10. Wobus, A.M., Boheler, K.R.: Embryonic stem cells: prospects for developmental biology and cell therapy. Physiol. Rev. **85**(2), 635–678 (2005)
11. Burdick, J.A., Vunjak-Novakovic, G.: Engineered microenvironments for controlled stem cell differentiation. Tissue Eng. Part A **15**(2), 205–219 (2009)
12. Khademhosseini, A., et al.: Microscale technologies for tissue engineering and biology. Proc. Natl. Acad. Sci. USA **103**(8), 2480–2487 (2006)
13. Bonassar, L.J., Vacanti, C.A.: Tissue engineering: The first decade and beyond. J. Cell. Biochem. **30–31**, 297–303 (1998)
14. Atala, A.: Tissue engineering of artificial organs. J. Endourol. **14**(1), 49–57 (2000)
15. Ren, D.F., et al.: Evaluation of RGD modification on collagen matrix. Artif. Cells Blood Subst. Biotechnol. **34**(3), 293–303 (2006)
16. Shin, H.: Fabrication methods of an engineered microenvironment for analysis of cell-biomaterial interactions. Biomaterials **28**(2), 126–133 (2007)
17. Langer, R., Vacanti, J.P.: Tissue engineering. Science **260**(5110), 920–926 (1993)
18. Murtuza, B., Nichol, J.W., Khademhosseini, A.: Micro- and nanoscale control of the cardiac stem cell niche for tissue fabrication. Tissue Eng. Part B Rev. **15**(4), 443–454 (2009)
19. Alsberg, E., von Recum, H.A., Mahoney, M.J.: Environmental cues to guide stem cell fate decision for tissue engineering applications. Exp. Opin. Biol. Ther. **6**(9), 847–866 (2006)
20. Lensch, M.W., Daheron, L., Schlaeger, T.M.: Pluripotent stem cells and their niches. Stem Cell Rev. **2**(3), 185–201 (2006)
21. Metallo, et al. C.M.: Engineering the stem cell microenvironment. Biotechnol. Prog. **23**(1), 18–23 (2007)
22. Tsai, R.Y., McKay, R.D.: Cell contact regulates fate choice by cortical stem cells. J. Neurosci. **20**(10), 3725–3735 (2000)
23. Gavard, J., et al.: N-cadherin activation substitutes for the cell contact control in cell cycle arrest and myogenic differentiation: involvement of p120 and beta-catenin. J. Biol. Chem. **279**(35), 36795–36802 (2004)

24. Nose, A., Tsuji, K., Takeichi, M.: Localization of specificity determining sites in cadherin cell adhesion molecules. Cell **61**(1), 147–155 (1990)
25. Niessen, C.M., Gumbiner, B.M.: Cadherin-mediated cell sorting not determined by binding or adhesion specificity. J. Cell Biol. **156**(2), 389–399 (2002)
26. Volk, T., et al.: Formation of heterotypic adherens-type junctions between L-cam-containing liver-cells and A-cam-containing lens cells. Cell **50**(6), 987–994 (1987)
27. Halbleib, J.M., Nelson, W.J.: Cadherins in development: cell adhesion, sorting, and tissue morphogenesis. Genes Dev. **20**(23), 3199–3214 (2006)
28. Takeichi, M.: The cadherins: cell-cell adhesion molecules controlling animal morphogenesis. Development **102**, 639–655 (1988)
29. Aberle, H., et al.: Cadherin-catenin complex: protein interactions and their implications for cadherin function. J. Cell Biochem. **61**(4), 514–523 (1996)
30. Qureshi, H.S., et al.: E-cadherin status in breast cancer correlates with histologic type but does not correlate with established prognostic parameters. Am. J. Clin. Pathol. **125**(3), 377–385 (2006)
31. Tlsty, T.D.: Cell-adhesion-dependent influences on genomic instability and carcinogenesis. Curr. Opin. Cell Biol. **10**(5), 647–653 (1998)
32. Cowin, P., et al.: Cadherins and catenins in breast cancer. Curr. Opin. Cell Biol. **17**(5), 499–508 (2005)
33. Chen, W.C., Obrink, B.: Cell-cell contacts mediated by E-cadherin (uvomorulin) restrict invasive behavior of L-cells. J. Cell Biol. **114**(2), 319–327 (1991)
34. Frixen, U.H., et al.: E-cadherin-mediated cell-cell adhesion prevents invasiveness of human carcinoma cells. J. Cell Biol. **113**(1), 173–185 (1991)
35. Silvestre, J., et al.: Cadherin and integrin regulation of epithelial cell migration. Langmuir **25**(17), 10092–10099 (2009)
36. Potthoff, S., et al.: N-cadherin engagement provides a dominant stop signal for the migration of MDA-MB-468 breast carcinoma cells. Breast Cancer Res. Treat. **105**(3), 287–295 (2007)
37. Bloor, D.J., et al.: Expression of cell adhesion molecules during human preimplantation embryo development. Mol. Hum. Reprod. **8**, 237–245 (2002)
38. Dang, S.M., et al.: Controlled, scalable embryonic stem cell differentiation culture. Stem Cells **22**, 275–282 (2004)
39. Huber, B.T.L., et al.: Cooperative effects of growth factors involved in the induction of hematopoietic mesoderm. Blood **92**(11), 4128–4137 (1998)
40. Johansson, B.M., Wiles, M.V.: Evidence for involvement of activin A and bone morphogenetic protein 4 in mammalian mesoderm and hematopoietic development. Mol. Cell Biol. **15**(1), 141–151 (1995)
41. Valdimarsdotter, G., Mummery, C.: Functions of the TGF-b superfamily in human embryonic stem cells. APMIS **113**, 773–789 (2005)
42. Morali, O.G. et al.: IGF-II promotes mesoderm formation. Dev. Biol. **227,** 133–145 (2000)
43. M. Schuldiner, et al., Effect of eight growth factors on the differentiation of cells derived from human embryonic stem cells. Proc. Natl. Acad. Sci. **97**(21), 11307–11312 (2000)
44. Dutta, R.C., Dutta, A.K.: Cell-interactive 3D-scaffold; advances and applications. Biotechnol. Adv. **27**(4), 334–339 (2009)
45. Weisenberg, E.: Pocket companion to robbins pathologic basis of disease. Arch. Pathol. Lab. Med. **124**(10), 1566 (2000)
46. Engler, A.J., et al.: Matrix elasticity directs stem cell lineage specification. Cell **126**(4), 677–689 (2006)
47. Handschel, J., et al.: Compatibility of embryonic stem cells with biomaterials. J. Biomater. Appl. **23**, 549–560 (2009)
48. Levenberg, S., et al.: Differentiation of human embryonic stem cells on three-dimensional polymer scaffolds. Proc. Natl. Acad. Sci. USA **100**(22), 12741–12746 (2003)
49. Battista, S., et al.: The effect of matrix composition of 3D constructs on embryonic stem cell differentiation. Biomaterials **26**(31), 6194–6207 (2005)

50. Fukuda, J., et al.: Micromolding of photocrosslinkable chitosan hydrogel for spheroid microarray and co-cultures. Biomaterials **27**, 5259–5267 (2006)
51. Tien, J., Chen, C.S.: Patterning the cellular microenvironment. IEEE Eng. Med. Biol. Mag. **21**(1), 95–98 (2002)
52. Tien, J., et al.: Fabrication of aligned microstructures with a single elastomeric stamp. Proc. Natl. Acad. Sci. USA **99**(4), 1758–1762 (2002)
53. Nelson, C.M., Chen, C.S.: VE-cadherin simultaneously stimulates and inhibits cell proliferation by altering cytoskeletal structure and tension. J. Cell Sci. **Sep 1**(116), 3571–3581 (2003)
54. Khademhosseini, A., et al.: Molded polyethylene glycol microstructures for capturing cells within microfluidic channels. Lab Chip **4**(5), 425–430 (2004)
55. Karp, J.M., et al.: Controlling size, shape and homogeneity of embryoid bodies using poly(ethylene glycol) microwells. Lab Chip **7**, 786–794 (2007)
56. Khademhosseini, A., et al.: Micromolding of photocrosslinkable hyaluronic acid for cell encapsulation and entrapment. J. Biomed. Mater. Res. A **79**(3), 522–532 (2006)
57. Napolitano, A.P., et al.: Dynamics of the self-assembly of complex cellular aggregates on micromolded nonadhesive hydrogels. Tissue Eng. Part A **13**(8), 2087–2094 (2007)
58. Nichol, J.W., Khademhosseini, A.: Modular tissue engineering: engineering biological tissues from the bottom up. Soft Matter **5**, 1312–1319 (2009)
59. Peppas, N.A., et al.: Hydrogels in biology and medicine: from molecular principles to bionanotechnology. Adv. Mater. **18**(11), 1345–1360 (2006)
60. Whitesides, G.M.: The origins and the future of microfluidics. Nature **442**(7101), 368–373 (2006)
61. Purcell, E.M.: Life at low reynolds-number. Am. J. Phys. **45**(1), 3–11 (1977)
62. McDonald, J.C., et al.: Fabrication of microfluidic systems in poly(dimethylsiloxane). Electrophoresis **21**(1), 27–40 (2000)
63. Park, T.H., Shuler, M.L.: Integration of cell culture and microfabrication technology. Biotechnol. Prog. **19**(2), 243–253 (2003)
64. Rhee, S.W., et al.: Patterned cell culture inside microfluidic devices. Lab Chip **5**(1), 102–107 (2005)
65. Tan, W., Desai, T.A.: Microfluidic patterning of cells in extracellular matrix biopolymers: effects of channel size, cell type, and matrix composition on pattern integrity. Tissue Eng. **9**(2), 255–267 (2003)
66. Chiu, D.T., et al.: Patterned deposition of cells and proteins onto surfaces by using three-dimensional microfluidic systems. Proc. Natl. Acad. Sci. USA **97**(6), 2408–2413 (2000)
67. Takayama, S., et al.: Subcellular positioning of small molecules. Nature **411**(6841), 1016 (2001)
68. Kim, L., et al.: Microfluidic arrays for logarithmically perfused embryonic stem cell culture. Lab Chip **6**(3), 394–406 (2006)
69. Meinel, L., et al.: Bone tissue engineering using human mesenchymal stem cells: effects of scaffold material and medium flow. Ann. Biomed. Eng. **32**(1), 112–122 (2004)
70. Vollmer, A.P., et al.: Development of an integrated microfluidic platform for dynamic oxygen sensing and delivery in a flowing medium. Lab Chip **5**(10), 1059–1066 (2005)
71. Park, S.H., et al.: An electromagnetic compressive force by cell exciter stimulates chondrogenic differentiation of bone marrow-derived mesenchymal stem cells. Tissue Eng. **12**(11), 3107–3117 (2006)
72. Metallo, C.M., et al.: The response of human embryonic stem cell-derived endothelial cells to shear stress. Biotechnol. Bioeng. **100**(4), 830–837 (2008)
73. Tourovskaia, A., et al.: Differentiation-on-a-chip: a microfluidic platform for long-term cell culture studies. Lab Chip **5**(1), 14–19 (2005)
74. Zhong, J.F., et al.: A microfluidic processor for gene expression profiling of single human embryonic stem cells. Lab Chip **8**(1), 68–74 (2008)

75. Davey, R.E., Zandstra, P.W.: Spatial organization of embryonic stem cell responsiveness to autocrine Gp130 ligands reveals an autoregulatory stem cell niche. Stem Cells **24**(11), 2538–2548 (2006)
76. Kirmizidis, G., Birch, M.A.: Microfabricated grooved substrates influence cell-cell communication and osteoblast differentiation in vitro. Tissue Eng. Part A. **15**(6), 1427–1436 (2009)
77. Abbott, A.: Cell culture: biology's new dimension. Nature **424**(6951), 870–872 (2003)
78. Zhang, S.: Beyond the petri dish. Nat. Biotech. **22**(2), 151–152 (2004)
79. Freed, L.E., et al.: Tissue engineering of cartilage in space. Proc. Natl. Acad. Sci. USA **94**(25), 13885–13890 (1997)
80. Radisic, M., et al.: Medium perfusion enables engineering of compact and contractile cardiac tissue. Am. J. Physiol. Heart Circ. Physiol. **286**(2), H507–516 (2004)
81. Radisic, M., et al.: Oxygen gradients correlate with cell density and cell viability in engineered cardiac tissue. Biotechnol. Bioeng. **93**(2), 332–343 (2006)
82. Chung, B.G., et al.: Micro- and nanoscale approaches for tissue engineering and drug discovery. Exp. Opin. Drug Dis. **2**(12), 1653–1668 (2007)
83. Moeller, H.-C., et al.: A microwell array system for stem cell culture. Biomaterials **29**(6), 752–763 (2008)
84. Bauwens, C.L., et al.: Control of human embryonic stem cell colony and aggregate size heterogeneity influences differentiation trajectories. Stem Cells **26**(9), 2300–2310 (2008)
85. Khademhosseini, A., et al.: Layer-by-layer deposition of hyaluronic acid and poly-L-lysine for patterned cell co-cultures. Biomaterials **25**(17), 3583–3592 (2004)
86. Mohr, J.C., et al.: 3-D microwell culture of human embryonic stem cells. Biomaterials **27**(36), 6032–6042 (2006)
87. Kang, L., et al.: Cell confinement in patterned nanoliter droplets in a microwell array by wiping. J. Biomed. Mater. Res. A **93**, 547–557 (2010)
88. Ungrin, M.D., et al.: Reproducible, ultra high-throughput formation of multicellular organization from single cell suspension-derived human embryonic stem cell aggregates. PLoS ONE **3**(2), e1565 (2008)
89. Rosenthal, A., Macdonald, A., Voldman, J.: Cell patterning chip for controlling the stem cell microenvironment. Biomaterials **28**(21), 3208–3216 (2007)
90. Wright, D., et al.: Generation of static and dynamic patterned co-cultures using microfabricated parylene-C stencils. Lab Chip **7**(10), 1272–1279 (2007)
91. Wu, X., et al.: Small molecules that induce cardiomyogenesis in embryonic stem cells. J. Am. Chem. Soc. **126**(6), 1590–1591 (2004)
92. Xu, Y., Shi, Y., Ding, S.: A chemical approach to stem-cell biology and regenerative medicine. Nature **453**(7193), 338–344 (2008)
93. Ding, S., Schultz, P.G.: A role for chemistry in stem cell biology. Nat. Biotechnol. **22**(7), 833–840 (2004)
94. Urquhart, A.J., et al.: TOF-SIMS analysis of a 576 micropatterned copolymer array to reveal surface moieties that control wettability. Anal. Chem. **80**(1), 135–142 (2008)
95. Levenberg, S., et al.: Differentiation of human embryonic stem cells on three-dimensional polymer scaffolds. Proc. Natl. Acad. Sci. U.S.A **100**(22), 12741–12746 (2003)
96. Chaudhry, G.R., et al.: Osteogenic cells derived from embryonic stem cells produced bone nodules in three-dimensional scaffolds. J. Biomed. Biotechnol. **4**, 203–210 (2004)
97. Levenberg, S., et al.: Neurotrophin-induced differentiation of human embryonic stem cells on three-dimensional polymeric scaffolds. Tissue Eng. **11**(3–4), 506–512 (2005)
98. Tsang, V.L., Bhatia, S.N.: Fabrication of three-dimensional tissues. Adv. Biochem. Eng. Biotechnol. **103**, 189–205 (2007)
99. Tsang, V.L., Bhatia, S.N.: Three-dimensional tissue fabrication. Adv. Drug. Deliv. Rev. **56**(11), 1635–1647 (2004)
100. Tsang, V.L., et al.: Fabrication of 3D hepatic tissues by additive photopatterning of cellular hydrogels. FASEB J. **21**(3), 790–801 (2007)

101. McGuigan, A.P., Sefton, M.V.: Vascularized organoid engineered by modular assembly enables blood perfusion. Proc. Natl. Acad. Sci. USA **103**(31), 11461–11466 (2006)
102. Du, Y., et al.: Directed assembly of cell-laden microgels for fabrication of 3D tissue constructs. Proc. Natl. Acad. Sci. USA **105**(28), 9522–9527 (2008)

Biomaterials as Stem Cell Niche: Cardiovascular Stem Cells

Ge Zhang and Laura J. Suggs

Abstract A tissue-specific stem cell niche functions to direct either self-renewal or differentiation. The niche comprises all local cues that can be sensed by the cell including soluble and insoluble signals, physical forces and cell–cell contacts. Approximating the stem cell niche through the utilization of biomaterials may give rise to a greater understanding of the biology of the stem cell niche as well as potential in vitro culture systems and translatable avenues for stem cell therapy, tissue engineering and regenerative medicine. Stem cell niches within the cardiovascular system have been described within the heart, the bone marrow compartment and in vascular beds within various tissues. Progenitor cell populations have been characterized which can give rise to all the major cells of the cardiovascular system including cardiomyocytes, endothelial cells, mural cells and fibroblasts. The extent to which these progenitor populations can be identified and isolated; however, is variable. Biomaterials have an important role in the development of artificial stem cell niches for in vitro culture or in vivo therapy. Biomaterials can be controlled in order to provide insoluble matrix signals, present soluble signals, control cell–cell contacts and transmit or augment physical signals all of which can contribute towards enhancing cell function or directing cell phenotype. This chapter focuses specifically on how biomaterials can be used within the context of a stem cell niche to direct and maintain differentiation towards cardiovascular cell types.

G. Zhang
Department of Biomedical Engineering, University of Akron,
Olson Research Center Room 301H, Akron OH, 44325-0302, USA

L. J. Suggs (✉)
Department of Biomedical Engineering, University of Texas at Austin,
1 University Station, C0800, Austin TX, 78712, USA
e-mail: laura.suggs@mail.utexas.edu

Stud Mechanobiol Tissue Eng Biomater (2010) 2: 173–193

DOI: 10.1007/8415_2010_4

Published online: 12 June 2010

1 Introduction

The stem cell niche comprises the local microenvironment of a stem cell residing in a particular tissue and encompasses all the factors that may direct cell function. In general, the stem cell niche serves to direct either self-renewal or differentiation towards relevant daughter cell types. Components of the stem cell niche include the insoluble matrix components, soluble factors, mechanical and other physical forces as well as direct cell–cell contact. Each of these factors has been manipulated in various ways to try and approximate the stem cell niche either in an in vitro culture system or in an in vivo system used for therapy. In particular, biomaterials have been use to control the microenvironment around the stem cell in order to direct stem cell differentiation and function. Biomaterials of interest include both natural and synthetic materials and span the spectrum from preformed hydrophobic materials to in situ polymerizable hydrogels. Biomaterials can not only be used to provide three-dimensional structure to the artificial niche but also provide insoluble matrix signals, present soluble signals, control cell–cell contacts and transmit or augment physical signals. Additionally, biomaterials can be used in combination with a cell therapy strategy to control cell phenotype and function following implantation. This chapter specifically addresses how biomaterials can function as a cardiovascular stem cell niche.

The content within this chapter begins with a discussion of stem cell niches within the adult organism that can give rise to three main cell types of interest for the cardiovascular system including cardiomyotyes, endothelial cells, and mural cells. Special emphasis is placed on identifying markers for these stem cell populations. The second section within this chapter addressed the use of biomaterials to approximate stem cell niches for the purpose of in vitro culture. Biomaterial systems can be used to either encourage self-renewal or promote differentiation towards specific lineages in an analogous manner to the native stem cell niche. This chapter focuses on cell culture systems to enhance differentiation towards cardiovascular lineages. Finally, the last section addresses how biomaterials can be used in conjunction with in vivo cell therapies to enhance differentiation, viability and cell delivery.

2 Adult Cardiovascular Stem Cells and Their Niches

2.1 Cardiac Stem Cells

The heart has traditionally been considered a postmitotic organ because mature cardiomyocytes cease to proliferate around the time of birth. However, this longstanding dogma has been challenged by the accumulating contradictory data that demonstrated the existence of cardiomyocyte proliferation under certain conditions such as ischemia and hypotension. The evidence subsequently drove the discovery of cardiac stem cells. Several types of cells have been found in adult hearts that possess stem cell characteristics, which include: c-kit$^+$ cardiac cells, sca-1$^+$ cardiac cells, side population cardiac progenitor cells, and cardiosphere-derived cells. C-kit is a transmembrane tyrosine kinase receptor for stem cell factor (SCF) and was initially used to identify hematopoietic stem cells. In 2003, c-kit$^+$ cells were isolated from rat adult hearts and they demonstrated the ability to replicate and differentiate into cardiomyocytes, endothelial and smooth muscle cells [1] (Fig. 1). C-kit$^+$ cell transplantation after ischemic injury leads to significant improvement in ventricular function [2–4]. Studies have also shown that c-kit has the ability to promote cardiac stem cell differentiation and regulate terminal cardiomyocyte differentiation [5].

While a variety of studies have focused on the isolation and expansion of cardiac stem cells based on c-kit expression, another cardiac progenitor cell marker, sca-1 (stem cell antigen-1), has been described by several other groups [6–8]. Upon

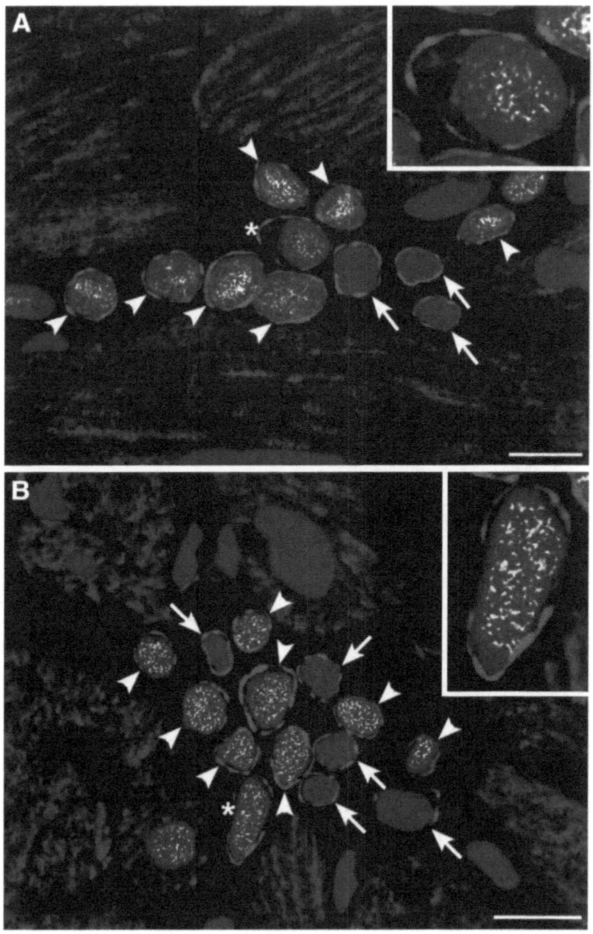

Fig. 1 Clusters of primitive and early committed cells in the heart. **a** Cluster of 11 c-kit^{POS} cells with three expressing c-kit only (*arrows*), seven expressing Nkx2.5 (*white dots*; *arrowheads*) in nuclei (propidium iodide, PI), and one Nkx2.5 and α-sarcomeric actin in the cytoplasm (*asterisk*, see *inset*). **b** Cluster of 15 c-kit^{POS} cells with five c-kit^{POS} cells only (*arrows*), eight expressing MEF2C (*arrowheads*), and one expressing MEF2C and α-sarcomeric actin (*asterisk*, see *inset*). *Bars*, 10 μm (reprinted with permission [1])

oxytocin stimulation, murine cardiac sca-1$^+$ cells are able to differentiate into spontaneously beating cardiomyocytes. Given intravenously, sca-1$^+$ cells were able to home to injured myocardium and improve cardiac function as well as promote new blood vessel formation [9]. Another cardiac progenitor cell population, termed side population (SP) cells, has been identified in various organs including bone marrow, skeletal muscle and adipose tissue [10]. A resident pool of SP cells has also been revealed in the heart [11]. It has been suggested that cardiac SP cells are capable of cardiomyogenic differentiation into mature cardiomyocytes and represent a distinct cardiac progenitor cell [9]. Most recently, cardiosphere-derived cells (CDCs) were isolated from cardiac tissues [12]. Isolated from postnatal atrial or ventricular human biopsy specimens and from murine hearts, CDCs grow as self-adherent clusters. These cells are clonogenic and express stem and progenitor cell markers. They are capable of long-term self-renewal and can differentiate to major specialized cell types of the heart; myocytes and vascular cells.

2.2 Endothelial Progenitor Cells

Vascular progenitor cells including endothelial progenitor cells (EPCs) have been isolated from peripheral blood and are thought to originate from bone marrow [13]. These progenitor cells share certain cell surface markers with hematopoietic (blood) stem cells (HSCs) which may include: Flk-1, Tek (Tie-2), c-kit, Sca-1, CD133, and CD34 [14]. The preponderance of evidence suggests that these two cell types arise from a common hemangioblastic precursor that exists within the bone marrow niche. Both EPCs and HSCs have been shown to mobilize from the bone marrow stroma into peripheral blood in response to certain cytokines including granulocyte macrophage-colony stimulating factor (GM-CSF) and EPCs in particular to vascular endothelial growth factor (VEGF) and stromal cell-derived factor 1 alpha (SDF-1α) [15]. An important mediator of EPC mobilization in response to ischemic injury is hypoxia inducible factor (HIF-1) [16].

Once mobilized from the bone marrow compartment, EPCs are typically classified as either "early" or "late." Early EPCs are characterized by the presence of markers CD45 and CD14, in addition to CD34, which are markers characteristic of hematopoietic and monocyte lineages, respectively. Late outgrowth EPCs lose CD45 and CD14 expression and assume markers that overlap with mature endothelial cells (ECs) including CD31, VE-cadherin, and vWF. EPCs have been shown to differentiate towards mature endothelial cells in culture both with and without VEGF [13, 17]. It is clear that EPCs are not a fixed cell phenotype, but exhibit different characteristics dependent on the local microenvironment and growth factor signaling.

Data from recent studies suggests that endothelial and vascular smooth muscle cells may arise developmentally from a common progenitor [18]. Yamashita et al. demonstrated that embryonic stem cells (ESCs) that are Flk-1 positive can give rise to both lineages in culture and in adult animals. SP cells from skeletal muscle are $CD45^{+}$ and sca-1^{+} and have been shown to differentiate towards both endothelial cells and smooth muscle cell lineages in vivo [19–21].

2.3 Mural Cell Progenitors/Mesenchymal Stem Cells

The phenotypic relationship among smooth muscle cells, pericytes and mesenchymal stem cells is still poorly understood. Vascular smooth muscle cells and pericytes may represent variations in phenotype of a mural cell lineage continuum. Pericytes may give rise to vascular smooth muscle cells given the appropriate stimulus. There is also evidence for the reverse phenomenon [22, 23]. It has also been suggested that pericytes provide a cellular niche existing as an intermediate between vascular smooth muscle cells and fibroblasts [24].

Populations of smooth muscle progenitor cells (SPCs) have been isolated from peripheral blood. SPCs grown in culture in the presence of PDGF-BB demonstrated markers specific for smooth muscle cells including: α-smooth muscle actin, myosin heavy chain, and calponin. However, they were also positive for bone marrow angioblastic markers such as: CD34, Flk-1, Flt-1, but not Tek(Tie-2), potentially representing a progenitor cell with both angioblastic and mesenchymal potential [25]. In related work, smooth muscle cell phenotype has been demonstrated as a result of transforming growth factor-beta (TGF-β) induction of bone marrow-derived mesenchymal stem cells (MSCs) [26].

Pericytes are polymorphic, elongated, multibranched cells of the mesodermal origin, which partially surround the endothelial cells of small vessels (e.g., capillaries, postcapillary venules, etc.) [27]. As pericytes may contain contractile muscle filaments on their endothelial cell side, they are occasionally regarded as microvascular counterparts of smooth muscle cells. Pericytes are in direct apposition with endothelial cells whereas vascular smooth muscle cells never directly contact endothelial cells [24]. Further, unlike smooth muscle cells, pericytes are covered by the same basement membrane as endothelial cells except where two mural cells are in contact [27].

The mesenchymal stem cell (MSC) is an adult stem cell with multilineage potential derived from bone marrow, adipose tissue or peripheral blood. These cells can differentiate readily into terminal cells of the mesenchyme, and recent evidence has indicated that MSCs can express phenotypic characteristics of endothelial cells, neural cells, smooth muscle cells, skeletal myoblasts, and cardiomyocytes [28–30]. Recent work has described a population of cells that occupy a perivascular niche with markers of both pericytes and MSCs (CD146, NG-2 and PDGF-Rβ) [31, 32]. Cells that were isolated from this niche could be clonally differentiated towards osteoblastic, chondrogenic and adipocytic phenotypes. It has been noted that this result established the overlap in adult tissues of pericytes and MSCs. Isolated, autologous bone marrow stem cells have been shown to contribute to tissue repair and formation of blood vessels following tissue ischemia, based on localized genetic markers.

Of particular interest is the use of MSCs or bone marrow-derived mononuclear cells for cell-based therapy [29, 33]. MSCs represent a more clearly defined stem cell population within a heterogeneous bone marrow cell adherent fraction. The differentiation potential of MSCs towards endothelial cells, pericytes, smooth muscle cells, myofibroblasts and cardiomyocytes has been theorized as a possible therapy for cardiac repair following myocardial infarction (MI), despite the literature differing widely on this possibility. Early reports directly injecting bone marrow cells in the margin bordering the infarct area of the murine left ventricle demonstrated that they were capable of repairing myocardium [34]. Subsequent clinical trials resulted in significant but modest improvements in cardiac function. These outcomes may be limited by the lack of sustained cell engraftment as discussed below. Nevertheless, implanted mesenchymal progenitors have been shown to express von Willebrand factor (vWF), vascular endothelial growth factor (VEGF), and enhance vasculogenesis after several weeks in the cardiac environment [35]. Since myocardial perfusion is critical to restoration of cardiac function after MI, involvement in vascular repair or regeneration may be an important role for MSCs in the ischemic heart. Furthermore, the vasculogenic potential of MSCs could be extended to other clinical settings such as peripheral ischemia or increasing perfusion in large tissue-engineered constructs.

2.4 Adult Cardiovascular Stem Cell Niches

Stem cells reside in specialized microenvironments created by supporting cells that promote stem cell maintenance through the production of factors that regulate self-renewal and differentiation. In adult bone marrow, hematopoietic stem cells are located in the trabecular endosteum (osteoblastic niche) or sinusoidal perivascular areas (vascular niche). A function for the vascular niche is to assist hematopoietic stem cells in transendothelial migration, which is important during both homing and mobilization.

Growth factor expression and chemokine crosstalk within the vascular niche is critical in tissue development and regeneration. VEGF is one of the most common direct angiogenic factors [36] and is likely the most well-characterized. VEGF acts directly on EPCs to modulate migration, proliferation and differentiation [37]. VEGF has also demonstrated the potential to upregulate phenotypic markers for ECs from MSCs [38]. Another important growth factor active within the vascular microenvironment is basic fibroblast growth factor (bFGF) which recruits supporting pericytes [39] and also stimulates EC and EPC proliferation and migration [40].

Similarly, platelet-derived growth factor-BB (PDGF-BB) recruits pericytes [37], SMCs, and fibroblasts [41] to facilitate maturation and stabilization of newly formed capillaries. Additionally, PDGF-BB indirectly acts on the angiogenesis cascade by stimulating release of VEGF and bFGF from SMCs [41]. PDGF-BB has also demonstrated the ability to upregulate phenotypic markers of both SMCs and ECs from progenitor cells [37]. TGF-β, as noted above, has the potential to direct SMC differentiation. Additionally, it has an effect on the stabilizing vascular pericyte population as evidenced by capillary structure stabilization in a co-culture of embryonic stem cells and ECs formed in vitro within Matrigel™ [42]. TGF-β; however, is a pleiotropic growth factor that has been shown to upregulate production of VEGF and bFGF from SMCs [41].

3 Biomaterials as Stem Cell Niches for 3D Cell Culture

3.1 3D Cell Culture Systems for Pluripotent Stem Cells

Pluripotent stem cells including embryonic and induced pluripotent stem cells will spontaneously differentiate under the appropriate conditions. Cell–cell contact in the absence of leukemia inhibitory factor (LIF) encourages mouse embryonic stem (ES) cells to differentiate and partition towards the three germ layers: endoderm, mesoderm, and ectoderm in the resulting cystic EB with endoderm on the outer layer. Specific induction can be enhanced through the use of soluble chemokines, co-culture and gene manipulation on the initial cell population. Desired cells can also be selected following differentiation based on FACS for certain cell markers of interest. Mouse ES cells have given rise to multiple mesodermally derived cell types including osteoblasts, chondrocytes, cardiomyocytes, endothelial cells, smooth muscle cells, and hematopoietic (blood) cells [43–47]. An important challenge in the development of pluripotent cell culture systems is the ability to drive specific cell types. Currently pluripotent cell culture is inefficient in both overall yield and the yield of desired cell populations.

Researchers have used 3D systems to culture ES cells in large aggregates termed “embryoid bodies” (EBs). A 3D environment serves to increase the space available for cell proliferation and to prevent agglomeration of EBs. Work from Zandstra’s group has described low temperature gelling agarose for EB encapsulation [48]. In this method, EBs remain encapsulated for 4 days, during which time cell–cell adhesion molecules are expressed. Concurrently with downregulation of these molecules, the agarose gel spontaneously collapses. The freed EBs can then be cultured in suspension under standard stirred culture conditions. Stirred suspension culture allows for high efficiency expansion of a cell population. In contrast, attached culture is limited by the large media volumes necessary to cover the culture surface.

Other techniques to increase the surface area to volume ratio of attached culture include the use of porous polymeric microbeads [49]. Results from Roy’s group demonstrate the use of porous tantalum constructs for ES cell seeding and subsequent differentiation toward hematopoietic cell populations [50]. Work from Langer’s group describe the use of porous poly(alpha hydroxyester) scaffolds to drive differentiation towards neural, chondrocyte and hepatocyte lineages along with growth factor supplementation [51]. An example of the matrix itself driving differentiation is research from Netti's group where EBs were cultured in collagen gels that had been loaded with either fibronectin or laminin. They demonstrated that cellular outgrowths were greater in culture with fibronectin. Immunohistochemical staining against CD31 demonstrated that the outgrowths expressed characteristic markers of endothelial cells and their progenitors. On gels with laminin as the supplementary protein, cardiomyocyte differentiation was enhanced. This result indicated the important role of the extracellular matrix in directing differentiation [52].

Differentiation of ES cells towards cardiomyocytes has been described spontaneously in culture as well as through the use of numerous induction agents including various growth factors, retinoic acid, DMSO, erythropoietin, nitric oxide and 5-azacytadine. The bulk of these studies, however, demonstrate low differentiation efficiency, typically less than half of EBs demonstrating beating areas with spontaneous differentiation resulting in 10–25% of EBs demonstrating beating areas [53–55]. It has also been proposed that cardiomyocyte differentiation can be enhanced through the use of mechanical stretch as a signaling tool under physiologically relevant conditions [56]. In general, the in vitro culture of cardiac muscle is challenged by the necessity for driving lineage differentiation of multiple cell types including cardiomyocytes, endothelial cells, smooth muscle cells and fibroblasts as well as the high metabolic demand of this tissue and the subsequent need for a well-vascularized tissue bed.

It is important to note that in the developing embryo cardiac cells arise from mesodermal origin from a population of cells within the cardiac crescent. As with the hemangioblast, differentiation of cardiac progenitors appears to be controlled in large part by an adjacent tissue, the anterior endoderm. In the amphibian model, selective removal of visceral endoderm blocks myocardial differentiation. In the chick model, mesoderm committed to a hematopoietic lineage can be reprogrammed towards a cardiac lineage, indicating that the VE serves an instructive function. As a result, Mummery's group has developed a co-culture system for enhancing cardiomyocyte differentiation from ES cells. A starting ES cell population has been induced to differentiate towards cardiomyocytes through co-culture with a visceral endoderm-like cell (END-2) [57, 58].

Endothelial cell differentiation has been demonstrated by Huang et al. where mouse ES cells were cultured on collagen IV well plates and the level of Flk1 positive cells was increased to 30%. Subsequent culture in polyurethane vascular grafts under both shear and circumferential strain in the presence of soluble VEGF demonstrated a cell layer with the morphology of endothelium and positive staining for CD31, a characteristic marker [59]. Limitations to this work include the relatively low levels of Flk1 positive cells that serve as the starting population. This low value may limit the VEGF response. Similarly, Langer's group used and immobilized RGD peptide in combination with microencapsulated VEGF in order to encourage commitment of vascular cell types from human ES cells [60].

Ando's group demonstrated that both ES cells and endothelial progenitors from circulating blood were able to differentiate towards mature endothelial cells under mechanical conditioning. The ES cell population was selected for Flk1 and upon exposure to shear in the physiologic range of 1.5–10 dyne/cm^2, expression of the endothelial markers, Flk1, Flt1, VE-cadherin, and CD31 were all increased over static controls. No change was seen in markers for smooth muscle cells [61]. This work was primarily focused on reporting the molecular mechanisms that regulate ES cell differentiation, including mechanical conditioning, and did not seek to develop robust culture systems.

Preliminary work in our laboratory has described the use of fibrin gels for the culture of mouse R1 ES cells [62]. The purpose of our work was to increase cell expansion within a 3D culture environment and potentially mediate cell differentiation. We compared culture in fibrin gels to standard 2D attached culture and culture in methylcellulose. It was hypothesized that fibrin has specific cell recognition sites that could participate in cell–matrix interactions and subsequently effect lineage specification. The intent of this study was to address two major problems within ES cell culture; the inability to culture at maximum yield due to agglomeration events as well as the inability to efficiency drive differentiation towards uniform cell populations.

Results from our work concluded that cell yield can be increased in 3D fibrin gels relative to attached culture or culture in methylcellulose (MC). In comparison, Zandstra's group has examined both attached and MC culture with similar results for cell proliferation after 12 days in culture [63]. In addition, EB culture in fibrin is similar to culture in agarose gels. Culture in agarose results in a two-fold expansion at 4 days compared to unencapsulated culture [48]. Similar results have also been demonstrated in alginate culture systems [64]. Our work demonstrated a population doubling at day 6 in fibrin relative to attached culture Our study demonstrated a two-fold expansion 6 days in fibrin and two and a half times greater cell expansion at day 12 when compared to attached culture.

Our work also demonstrated that differentiation could be enhanced in fibrin culture. We demonstrated the relative degree of differentiation using FACS and quantitative RT-PCR. Oct-4 gene expression and protein expression of both SSEA-1 and Oct-4 were as low as or lower than either attached culture or culture in methylcellulose. Methylcellulose culture demonstrated high levels of Oct-4 gene expression and high levels of surface expression of SSEA-1 indicating that differentiation was more limited in the methylcellulose culture. MC culture is typically employed for hematopoiesis and prior reports from other groups have not demonstrated differences between MC and attached culture in relative efficiency of hematopoietic differentiation. It was noted; however, that limited serum exposure is likely in the MC culture due to the lack of complete medium replacement. Other research has demonstrated that limited serum exposure can enhance endodermal differentiation and limit expression of hematopoietic phenotype [65].

Additionally, we demonstrated that fibrin culture of EBs resulted in greater expression of VE-cadherin, a marker of endothelial and hemangioblastic progenitors. This result serves as additional evidence that the cell's ability to form cell–matrix attachments may significantly affect the resulting lineage commitment [66].

Known cell attachment sites on fibrin include both integrin and cadherins binding sites including VE-cadherin. The N-terminus of the fibrin beta chain is known to contain this binding site [67]. Our work demonstrated that the fibrin cultures maintained the highest level of VE-cadherin expression as a measure of angioblastic progenitors following the 12-day EB differentiation [68].

3.2 3D Cell Culture Systems for Adult Stem Cells

Evidence suggests that adult mesenchymal stem cells (MSCs) can express phenotypic characteristics of vascular cell types including endothelial cells, smooth muscle cells and cardiomyocytes both in vitro and in vivo. While delivering isolated MSCs or bone marrow-derived cells to the infarcted site has become a potential therapy for MI and is now being evaluated clinically, the extent of stem cell differentiation towards cardiomyocytes in the infarcted heart has now been shown to be quite low [69]. Other groups also demonstrated short term improvement in cardiac function and resting blood flow in infarcted myocardium with injection of human bone marrow stem cells; however, the mechanism for this improvement has not been conclusively demonstrated but the likelihood that significant cardiomyocyte differentiation occurs is minimal [70, 71].

The differentiation of MSCs towards smooth muscle cells in vitro may be affected by numerous factors including soluble factors such as PDGF-BB, TGF-β1, and bFGF; extracellular matrix proteins; as well as mechanical forces such as cyclic strain or shear. Niklason's group has explored the effects of various parameters for the purpose of culturing engineered blood vessel replacements. In their work, human bone marrow-derived MSCs were cultured in a biomimetic system consisting of a fibronectin-coated polyglycolic acid scaffold. Their suggested culture protocol consisted of a two-phase timecourse where the first four weeks enhanced proliferation while the last four weeks optimized differentiation. The proliferative phase was enhanced by the addition of PDGF-BB and bFGF, while the differentiation phase consisted of removal of proliferative factors and media supplementation with 1 ng/ml of TGF-β1. Cyclic strain on the cell-seeded conduits was also imposed at this time. Markers for smooth muscle cell phenotype including α-smooth muscle actin and myosin heavy chain increased with these culture conditions.

In driving the differentiation of MSCs toward endothelial cells, the matrix microenvironment may play a significant role. MSCs have been shown form tubes in vitro on ECM matrices (Matrigel^{TM}) under a number of different conditions including growth factor stimulation and hypoxia [72, 73]. MSCs have also been shown to stimulate vascular ingrowth into plugs in vivo in the absence of exogenous stimulation [72]. To date, in vitro studies have not demonstrated spontaneous tube formation in fibrin gels by MSCs. There is a large body of literature, however, demonstrating that microvascular endothelial cells are able to form tubes within 3D fibrin gels, and this process can be augmented through the addition of various stimuli [74, 75]. Stimulation may include hypoxia, the addition of MSCs in co-culture, the addition of angiogenic growth factors, as well as the alteration of various matrix parameters [76–78].

Work in our laboratory has demonstrated that an adult mesodermal progenitor cell, specifically porcine mesenchymal stem cells (MSCs), seeded in a PEGylated fibrin gel within 48 h in vitro began to form vascular tube-like networks, in contrast to controls of unreactive PEG mixed with fibrinogen or fibrin alone [79]. These tubes stained positive for mature endothelial cell specific markers like CD31 and vWF (Fig. 2a,b). Our RT-PCR data also demonstrated that CD31 and vWF as well as VEGF mRNA was expressed in MSCs (Fig. 2c). Results from MSC seeded gel plugs comparing PEGylated fibrin and fibrin alone are shown in Fig. 2d,e. Gross images demonstrate visible perfused vessels in 1 ml preformed PEGylated fibrin gels after 7 days in vivo in a subcutaneous implantation site in Lewis rats. From our previous work and that of other groups, it has been demonstrated that MSCs by themselves cannot differentiate into cells expressing EC phenotype without induction by the appropriate cytokines [28, 80]. Even when cultured in Matrigel™ MSCs do not demonstrate vascular tube formation without the addition of VEGF. Endothelial cell tubes in Matrigel™ typically disappear rapidly; however, in our PEGylated fibrin gel, MSCs maintained vascular tube-like networks until the gel degraded, approximately 7–10 days.

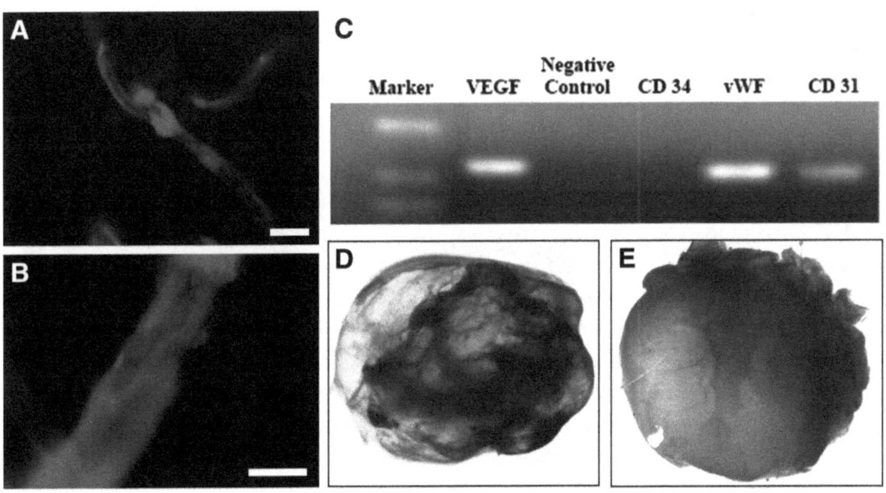

Fig. 2 Immunochemistry against vWF of porcine MSCs seeded in PEGylated fibrin with DAPI nuclear counterstain **a** ×40, **b** ×60 (48 h, *scale bar* = 10 μm). **c** Endothelial cell gene expression by porcine MSCs. PCR demonstrated that entrapped MSCs highly expressed VEGF and vWF (*lanes 2* and *5*), expressed CD31 (*lane 6*), but did not express CD34 (*lane 4*). *Lane 1* was marker and *lane 3* is the negative control (reprinted with permission from [79]). Seven day gel plug explants from Lewis rats of syngeneic MSCs seeded at 50,000 cells per 1 ml gel in **d** PEGylated fibrin and **e** fibrin only (unpublished data)

4 Biomaterials as Stem Cell Niches for Cardiac Cell Therapy

4.1 Cardiac Cell Therapy

In recent years, advances in developmental and cellular biology have provided powerful techniques that have offered hope for future therapeutic modalities for the treatment of end-stage cardiac failure. One of the most appealing and promising fields in the therapy for myocardial infarction and heart failure is cardiac cell therapy [81, 82]. The aim of cardiac cell therapy is to restore at least in part the functionality of the diseased or injured myocardium by the use of stem/progenitor cells. Cardiac cell therapy has progressed rapidly in the last decade. The key elements of cell therapy for myocardial repair are: the choice of donor cell, the strategy of delivery and the understanding of the mechanisms by which these cells improve cardiac function.

A variety of cell types have been tested in cardiology and there is emerging preclinical and clinical data on the feasibility and safety of different cell lines in the setting of acute myocardial infarction and chronic heart failure. Although human embryonic stem cells indisputably differentiate into cardiomyocytes they are not currently prime candidates for cell therapy due to issues with their immunogenicity, tumorigenesis and ethical challenges. So far, adult bone marrow-derived stem cells and skeletal-muscle derived myoblasts are the most frequently used autologous cell sources in cardiac cell therapy. Since the occurrence of ventricular arrhythmias after skeletal myoblast transplantation has raised a major safety issue, bone marrow-derived stem cells became the focus of stem cell-based therapies [83]. Experimental studies have suggested that the infusion of different subsets of bone marrow-derived progenitor cells, circulating endothelial progenitor cells, or tissue-residing stem cells improved neovascularization and cardiac function.

Clinical studies at present predominantly use bone marrow mononuclear cells (BMNC) isolated from bone marrow aspirates by density gradient centrifugation. Intracoronary infusion of BMNC significantly increased global or regional ejection fraction and/or reduced infarct size and end systolic volumes in patients with acute myocardial infarction as demonstrated in initial pilot trials and in randomized studies, although there is still opportunity to improve functional outcomes. Bone marrow-derived stem cells are a desirable cell source for cardiac muscle regeneration, as they reside in the bone marrow of patients and can be used for autologous transplantation, thus eliminating the need for immunosuppression. An additional capability of bone marrow-derived mesenchymal stem cells is that they can induce therapeutic vasculogenesis in the remaining viable myocardium, therefore increasing viability and restoring heart function [15, 84, 85]. Other sources of stem cells for myocardial regeneration are also currently under investigation. These include endothelial progenitor cells (EPCs), adipose tissue-derived stem cells (ASCs) and cardiac stem cells (CSCs).

4.2 Biomaterial Scaffolds for Cardiac Cell Therapy

It is essential to deliver stem cells to the site of injured hearts to maximize the therapeutic effects of cell therapy. Delivery strategies widely used in animal studies or clinical trials include: (1) Intramyocardial injection; Orlic and colleagues isolated bone marrow stem cells and directly injected them in the margin bordering the infarct of the left ventricle of mice [34]. This kind of direct intramyocardial injection may require fewer cells to achieve engraftment compared with intracoronary or intravenous injection. Also the injection process is simple and can be performed by direct inspection of the potential target zones [33]. However, this invasive delivery is associated with intraoperative and postoperative risks. An improved approach for intramyocardial injection is to implant stem cells through a percutaneous catheter guided by left ventricular electromechanical mapping with the NOGA™ system [33]. This system allows for injection with high precision into nonviable areas of the myocardium with an injection-needle catheter, which offers an advantage over the more invasive surgical approach and its associated risks. (2) Intracoronary injection; a percutaneous transluminal coronary catheter can be used for intracoronary delivery of bone marrow-derived stem cells after myocardial infarction [86]. It can deliver the maximum concentration of cells to the site of infarct and peri-infarct tissue, which is advantageous over intravenous injection. Intracoronary administration into the infarct artery allows the stem cells to "home to" and incorporate in the areas bordering the infarct zone in a homogenous manner. This is in contrast to direct myocardial injection, which may lead to 'islands' of cells in the infarcted myocardium, providing a possibility for electrical instability and ventricular tachyarrhythmias [87]. (3) Intravenous injection; intravenous injection obviates the need for cardiac surgery or cardiac catheterization [86]. If stem cells have an effective cellular homing mechanism to localize in the infarcted myocardium, intravenous injection will be an attractive and practical mode of stem cell delivery. Microenvironmental factors, expression of matrix and adhesion molecules by injured tissue, homing receptors and various factors relating to migration are believed to be involved in the homing process of stem cells [87]. However, because of the long circulation time, cells could be lost by extraction towards noncardiac organs and fail to home to the area of infarct. Consequently, a large dose of stem cells may be needed to get enough cells reach to the heart compared with other delivery routes.

Besides the above-mentioned relatively direct method, another option is chemotactic mobilization of stem cells from both heart and bone marrow (BM) to the desired ischemic area to assist repair. The homing of stem cells has been performed via local delivery of various chemotactic factors. Our group has used a controlled release strategy based on the stem cell homing factor, stromal derived factor 1 alpha (SDF-1α) in an injectable biomaterial in order to enhance autologous stem cell repair [88]. Our results demonstrated enhanced recruitment of c-kit$^+$ stem cells to myocardium in a mouse infarct model. Stem cell recruitment was correlated with increased ejection fraction and fractional shortening as determined from echocardiography.

Unfortunately, regardless of the application method the majority of transplanted cells die within the first days post-transplantation and the long-term engraftment rate is very low [89–91]. The main factor contributing to cell death is that the areas where stem cells are transplanted to are ischemic regions. They are devoid of blood flow and exhibit the toxic effects of inflammation and tissue death. Consequently, the development of strategies designed to enhance cell survival is a major research area for optimizing the functional benefits of cardiac cell therapy. More recently, some groups have used a tissue-engineering approach to achieve myocardial regeneration by implanting cells into a scaffold onto the surface of the heart [92, 93]. The scaffold served as a temporary stem cell niche to maintain the proliferation and differentiation of the transplanted stem cells. Recent results from the MAGNUM clinical trial used a porous collagen type I matrix upon which autologous bone marrow mononuclear cells were seeded [94]. This patch was grafted onto the surface of injured myocardium and cardiac function was assessed with echocardiography. Feasibility of this strategy was established; however, direct bone marrow cell injection resulted in a maximum increase in ejection fraction of 7.4% with no difference between groups receiving a cell-loaded patch and those that did not.

Our group performed in vivo experiments by applying a fibrin patch with porcine mesenchymal stem cells to the LV anterior wall of swine LV myocardial infarction models which were produced by left circumflex coronary occlusion [80]. The results indicated this fibrin-MSC patch may prevent LV wall thinning and rescue myocardial function. Simultaneously, it was reported by Karen and colleagues that either a patch alone or skeletal myoblasts in a fibrin patch is capable of preserving infarct wall thickness and cardiac function after a myocardial infarction in rats [90]. These data demonstrate that a fibrin patch can be used as a biomaterial scaffold for myocardial cell transplantation. The main concern with fibrin patch application is the trade-off between mechanical properties and cell viability. High concentrations of fibrinogen and thrombin facilitate fibrin gelation and lead to a high strength product in a few seconds, which is very suitable to surgical placement but does not allow for cell viability and growth.

We have developed a novel PEGylated fibrin biomatrix using a difunctional, amine reactive PEG [79]. This system was designed to realize a combination strategy which aims to facilitate myocardium regeneration after myocardial infarction by delivering both stem cells and growth factors to the injured myocardium. The efficacy of our system has been evidenced both in vitro and in vivo [95]. Many other natural and synthetic biomaterials besides fibrin have been used as scaffolds to repair ischemic heart as well (see Table 1). A very recent development that aims to bring a higher degree of control to tissue engineering is called micro-patterning or bioprinting [96, 97]. Instead of seeding cells into a scaffold with random attachment, bioprinting can transfer cells to a culture substrate in a precise spatial order and distribution. In this way, multiple layers of cells and biopolymers can be printed and theoretically be used to replace the infarcted wall as a cardiac patch. This technique is still in the early stages of development and will likely require additional optimization steps before it can be applied to translational research.

Table 1 Scaffolds for cardiac cell therapy

Scaffold	Cell type	Application method	Animal model	Outcome benefits	References
Collagen	MSC	Intramyocardial injection	Fischer rat LAD model	Reduced the relocation of transplanted MSCs to remote organs and non-infarcted myocardium	[98]
Matrigel	Skeletal muscle cells	Fixed over the infarct area	Lewis rat LAD model	Significantly increased LVFS (41.6 ± 5.8 vs. 31.8 ± 5.2%)	[99]
Hyaluronic acid-based hydrogel	N/A	Intramyocardial injection	Rat LAD model	Increased angiogenesis and improved heart function	[100]
Alginate	N/A	Intramyocardial injection	Rat LAD model	Increased arteriole density and LVFS	[101]
Decellularized ECM	N/A	Intramyocardial injection	Normal rat	Increased arteriole density	[102]
PGAC	BMC	Sutured over the infarct area	Rat LAD model	Increased angiogenesis	[103]
PU	Skeletal myoblasts	Sutured over the infarct area	Rat LAD model	Increased cell survival rate after transplantation	[104]

LAD ligation of the left anterior descending coronary artery, *LVFS* left ventricle fractional shortening, *ECM* extracellular matrix, *PGAC* polyglycolic acid cloth, *BMC* bone marrow cells, *PU* polyurethane

It is reasonable to anticipate that the optimal cell therapy strategies for myocardial regeneration will need to incorporate a renewable source of proliferating, functional cardiomyocytes and a network of capillaries and blood vessels for the supply of oxygen and nutrients to both the ischemic endogenous myocardium and the newly implanted cells. In addition, the cardiomyocytes lost during myocardial infarction and subsequent heart failure may be comprised not only of transplanted stem cells with cardiomyocyte potential but also by endogenous mature cardiomyocytes that have been stimulated to proliferate, resident cardiac stem cells that have been induced to differentiate, or resident bone marrow stem cells that have been mobilized to the site of injury through local treatment with cytokines. A combination of these strategies may be required for full realization of cardiac repair.

5 Conclusions

Although multiple preclinical and clinical studies have provided evidence that stem cell therapy is efficacious, the underlying mechanism remains unclear and there is significant opportunity to improve clinical outcomes. Stem cells may improve cardiac function by trans-differentiating into endothelial cells or cardiac myocytes, by promoting angiogenesis and improving myocardial blood flow, or by paracrine effects. Studies using MSCs for cardiac damage have suggested that the degree of stem cell contribution through differentiation into the transplanted tissue has turned out to be generally low [105, 106]. Increasing evidence indicates that improved ventricular function and enhanced myocardial perfusion are mainly caused by angiogenesis, which is induced by multiple angiogenic factors secreted by the transplanted cells [107, 108]. Therefore, lessons learned from stem cell therapy in the cardiac environment may also be used to treat vascular ischemia, enhance wound healing and make progress towards the development of prevascularized tissue constructs. It is also likely that the progress in culture and delivery strategies made during development of cardiac cell therapies will contribute towards a greater understand of the cardiovascular stem cell niche and the role that it plays in tissue development and regeneration.

It is clear from the body of work cited here that biomaterials can have a profound effect on stem cell phenotype and function. The development of in vitro culture systems to direct both pluripotent and adult stem cell differentiation employ a host of strategies such as: providing insoluble matrix signals, presenting soluble signals, controlling cell–cell contacts and transmitting or augmenting physical signals all of which can contribute towards enhancing cell function. As our understanding of stem cells and their native niches develops, it will be important to engineer the appropriate interaction between cells and biomaterials. The ability to control this interaction will lead to culture systems that enhance cardiovascular differentiation and advances in cell therapy and regenerative medicine that can be used in the clinical setting to treat a broad range of cardiovascular diseases.

References

1. Beltrami, A.P., et al.: Adult cardiac stem cells are multipotent and support myocardial regeneration. Cell **114**(6), 763–776 (2003)
2. Dawn, B., et al.: Cardiac stem cells delivered intravascularly traverse the vessel barrier, regenerate infarcted myocardium, and improve cardiac function. Proc. Natl. Acad. Sci. U.S.A. **102**(10), 3766–3771 (2005)
3. Bearzi, C., et al.: Human cardiac stem cells. Proc. Natl. Acad. Sci. U.S.A. **104**(35), 14068–14073 (2007)
4. Rota, M., et al.: Local activation or implantation of cardiac progenitor cells rescues scarred infarcted myocardium improving cardiac function. Circ. Res. **103**(1), 107–116 (2008)
5. Li, M., et al.: c-kit is required for cardiomyocyte terminal differentiation. Circ. Res. **102**(6), 677–685 (2008)
6. Oh, H., et al.: Cardiac progenitor cells from adult myocardium: homing, differentiation, and fusion after infarction. Proc. Natl. Acad. Sci. U.S.A. **100**(21), 12313–12318 (2003)
7. Lu, G., et al.: Sca-1^+ stem cell survival and engraftment in the infarcted heart: dual role for preconditioning-induced connexin-43. Circulation **119**(19), 2587–2596 (2009)

8. Wang, X., et al.: The role of the sca-1^{+}/CD31^{-} cardiac progenitor cell population in postinfarction left ventricular remodeling. Stem Cells **24**(7), 1779–1788 (2006)
9. Matsuura, K., et al.: Adult cardiac Sca-1-positive cells differentiate into beating cardiomyocytes. J. Biol. Chem. **279**(12), 11384–11391 (2004)
10. Challen, G.A., Little, M.H.: A side order of stem cells: the SP phenotype. Stem Cells **24**(1), 3–12 (2006)
11. Hierlihy, A.M., et al.: The post-natal heart contains a myocardial stem cell population. FEBS Lett. **530**(1–3), 239–243 (2002)
12. Messina, E., et al.: Isolation and expansion of adult cardiac stem cells from human and murine heart. Circ. Res. **95**(9), 911–921 (2004)
13. Boyer, M., et al.: Isolation of endothelial cells and their progenitor cells from human peripheral blood. J. Vasc. Surg. **31**(1 Pt 1), 181–189 (2000)
14. Asahara, T., Isner, J.M.: Endothelial progenitor cells for vascular regeneration. J. Hematother. Stem Cell Res. **11**(2), 171–178 (2002)
15. Takahashi, T., et al.: Ischemia- and cytokine-induced mobilization of bone marrow-derived endothelial progenitor cells for neovascularization. Nat. Med. **5**(4), 434–438 (1999)
16. Fadini, G.P., et al.: Diabetes impairs progenitor cell mobilisation after hindlimb ischaemia-reperfusion injury in rats. Diabetologia **49**(12), 3075–3084 (2006)
17. Gehling, U.M., et al.: In vitro differentiation of endothelial cells from AC133-positive progenitor cells. Blood **95**(10), 3106–3112 (2000)
18. Hirschi, K., Goodell, M.: Common origins of blood and blood vessels in adults? Differentiation **68**(4–5), 186–192 (2001)
19. Majka, S.M., et al.: Distinct progenitor populations in skeletal muscle are bone marrow derived and exhibit different cell fates during vascular regeneration. J. Clin. Invest. **111**(1), 71–79 (2003)
20. Nadin, B.M., Goodell, M.A., Hirschi, K.K.: Phenotype and hematopoietic potential of side population cells throughout embryonic development. Blood **102**(7), 2436–2443 (2003)
21. Olmsted-Davis, E.A., et al.: Primitive adult hematopoietic stem cells can function as osteoblast precursors. Proc. Natl. Acad. Sci. U.S.A. **100**(26), 15877–15882 (2003)
22. Nehls, V., Denzer, K., Drenckhahn, D.: Pericyte involvement in capillary sprouting during angiogenesis in situ. Cell Tissue Res. **270**(3), 469–474 (1992)
23. Nehls, V., Schuchardt, E., Drenckhahn, D.: The effect of fibroblasts, vascular smooth muscle cells, and pericytes on sprout formation of endothelial cells in a fibrin gel angiogenesis system. Microvasc. Res. **48**(3), 349–363 (1994)
24. Lindblom, P., et al.: Endothelial PDGF-B retention is required for proper investment of pericytes in the microvessel wall. Genes Dev. **17**(15), 1835–1840 (2003)
25. Simper, D., et al.: Smooth muscle progenitor cells in human blood. Circulation **106**(10), 1199–1204 (2002)
26. Seruya, M., et al.: Clonal population of adult stem cells: life span and differentiation potential. Cell Transplant. **13**(2), 93–101 (2004)
27. Shepro, D., Morel, N.M.L.: Pericyte physiology. FASEB J. **7**, 1031–1038 (1993)
28. Pittenger, M.F., et al.: Multilineage potential of adult human mesenchymal stem cells. Science **284**(5411), 143–147 (1999)
29. Toma, C., et al.: Human mesenchymal stem cells differentiate to a cardiomyocyte phenotype in the adult murine heart. Circulation **105**(1), 93–98 (2002)
30. Makino, S., et al.: Cardiomyocytes can be generated from marrow stromal cells in vitro. J. Clin. Invest. **103**(5), 697–705 (1999)
31. Abedin, M., Tintut, Y., Demer, L.L.: Mesenchymal stem cells and the artery wall. Circ. Res. **95**(7), 671–676 (2004)
32. Tintut, Y., et al.: Multilineage potential of cells from the artery wall. Circulation **108**(20), 2505–2510 (2003)
33. Strauer, B.E., et al.: Repair of infarcted myocardium by autologous intracoronary mononuclear bone marrow cell transplantation in humans. Circulation **106**(15), 1913–1918 (2002)

34. Orlic, D., et al.: Bone marrow cells regenerate infarcted myocardium. Nature **410**(6829), 701–705 (2001)
35. Davani, S., et al.: Mesenchymal progenitor cells differentiate into an endothelial phenotype, enhance vascular density, and improve heart function in a rat cellular cardiomyoplasty model. Circulation **108** (Suppl 1), II253–II258 (2003)
36. Visconti, R.P., Richardson, C.D., Sato, T.N.: Orchestration of angiogenesis and arteriovenous contribution by angiopoietins and vascular endothelial growth factor (VEGF). Proc. Natl. Acad. Sci. U.S.A. **99**(12), 8219–8224 (2002)
37. Ahrendt, G., Chickering, D.E., Ranieri, J.P.: Angiogenic growth factors: A review for tissue engineering. Tissue Eng. **4**(2), 117–130 (1998)
38. Oswald, J., et al.: Mesenchymal stem cells can be differentiated into endothelial cells in vitro. Stem Cells **22**(3), 377–384 (2004)
39. Nugent, M.A., Iozzo, R.V.: Fibroblast growth factor-2. Int. J. Biochem. Cell Biol. **32**(2), 115–120 (2000)
40. Conklin, B.S., et al.: Basic fibroblast growth factor coating and endothelial cell seeding of a decellularized heparin-coated vascular graft. Artif. Organs **28**(7), 668–675 (2004)
41. Isner, J.M., Takayuki, A.: Therapeutic angiogenesis. Front Biosci. **3,** e49–e69 (1998)
42. Darland, D.C., D'Amore, P.A.: TGF beta is required for the formation of capillary-like structures in three-dimensional cocultures of 10T1/2 and endothelial cells. Angiogenesis **4**(1), 11–20 (2001)
43. Boheler, K.R., et al.: Differentiation of pluripotent embryonic stem cells into cardiomyocytes. Circ. Res. **91**(3), 189–201 (2002)
44. Hegert, C., et al.: Differentiation plasticity of chondrocytes derived from mouse embryonic stem cells. J. Cell Sci. **115**(Pt 23), 4617–4628 (2002)
45. Kramer, J., Hegert, C., Rohwedel, J.: In vitro differentiation of mouse ES cells: bone and cartilage. Methods Enzymol. **365,** 251–268 (2003)
46. Levenberg, S., et al.: Endothelial cells derived from human embryonic stem cells. Proc. Natl. Acad. Sci. U.S.A. **99**(7), 4391–4396 (2002)
47. Wobus, A.M., et al.: Embryonic stem cells as a model to study cardiac, skeletal muscle, and vascular smooth muscle cell differentiation. Methods Mol. Biol. **185,** 127–156 (2002)
48. Dang, S.M., et al.: Controlled, scalable embryonic stem cell differentiation culture. Stem Cells **22**(3), 275–282 (2004)
49. Oh, S.K., et al.: High density cultures of embryonic stem cells. Biotechnol. Bioeng. **91**(5), 523–533 (2005)
50. Liu, H., Roy, K.: Biomimetic three-dimensional cultures significantly increase hematopoietic differentiation efficacy of embryonic stem cells. Tissue Eng. **11**(1–2), 319–330 (2005)
51. Levenberg, S., et al.: Differentiation of human embryonic stem cells on three-dimensional polymer scaffolds. Proc. Natl. Acad. Sci. U.S.A. **100**(22), 12741–12746 (2003)
52. Battista, S., et al.: The effect of matrix composition of 3D constructs on embryonic stem cell differentiation. Biomaterials **26**(31), 6194–6207 (2005)
53. He, J.Q., et al.: Human embryonic stem cells develop into multiple types of cardiac myocytes: action potential characterization. Circ. Res. **93**(1), 32–39 (2003)
54. Kehat, I., et al.: Human embryonic stem cells can differentiate into myocytes with structural and functional properties of cardiomyocytes. J. Clin. Invest. **108**(3), 407–414 (2001)
55. Xu, C., et al.: Characterization and enrichment of cardiomyocytes derived from human embryonic stem cells. Circ. Res. **91**(6), 501–508 (2002)
56. Zimmermann, W.H., Eschenhagen, T.: Embryonic stem cells for cardiac muscle engineering. Trends Cardiovasc. Med. **17**(4), 134–140 (2007)
57. Mummery, C., et al.: Differentiation of human embryonic stem cells to cardiomyocytes: role of coculture with visceral endoderm-like cells. Circulation **107**(21), 2733–2740 (2003)
58. Mummery, C.L., et al.: Visceral-endoderm-like cell lines induce differentiation of murine P19 embryonal carcinoma cells. Differentiation **46**(1), 51–60 (1991)
59. Huang, H., et al.: Differentiation from embryonic stem cells to vascular wall cells under in vitro pulsatile flow loading. J. Artif. Organs **8**(2), 110–118 (2005)

60. Ferreira, L.S., et al.: Bioactive hydrogel scaffolds for controllable vascular differentiation of human embryonic stem cells. Biomaterials **28**(17), 2706–2717 (2007)
61. Yamamoto, K., et al.: Fluid shear stress induces differentiation of Flk-1-positive embryonic stem cells into vascular endothelial cells in vitro. Am. J. Physiol. Heart Circ. Physiol. **288**(4), H1915–H1924 (2005)
62. Liu, H., Collins, S.F., Suggs, L.J.: Three-dimensional culture for expansion and differentiation of mouse embryonic stem cells. Biomaterials **27**(36), 6004–6014 (2006)
63. Dang, S.M., et al.: Efficiency of embryoid body formation and hematopoietic development from embryonic stem cells in different culture systems. Biotechnol. Bioeng. **78**(4), 442–453 (2002)
64. Gerecht-Nir, S., et al.: Three-dimensional porous alginate scaffolds provide a conducive environment for generation of well-vascularized embryoid bodies from human embryonic stem cells. Biotechnol. Bioeng. **88**(3), 313–320 (2004)
65. Kubo, A., et al.: Development of definitive endoderm from embryonic stem cells in culture. Development **131**(7), 1651–1662 (2004)
66. Francis, S.E., et al.: Central roles of alpha5beta1 integrin and fibronectin in vascular development in mouse embryos and embryoid bodies. Arterioscler. Thromb. Vasc. Biol. **22**(6), 927–933 (2002)
67. Mosesson, M.W., Siebenlist, K.R., Meh, D.A.: The structure and biological features of fibrinogen and fibrin. Ann. N Y Acad. Sci. **936,** 11–30 (2001)
68. Zhang, W.J., et al.: Modulation of hematopoietic and endothelial cell differentiation from mouse embryonic stem cells by different culture conditions. Blood **105**(1), 111–114 (2005)
69. Orlic, D.: Adult bone marrow stem cells regenerate myocardium in ischemic heart disease. Ann. N Y Acad. Sci. **996,** 152–157 (2003)
70. Schachinger, V., et al.: Intracoronary bone marrow-derived progenitor cells in acute myocardial infarction. N. Engl. J. Med. **355**(12), 1210–1221 (2006)
71. Schachinger, V., et al.: Improved clinical outcome after intracoronary administration of bone-marrow-derived progenitor cells in acute myocardial infarction: final 1-year results of the REPAIR-AMI trial. Eur. Heart J. **27**(23), 2775–2783 (2006)
72. Al-Khaldi, A., et al.: Postnatal bone marrow stromal cells elicit a potent VEGF-dependent neoangiogenic response in vivo. Gene Ther. **10**(8), 621–629 (2003)
73. Annabi, B., et al.: Hypoxia promotes murine bone-marrow-derived stromal cell migration and tube formation. Stem Cells **21**(3), 337–347 (2003)
74. Mosesson, M.W.: Fibrinogen and fibrin structure and functions. J. Thromb. Haemost. **3**(8), 1894–1904 (2005)
75. van Hinsbergh, V.W., Collen, A., Koolwijk, P.: Role of fibrin matrix in angiogenesis. Ann. N Y Acad. Sci. **936,** 426–437 (2001)
76. Collen, A., et al.: Influence of fibrin structure on the formation and maintenance of capillary-like tubules by human microvascular endothelial cells. Angiogenesis **2**(2), 153–165 (1998)
77. Vailhe, B., et al.: The formation of tubular structures by endothelial cells is under the control of fibrinolysis and mechanical factors. Angiogenesis **2**(4), 331–344 (1998)
78. Vailhe, B., et al.: In vitro angiogenesis is modulated by the mechanical properties of fibrin gels and is related to alpha(v)beta3 integrin localization. In Vitro Cell Dev. Biol. Anim. **33**(10), 763–773 (1997)
79. Zhang, G., et al.: A PEGylated fibrin patch for mesenchymal stem cell delivery. Tissue Eng. **12**(1), 9–19 (2006)
80. Liu, J., et al.: Autologous stem cell transplantation for myocardial repair. Am. J. Physiol. Heart Circ. Physiol. **287**(2), H501–H511 (2004)
81. Joggerst, S.J., Hatzopoulos, A.K.: Stem cell therapy for cardiac repair: benefits and barriers. Expert Rev. Mol. Med. **11,** e20 (2009)
82. LaPar, D.J., Kron, I.L., Yang, Z.: Stem cell therapy for ischemic heart disease: where are we? Curr. Opin. Organ Transplant. **14**(1), 79–84 (2009)

83. Menasche, P.: Stem cell therapy for heart failure: are arrhythmias a real safety concern? Circulation **119**(20), 2735–2740 (2009)
84. Shintani, S., et al.: Augmentation of postnatal neovascularization with autologous bone marrow transplantation. Circulation **103**(6), 897–903 (2001)
85. Tse, H.F., et al.: Angiogenesis in ischaemic myocardium by intramyocardial autologous bone marrow mononuclear cell implantation. Lancet **361**(9351), 47–49 (2003)
86. Assmus, B., et al.: Transplantation of progenitor cells and regeneration enhancement in acute myocardial infarction (TOPCARE-AMI). Circulation **106**(24), 3009–3017 (2002)
87. Kocher, A.A., et al.: Neovascularization of ischemic myocardium by human bone-marrow-derived angioblasts prevents cardiomyocyte apoptosis, reduces remodeling and improves cardiac function. Nat. Med. **7**(4), 430–436 (2001)
88. Zhang, G., et al.: Controlled release of stromal cell-derived factor-1 alpha in situ increases c-kit$^+$ cell homing to the infarcted heart. Tissue Eng. **13**(8), 2063–2071 (2007)
89. Smythe, G.M., Hodgetts, S.I., Grounds, M.D.: Problems and solutions in myoblast transfer therapy. J. Cell Mol. Med. **5**(1), 33–47 (2001)
90. Christman, K.L., et al.: Fibrin glue alone and skeletal myoblasts in a fibrin scaffold preserve cardiac function after myocardial infarction. Tissue Eng. **10**(3, 4), 403–409 (2004)
91. Hodgetts, S.I., et al.: Why do cultured transplanted myoblasts die in vivo? DNA quantification shows enhanced survival of donor male myoblasts in host mice depleted of CD4$^+$ and CD8$^+$ cells or Nk1.1$^+$ cells. Cell Transplant. **9**(4), 489–502 (2000)
92. Leor, J., et al.: Bioengineered cardiac grafts: a new approach to repair the infarcted myocardium? Circulation **102**(19 Suppl 3), III56–III61 (2000)
93. Bensaid, W., et al.: A biodegradable fibrin scaffold for mesenchymal stem cell transplantation. Biomaterials **24**(14), 2497–2502 (2003)
94. Chachques, J.C., et al.: Myocardial assistance by grafting a new bioartificial upgraded myocardium (MAGNUM trial): clinical feasibility study. Ann. Thorac. Surg. **85**(3), 901–908 (2008)
95. Zhang, G., et al.: Enhancing efficacy of stem cell transplantation to the heart with a PEGylated fibrin biomatrix. Tissue Eng. Part A **14**(6), 1025–1036 (2008)
96. Murtuza, B., Nichol, J.W., Khademhosseini, A.: Micro- and nanoscale control of the cardiac stem cell niche for tissue fabrication. Tissue Eng. Part B Rev **15**(4), 443–454 (2009)
97. Phillippi, J.A., et al.: Microenvironments engineered by inkjet bioprinting spatially direct adult stem cells toward muscle- and bone-like subpopulations. Stem Cells **26**(1), 127–134 (2008)
98. Dai, W., et al.: Delivering stem cells to the heart in a collagen matrix reduces relocation of cells to other organs as assessed by nanoparticle technology. Regen. Med. **4**(3), 387–395 (2009)
99. Giraud, M.N., et al.: Hydrogel-based engineered skeletal muscle grafts normalize heart function early after myocardial infarction. Artif. Organs **32**(9), 692–700 (2008)
100. Yoon, S.J., et al.: Regeneration of ischemic heart using hyaluronic acid-based injectable hydrogel. J. Biomed. Mater. Res. B Appl. Biomater. **91**(1), 163–171 (2009)
101. Yu, J., et al.: The effect of injected RGD modified alginate on angiogenesis and left ventricular function in a chronic rat infarct model. Biomaterials **30**(5), 751–756 (2009)
102. Singelyn, J.M., et al.: Naturally derived myocardial matrix as an injectable scaffold for cardiac tissue engineering. Biomaterials **30**(29), 5409–5416 (2009)
103. Fukuhara, S., et al.: Bone marrow cell-seeded biodegradable polymeric scaffold enhances angiogenesis and improves function of the infarcted heart. Circ. J. **69**(7), 850–857 (2005)
104. Siepe, M., et al.: Construction of skeletal myoblast-based polyurethane scaffolds for myocardial repair. Artif. Organs **31**(6), 425–433 (2007)
105. Rose, R.A., et al.: Bone marrow-derived mesenchymal stromal cells express cardiac-specific markers, retain the stromal phenotype, and do not become functional cardiomyocytes in vitro. Stem Cells **26**(11), 2884–2892 (2008)
106. Marin-Garcia, J., Goldenthal, M.J.: Application of stem cells in cardiology: where we are and where we are going. Curr. Stem Cell Res. Ther. **1**(1), 1–11 (2006)

107. Chugh, A.R., Zuba-Surma, E.K., Dawn, B.: Bone marrow-derived mesenchymal stems cells and cardiac repair. Minerva Cardioangiol. **57**(2), 185–202 (2009)
108. Tang, J., et al.: Mesenchymal stem cells participate in angiogenesis and improve heart function in rat model of myocardial ischemia with reperfusion. Eur. J. Cardiothorac. Surg. **30**(2), 353–361 (2006)

The Integrated Role of Biomaterials and Stem Cells in Vascular Regeneration

Guoming Sun, Sravanti Kusuma and Sharon Gerecht

Abstract A healthy vascular system is essential for maintaining normal blood supply and circulation in the body, while ischemia can lead to limb amputation or even death. Vascular regeneration engineering holds the promise of permanent, effective treatments for many vascular diseases. However, many challenges also remain to bring the therapy to the clinic, as the formation of blood vessels is a complicated process. One major challenge facing vascular engineering is developing the ability to maintain large masses of viable and functional cells during in vitro culture and following their transfer from in vitro conditions into the patient. This chapter introduces the cells being studied for vascular differentiation and regeneration and introduces the biomaterials being investigated for vascular engineering, including their sources, properties, and different scaffold types. We then discuss recent approaches to engineering microenvironments, including proper signaling cues and biodegradable scaffolds that will guide the development of these cells into vessels suitable for cell-based vascular therapy. These functional biomaterials may be used as environments to stimulate the generation of blood vessels, to deliver cells to angiogenic areas of the vasculature, or to promote differentiation from progenitor cells into mature vascular cells.

G. Sun and S. Kusuma contributed equally.

G. Sun, S. Kusuma and S. Gerecht (✉)
Departments of Chemical and Biomolecular Engineering, Johns Hopkins University, Baltimore, USA
e-mail: gerecht@jhu.edu

S. Kusuma
Department of Biomedical Engineering, Johns Hopkins University, 3400 North Charles St, Baltimore, MD 21218, USA

Stud Mechanobiol Tissue Eng Biomater (2010) 2: 195–223
DOI: 10.1007/8415_2010_10

Published online: 15 June 2010

1 Introduction

Because an intact and functioning vasculature is critical to human health, diseases which affect the integrity of blood vessels lead to serious and often deadly outcomes. Vascular diseases encompass conditions that cause artery, vein, or lymph vessel damage and disorders that affect blood circulation, such as peripheral vascular disease, cardiovascular disease, and carotid artery disease [1]. Vascular diseases, such as coronary artery diseases, cerebrovascular diseases, and peripheral vascular diseases, account for more than half of all deaths each year [2]. As adults over the age of 50 are more susceptible than younger generations to certain vascular diseases—such as peripheral vascular disease, which leads to gangrene, heart attack, stoke, and amputation—improved vascular disease therapies are especially desirable with the growth of the senior population.

Upon injury or trauma, our bodies have limited inherent self-healing capabilities. However, vascular diseases are even more complicated due to their various causes and localized effects. When blood vessels become injured, for example, they cannot completely repair the injury without the help of progenitor cells, due to the low proliferative capacity of vascular cells. Current vascular disease treatments include both surgical approaches—such as heart bypass surgery and arterial replacement—and nonsurgical approaches—such as arterial clearance therapy. Although some of these methods have been well developed, they usually involve complicated procedures and unpleasant side effects. Furthermore, the current therapies cannot prevent or reduce damage to target organs and thus cripple many patients with heart failure, stroke, and limb ischemia syndromes; moreover, some of these therapies are only temporary. Vascular regeneration offers the hope of permanent and effective treatments for many vascular diseases [1].

Vascular regenerative engineering is a subfield of regenerative medicine that involves designing scaffolds, incorporating signaling molecules to the scaffolds, loading stem cells into the scaffolds, manipulating the encapsulation and release of growth factors, and transplanting functional vascular structures to the site of interest.

Several types of adult stem cells have been evaluated for their ability to differentiate into cells of the vasculature, including endothelial cells (ECs) of the luminal side of blood vessels and smooth muscle cells (SMCs), which surround and provide structure to the EC layer [3, 4]. Progenitor cells reside in the bone marrow and serve to regenerate cells upon trauma or injury. Various endogenous agents trigger these adult stem cells to differentiate upon injury. The study of these progenitor cells focuses on external cues and the signaling pathways that alert the progenitor cells to differentiate into a specific cell type. Exploiting these progenitor cells using functional biomaterials as scaffolds or coatings holds great promise for vascular regenerative therapies and could enhance the body's ability to repair itself.

The vascular extracellular matrix (ECM) is a natural scaffold that provides critical mechanical properties for the proper function of the adult vascular system [5]. Physiological and pathological vascular remodeling entails the production, degradation, and reorganization of the ECM of the vessel wall [6], which plays a critical role in the development, growth, and biomechanical properties of blood vessels. In vitro studies have demonstrated that the spatial presentation of specific ECM domains direct EC growth behavior, differentiation, and organization into blood vessels [7–11]. Specifically, ECM proteins, including collagen, fibronectin, and laminin, have been shown to provide adhesion signals, growth factor binding sites, and migration signal initiation sites that regulate neovascularization [12]. Therefore, ECM scaffolds could be used to generate in vitro microenvironments to engineer vascular structures or networks, to provide mechanical support during vascular or tissue regeneration, or as carriers to deliver cells to desired sites in the patient's body to repair damaged tissue or organs. For example, a recent study indicated that the utility of endothelial progenitor cells (EPCs) depended highly on the mode of delivery and that scaffold deployment improved the efficacy of these cells in restoring blood flow to an ischemic limb in mice [13]. The sources of natural ECMs and opportunities to manipulate their properties are limited, and scientists thus have turned to biodegradable polymers in order to engineer different ECM scaffolds.

However, the integration of biodegradable polymeric scaffolds with stem cells requires a full understanding of how each factor contributes to vasculogenesis, and many challenges must be overcome before therapies can be translated to the clinic, since the formation of blood vessels is a complicated process. One major challenge facing vascular engineering is maintaining large masses of viable and functional cells during in vitro culture and following their transfer from in vitro conditions into the patient. Therefore, a key issue is the development of biodegradable scaffolds that can provide an appropriate microenvironment to support cell attachment, migration and, ultimately, vascular structure formation and that will degrade and be replaced by the host's own tissues as new tissues form. Progress in developing stem cell culture and functional scaffolds is crucial for advancing vascular engineering theory from the bench to practical application.

2 Stem Cells for Vascular Regeneration

2.1 *Vascular Development of ECs and SMCs from Pluripotent Stem Cells*

The importance of an intact circulation is aptly illustrated by the fact that the cellular lineages of a rudimentary circulatory system are among the first cell types to develop during embryogenesis. Vasculogenesis, or the formation of blood vessels de novo, occurs early in the developing embryo and is facilitated by the arrangement of blood islands with angioblasts in the yolk sac [14], giving rise to a capillary network which further develops into an arteriovenous vascular system made of ECs enclosed by SMCs that protect and control blood flow [15].

The self-renewal capability of undifferentiated human embryonic stem cells (hESCs) in vitro makes them attractive for tissue engineering applications. ESCs are pluripotent cells isolated from the inner cell mass of the blastocyst. In vitro, these cells can be maintained as undifferentiated cells via culture on mouse embryonic fibroblasts (MEFs) or via culture in feeder-free conditions with the medium composition defined to supply essential cytokines and nutrients to promote the undifferentiated phenotype [16]. Upon removal from these conditions, hESCs spontaneously aggregate and form differentiated embryoid bodies (EBs) when cultured in suspension. The EBs contain cells of the ectoderm, endoderm, and mesoderm. Of particular interest for vascular regenerative therapies, cells of the mesoderm lineage contain a subpopulation of vascular progenitor cells with the ability to further differentiate into ECs and SMCs for angiogenesis and vasculogenesis [17]. Our previous studies examine the upregulation and localization of vasculogenic genes, such as *CD34*, *PECAM1*, *SMA*, *Ang1*, *Tie2*, *VEGFR2*, and *VEGFR3*, along this pathway (Fig. 1) [18].

ECs are characterized by their cobblestone morphology, capillary-like formation when cultured on Matrigel, incorporation of DiI-labeled acetylated low-density lipoprotein (DiI-Ac-LDL), and expression of endothelial-specific markers, such as platelet endothelial cell adhesion molecule-1 (PECAM1), CD34, vascular endothelial cadherin (VE-cad), vascular endothelial growth factor receptor 2 (VEGFR2), and von Willebrand factor (vWF) [19].

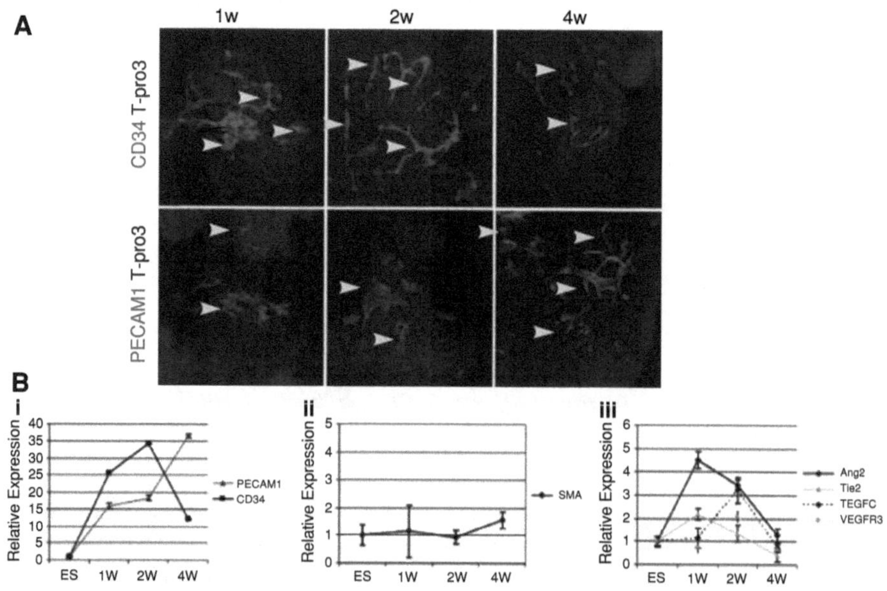

Fig. 1 Differential vasculogenic protein and gene expression. **a** Confocal images of differential CD34 and PECAM1 expression along EB development indicate primitive vascular structures (*arrow heads*). **b** RT-PCR analysis of vascular gene expression. Reprinted from [18] with permission from Elsevier

SMCs differ in their characterization; they have a spindle-like morphology, contract and relax in response to carbachol and atropine, rarely form capillary-like structures in Matrigel, and express SMC-specific markers, such as α-smooth muscle actin (α-SMA), smooth muscle myosin heavy chain (SM-MHC), and calponin [4].

Many studies, using gene knockouts or silencing, have examined important genes which direct pluripotent cells to the vascular lineage. Mice embryos deficient in Flk1, a tyrosine kinase receptor for VEGF, were unable to form blood vessels, and thus died in utero [20]. Further studies confirmed the potency of VEGF in the formation of vascular cells [21].

Vascular regenerative therapies look to the formation of new ECs and SMCs to repair damaged blood vessels or to vascularize engineered tissue. Current research has adopted this approach from two angles—via the derivation of vascular cells from stem cells and from adult stem cells, mainly endothelial progenitor cells, in vitro.

2.2 *Stem-cell-derived Vascular Cells*

ESCs have been extensively studied for their ability to differentiate into all cell types of the body. Of particular interest to regenerative medicine is the derivation and directed differentiation of vascular cells from hESCs. Elucidating these mechanisms and appropriate differentiation schemes could have immense implications for vascularizing engineered tissue, rebuilding blood vessels, and repairing ischemic wounds [19]. Moreover, the recent discovery of induced pluripotent stem cells (iPSCs) has raised the question of whether vascular cells may also be derived from this pluripotent cell source under the same conditions. The derivation of vascular cells from both pluripotent sources has provided insight into key molecular events involved in angiogenesis and vasculogenesis. This section discusses some of the key discoveries in the efficient derivation of ECs and SMCs from pluripotent stem cells as outlined in Table 1.

2.2.1 Stem-cell-derived ECs

Direct isolation of ECs from human tissue raises several issues. ECs from different parts of the vasculature have slight variations in morphology and protein expression; thus, obtaining a pure population of ECs remains a challenge. Furthermore, these cells are terminally differentiated, and therefore exhibit a low proliferative capacity. These issues have motivated the search for endogenous progenitor cells, which may provide an unlimited and self-renewing source of ECs, and for techniques to derive a pure population of ECs from ESCs and iPSCs.

Table 1 Pluripotent cells and their differentiation

References	Pluripotent cell type	Vascular cell derived	Differentiation scheme
Asahara et al. [3]	EPCs	ECs	Isolation of EPCs from systemic circulation and culture on fibronectin
Levenberg et al. [22]	hESCs	ECs	Isolation of PECAM+ cells from 13-day-old EBs
Ferreira et al. [4]	hESCs	VPC	CD34+ cells isolated from 10-day-old EBs
		ECs	VPCs cultured in media supplemented with 50 ng/ml VEGF
		SMCs	VPCs cultured in media supplemented with 50 ng/ml PDGF
Marchetti et al. [23]	Mouse ESCs	ECs	Isolation and expansion of EB cells transfected with *tie-1* promoter gene
Cao et al. [24]	Mouse ESCs	ECs	Isolation and expansion of EB cells transfected with *Ve-cad* promoter gene
Gerecht et al. [25]	hESCs	ECs	Single ESCs grown on Collagen IV, filtered through 40 μm strainer, and cultured in VEGF (for ECs), and PDGF-bb (for SMCs)
Kaufman et al. [26]	Monkey ESCs	ECs	Single ESCs grown in media supplemented with VEGF, bFGF, IGF, and EGF
Vodyanik et al. [27]	hESCs	ECs and hematopoietic cells	Co-culture on OP9 stromal cells
Yamashita et al. [28] and Narazaki et al. [29]	Mouse ESCs, mouse iPSCs	Mesoderm cells	Cultured on Collagen IV and Flk1+ cells purified for further differentiation
		SMCs and venous ECs	Flk1+ cells in media supplemented with VEGF
		Arterial ECs	Flk1+ cells in media supplemented with VEGF and cAMP analog
		Lymphatic ECs	Flk1+ cells co-cultured with OP9 stroma cells
		Cardiomyoctes	Flk1+ cells co-cultured with OP9 cells

(continued)

Table 1 (continued)

References	Pluripotent cell type	Vascular cell derived	Differentiation scheme
Sone et al. [30] and Taura et al. [31]	hESCs/iPSCs	SMCs	Cultured on OP9 feeder layer, sorted for VE-CAD-, and re-cultured on Collagen IV in media supplemented with 10 ng/ml PDGF-bb
Huang et al. [32]	hESCs	SMCs	All-trans RA treatment
Xie et al. [33]	Mouse ESCs and iPSCs	SMCs	RA treatment
Xiao et al. [34]	Mouse ESCs	SMCs	Transfection of Nox4
Sinha et al. [35]	Mouse ESCs	SMCs	Transfection of α-SMA or SM-MHC promoters
Riha et al. [36]	MSCs	SMCs	Cyclic strain

Endothelial Progenitor Cells

Asahara et al. [3] were the first group to isolate and characterize a vasculogenesis-contributing endogenous progenitor population, which they defined as endothelial progenitor cells (EPCs). These bone marrow-derived cells are adult stem cells that circulate in the bloodstream; when triggered, they form new vascular cells which contribute to angiogenesis and vasculogenesis. The research group isolated CD34+ Flk1+ cells from human peripheral blood using magnetic microbeads; these two markers are common to hematopoietic stem cells and ECs but are not present on differentiated hematopoietic cells. The study showed that these isolated cells can differentiate into ECs when plated on fibronectin in vitro and can home into sites of angiogenesis in hind limb ischemia models in vivo, demonstrating their progenitor-like capabilities and their critical presence in the circulation.

Though vasculogenesis had been well characterized in embryos, vasculogenesis in an adult was a novel concept, and its mechanism had yet to be elucidated. Characterization of EPCs by Asahara et al. [1] engendered new insights into processes involved in postnatal vasculogenesis. In a healthy human, the number of circulating EPCs remains low. However, the discovery and isolation of EPCs gave rise to a novel therapeutic strategy to treat vascular disease: by increasing their number in the circulation, EPCs could target sites of injury and differentiate into ECs as needed to heal the wound.

As part of the body's self-healing mechanisms, perturbations to the vessel wall—such as limb ischemia, burn injury, acute myocardial infarction, or coronary bypass surgery—greatly increase the number of EPCs in circulation [37, 38], further implicating them in endothelium regeneration. Unfortunately, this increase in circulating EPCs is not sufficient to compensate for all of the damage caused by these conditions, and certain conditions are associated with decreased EPC levels in the circulation. For example, patients with diabetes mellitus type II and Alzheimer's disease exhibit lower levels of circulating EPCs [39, 40]. Patients at high risk for ischemic cardiovascular disease also present reduced numbers of EPCs in the circulation [41, 42].

These conditions could greatly benefit from cell-based therapies that would allow a sufficient increase in EPCs to promote vessel wall repair. Although the number of EPCs in the circulation is low and thus more difficult to isolate and expand for therapeutic use, cord blood possesses higher quantities of EPCs than are present in the systemic circulation, and several groups have successfully isolated these progenitor cells from cord blood and demonstrated their differentiation into ECs in vitro and in vivo [43, 44].

ECs Derived from ESC and iPSC Populations

Deriving ECs from ESCs still remains the focus of much research because ESCs renew rapidly and thus could provide ECs en masse. An early study focused on determining the sequential expression of vascular genes during development into ECs, indicating first the presence of *Flk1*, then *PECAM* and *Tie-2*, and finally *VE-cad* and *Tie-1* [45]. Elucidation of this information aids the derivation of vascular cells from a pluripotent population.

Levenberg et al. [22] described an efficient derivation method for ECs from hESCs, determining an appropriate method and time point at which to isolate primitive ECs via EB formation (the process by which ESCs spontaneously differentiate). This study demonstrated that, on day 13 of EB formation, cells positive for PECAM, a cell surface receptor that mediates cell-to-cell adhesion, expressed CD34 and VE-cad and formed capillary-like structures on Matrigel after several passages [22]. Ferreira et al. [4] were the first group to isolate and identify a common human vascular progenitor cell (VPC) that differentiates into both ECs and SMCs from hESCs. CD34+ cells were isolated from EBs differentiating for 10 days and were labeled VPCs, because this subpopulation was able to differentiate into multiple vascular cells. Upon culture in media supplemented with 50 ng/ml VEGF or platelet-derived growth factor–BB (PDGF-BB), these progenitor cells differentiated into ECs and SMCs, respectively, as characterized by immunohistochemistry and functionality in in vivo models.

Other derivation methods via EB formation involved transfecting *Tie-1* promoter genes, combined with puromycin resistance, into EB cells and expanding the transfected population in VEGF-supplemented media [23]. These genetically selected cells were confirmed as ECs by their expression of endothelial markers. Cao et al. [24] also employed genetic selection by using a *VE-cad* promoter-driven system to isolate and expand ECs.

The identification of an induced differentiation procedure also benefits regenerative medicine, bypassing EB formation and thus bypassing the heterogeneous mix of cells that are incorporated into these bodies. We previously determined that undifferentiated hESCs grown on collagen IV-coated dishes expressed early endothelial markers, such as CD34, PECAM, and VE-CAD after 6 days [25]. ECs have also been derived from rhesus monkeys via single cells of ESCs cultured in a cocktail of growth factors—including VEGF, basic fibroblast growth factor (bFGF), insulin-like growth factor (IGF), and epidermal growth factor (EGF)—in the absence of a feeder layer [26]. Vodyanik and Slukvin characterized a differentiation method for the induction of ECs and hematopoietic cells from hESCs and iPSCs via coculture on mouse OP9 bone marrow stromal cells [27, 46]. They compared five hESC lines with seven iPSC lines and found that both populations generated similar subpopulations of hematopoietic cells and ECs under similar differentiation conditions.

Yamashita et al. [29] derived various forms of ECs from mouse ESCs (mESCs) and mouse iPSCs via isolation of Flk1+ cells, identifying this as an early vascular marker in mouse cells. Their novel differentiation system mimics in vitro what is expected to occur within the yolk sac: the induced differentiation of ESCs into mesoderm cells followed by further differentiation into cardiovascular cells, such as SMCs, ECs, and cardiomyocytes. Purified Flk1+ cells from mESCs and miPSCs gave rise to SMCs and ECs after culture in VEGF. Arterial ECs were induced by media supplemented with VEGF and a cAMP analog, whereas venous ECs were observed in media supplemented with VEGF alone. ECs of the lymphatic system were also derived from the Flk1+ subpopulation via culture on OP9 stroma cells. Furthermore, the purified Flk1+ population was able to differentiate into cardiomyocytes when cocultured with OP9 cells. The differences in these differentiation schemes highlight the complexities of endogenous signaling within the embryo microenvironment to produce these various vascular cell types.

2.2.2 Stem-cell-derived SMCs

The SMCs which surround the endothelium are also essential components of the vasculature, since they provide support, maintain vascular tone, and are specialized to respond to endogenous stimuli in the body [15]. SMC dysfunction or abnormal accumulation has been implicated in many life-threatening diseases, including atherosclerosis [47]. Many research groups have studied the differentiation of pluripotent cells into SMCs specifically, as these are a very specialized cell type suited to control transport through blood vessels in response to external cues.

Ferreira et al. [4] found that CD34+ cells isolated from day 10 EBs differentiated into SMCs via culture in media supplemented with PDGF. SMCs were also derived from hESCs and iPSCs via culture on an OP9 feeder layer and subsequent selection for VE-cad negative cells and culture on collagen IV in 10 ng/ml PDGF-BB [30, 31].

While most EC, and few SMC, derivation studies seem to focus on EB formation and/or marker-specific purification, a more convenient method exists for SMC derivation. Treatment with all-trans retinoic acid (RA) has been shown to induce SMCs from mESCs and hESCs, with up to approximately 93% efficiency in hESCs [32, 48].

In 2009, Xie et al. [33] used RA treatment to compare SMCs derived from miPSCs and mESCs. RA treatment successfully induced SMCs from mESCs.

Two iPSC lines were tested, O9 and TT025, though the differentiation treatment was not suitable for the latter. The O9 cells, however, differentiated into SMCs remarkably similar to those derived from ESCs. They expressed α-SMA and SM-MHC, markers specific to SMCs and contracted in response to treatment with carbachol, a drug that activates the AchR and stimulates contraction. However, fine nuances were observed between ESC-derived and iPSC-derived SMCs, reflecting inherent dissimilarities between the pluripotent cell types. SMC protein levels were higher in ESC derivatives than in iPSC derivatives, which could have an effect in vivo, though this effect was not tested in the study. Additionally, mRNA levels of myocardin and SM-MHC followed different patterns of expression and expression levels between the two forms of SMCs. A crucial difference, and a parameter needing critical assessment before clinical translation, is that the differentiation treatment was lethal to the TT025 line of iPSCs.

Other SMC derivation methods have induced expression of NADPH oxidase (Nox4) in mESCs via transfection and have shown that increased Nox4 expression resulted in the SMC phenotype [34]. Transforming growth factor β1 (TFG-β1), which is autosecreted, activated the Nox4, further driving SMC differentiation. Functional SMCs were also generated via transgenic ESC lines, with either α-SMA or SM-MHC promoters, and selected from EBs via puromycin resistance, generating a relatively pure population of SMCs [35].

Lastly, mesenchymal stem cells (MSCs) are endogenous progenitor cells which differentiate into SMCs. A large body of research has focused on methods to induce SMC differentiation from these adult stem cells in vitro. Cyclic strain was sufficient to differentiate murine embryonic mesenchymal cells into SMCs, as characterized by SMC protein and mRNA levels [36].

This section has focused on the derivation of vascular cells based on molecular factors and genes important to vascular development. Insight into these processes engenders new approaches in tissue engineering that would be conducive to promoting vascular differentiation. However, the successful transition from the benchtop to the clinic requires highly functional biomimetic scaffolds. An ideal scaffold should provide not just ECM support; it should also hold the necessary stimulation cues and growth factors needed for vascular formation. Next, we will discuss these biomimetic scaffolds for vascular regeneration.

3 Biomimetic Scaffolds for Vascular Regeneration

3.1 General Requirements for Biomimetic Scaffolds

A biomimetic ECM is pivotal for blood vessel development and growth. Much effort has been expended to design and develop such scaffolds for vascular regeneration. However, many challenges remain for designing a practical, feasible scaffold suitable for implantation. A key requirement is that the construct must not be toxic to the host. Its cytotoxicity must be thoroughly evaluated prior to clinical use, as any toxic components will harm the host. In addition to being nontoxic, the scaffold should be biocompatible. A biocompatible scaffold will permit cell adhesion, promote cell growth, and allow cells to maintain a differentiated state. Moreover, the host ECM should degrade and replace the scaffold; the degradation rate should match the tissue growth rate, and the byproduct should be nontoxic and capable of being eliminated from the body by either metabolic processes or kidney filtration. Such physical properties as porosity, pore size, mechanics, and shapability are also critical for developing a vascular engineering scaffold. A porous scaffold should have interconnected channels to make transport of chemicals, nutrients, and metabolic wastes possible. For blood vessel regeneration, the scaffold should support the presence of several different cell types of the vasculature (e.g., ECs, SMCs, and fibroblasts) and other cell types (i.e., organ-specific, such as hepatocytes, cardiomyocytes, etc.). The topography of the scaffold will affect cell attachment, migration, proliferation, differentiation, and thus will direct blood vessel formation. In addition, an ideal scaffold could be used as a carrier for loading and releasing growth factors or other molecules. Furthermore, the scaffold should have enough mechanical strength to support and protect cell growth and differentiation.

3.2 *Polymeric Biomimetic Scaffolds*

Naturally occurring ECMs are limited in both availability and manipulation for further formulation. Thus, synthetic polymeric scaffolds have been developed to overcome these limitations. Although new polymers are emerging every day, only a few polymers can meet the above requirements and be used for vascular scaffolds.

Generally, polymeric scaffolds for soft tissue engineering are made from either natural or synthetic biodegradable polymers. Natural polymers have gained interest over the past few decades because they regenerate and break down into small molecules without causing toxic effects and may contain biologically recognizable groups to support cellular activities [49]. Natural polymers used in vascular engineering can be roughly divided into two groups: protein/peptide-based and polysaccharide-based scaffolds.

Protein-based polymers are naturally suited for vascular engineering. A blood vessel scaffold fabricated from collagen demonstrated that ECs seeded onto the scaffold function much like the endothelium of normal blood vessels [50]. Tubular scaffolds electrospun from blends of proteins (e.g., gelatin and elastin) and polyglyconate were prepared for use as vascular tissue scaffolds and showed similar pore structure and porosity [51]. He et al. demonstrated that collagen-coated random and aligned poly(L-lactic acid)-co-poly(ε-caprolactone) [P(LLA-CL)] nanofiber meshes can maintain the phenotype and functions of human coronary artery endothelial cells (HCAECs) [52]. Fibrin gels, studied as an alternative to collagen scaffolds, could achieve high seeding efficiency and uniform cell distribution and exhibited remarkable remodeling, with considerable production of collagen and elastin and with a significant increase in mechanical strength [53]. Another approach incorporates protein or peptide moieties (e.g., fibronectin, RGD) into the scaffolds to guide cells toward vascular development. For example, Beamish et al. [54] demonstrated that the incorporation of arginine-glycine-aspartic acid (RGD), fibronectin, and laminin into polyethylene glycol diacrylate hydrogels, though not degradable, supported the redifferentiation of cultured vascular SMCs toward a contractile phenotype.

Polysaccharide-based scaffolds for vascular engineering include alginate, chitosan, dextran, hyaluronic acid (HA), pullulan, etc. Alginate, a negatively charged polysaccharide, can form physical hydrogels with the addition of anionic salts. Alginate hydrogels are very attractive as ECM scaffolds for vascular engineering. Freeman et al. [55] demonstrated that sequential delivery of three angiogenic factors (VEGF, PDGF-BB and tumor growth factor β1 [TGF-β1]) from an alginate/alginate-sulfate scaffold could be achieved via their differential affinity for binding to alginate-sulfate, leading to the formation of stable and mature blood vessels within the scaffold after implantation in rats. The release of VEGF from alginate scaffolds was also found to enhance vascularization and engraftment of hepatocytes transplanted on liver lobes [56]. A subcutaneous study using alginate-based macroporous hydrogel beads covalently coupled with RGD produced a vascular bed, minimal inflammation, capsule formation, and good tissue ingrowth [57]. When undifferentiated hESCs were removed from a feeder layer and seeded in a porous alginate scaffold, EB formation was observed with a relatively high degree of cell proliferation and differentiation [58].

Chitosan is a unique polymer that can be used in every field of tissue engineering. Chupa et al. [59] demonstrated that chitosan and chitosan complexes with glycosaminoglycans and dextran sulfate have significant potential for the design of new biologically active biomaterials which can modulate the activities of vascular ECs and SMCs in vitro and in vivo. Meanwhile, dextran has attracted increased interest because of its biodegradation, its biocompatibility and resistance to protein adsorption, and its easy modification with cell adhesive ligands [60, 61]. In a recent study [62], we demonstrated that dextran-based hydrogels are good candidates for vascular engineering. Ferreira et al. [17] recently demonstrated that bioactive dextran-based hydrogels are promising for vascular differentiation of hESCs. In this study, undifferentiated hESCs were encapsulated in RGD-modified dextran hydrogels preloaded with VEGF. The study found that the encapsulated hESCs differentiated within the dextran hydrogels while forming EBs. Upregulation of vascular markers and well-organized vasculature networks were observed in EBs encapsulated in the dextran hydrogels. Furthermore, when these cells were released from the dextran hydrogels and cultured on a Petri dish in vascular differentiating media (endothelial growth media-2 supplemented with VEGF), the vascular cells proliferated along a vascular lineage. HA-based hydrogels hold promise as scaffolds for vascular regeneration [63, 64]. HA was reported to stimulate EC proliferation, migration, and sprouting [65]. We have demonstrated that HA hydrogels can support long-term self-renewal of hESCs in the presence of conditioned medium from mouse embryonic fibroblast feeder layers and can direct cell differentiation [66]. When encapsulated in three-dimensional (3D) HA

Table 2 Natural polymer scaffolds for vascular regeneration

Polymers	Polymer source	Scaffold type	Applications	References
Collagen	Protein	Gel, electrospun fibers	Incorporation of VEGF significantly affects both the formation of microvessels and the tissue ingrowth into the implant	[50, 52, 68–71]
Fibrin	Protein	Gel	Supports vascular and tissue regeneration	[53, 71]
Silk	Protein	Electrospun fibers	These biomaterial matrices are potentially useful as scaffolds for tissue engineering	[72–74]
Alginate	Polysaccharide	Gel, porous scaffold	Attractive delivery, efficient hEB formation	[57, 58, 75]
Chitosan	Polysaccharide	Gel, smart scaffold	Good growth factor delivery, promotes neovascularization	[76, 77]
Dextran	Polysaccharide	Gel	Enhances the vascular differentiation of hESCs	[17, 61]
Hyaluronic acid	Polysaccharide	Gel	Stimulates EC proliferation, migration, and sprouting support long-term self-renewal of hESCs	[17, 63–66]
Pullulan	Polysaccharide	Gel	Supports SMC adhesion, spreading, and proliferation	[67]

hydrogels (but not within other hydrogels or in monolayer cultures on HA), hESCs maintained their undifferentiated state, preserved their normal karyotype, and maintained their full differentiation capacity, as indicated by EB formation. In addition, pullulan-based hydrogels support SMC adhesion, spreading, and proliferation and hold promise as scaffolds for vascular engineering [67]. Table 2 summarizes the natural polymeric scaffolds.

However, natural polymers have some limitations (e.g., mechanical properties) that they cannot overcome. To some extent, synthetic polymers can be chemically manipulated to have those desired properties. As synthetic scaffolds, these polymers should be biocompatible and biodegradable, and they should allow nutrients and oxygen to penetrate to nourish cells and tissue (e.g., have a porous structure); most natural polymers and their derivatives meet these requirements. As Table 3 indicates, polyester-based synthetic polymers are the major synthetic polymers used for vascular engineering. Polyester-based synthetic polymers, such as poly(ester amide) (PEA) [78, 79], poly(lactic-co-glycolic acid) (PLGA), poly-L-lactic acid (PLLA) [80, 81], polycaprolactone [82–85], and their copolymer poly(L-lactide-co-ε-caprolactone) [86] have been investigated thoroughly. As PEA can combine the properties of polyesters and polyamides with tunable properties, they have shown promising results. Hemmrich et al. [87] demonstrated that PEA prepared from ε-caprolactame, 1,4-butanediol, and adipic acid—and its 3D nonwoven scaffolds—support good adherence, proliferation, and differentiation of preadipocytes.

Table 3 Synthetic polymer scaffolds for vascular regeneration

Polymers	Polymer type	Scaffold type	Results	References
Poly(ester amide) (PEA)	Polyester polyamide	Film, electrospun fibers	Supports good adherence, proliferation, and differentiation	[78, 79, 87]
Poly(lactide-co-glycolide) (PLGA)	Polyester	Porous scaffold	Release of multiple growth factors promotes the vascular structure formation	[80, 81, 91, 92]
Poly-L-lactic acid (PLLA)	Polyester	Porous scaffold	Delivers growth factor, promotes cell proliferation and 3D vascularization	[80, 93]
Polycaprolactone (PCL)	Polyester	Gel, electrospun fibers	Promotes cell attachment, improves cell spreading and proliferation	[82–85, 93–95]
Poly(glycerolcosebacate) (PGS), acrylated-PGS (PGSA)	Polyester	Porous elastomer	Promotes tissue ingrowth and vascularization	[88, 89]
Poly(β-amino esters(PBAE)	Polyester	Nanoparticles	Enhances angiogenesis	[90]

The newly developed poly(glycerol-co-sebacate) (PGS) and its acrylated derivative poly(glycerol-co-sebacate) acrylate (PGSA) [88] have been studied for vascular regeneration because of their biocompatibility, biodegradation, and mechanical strength for 3D structure. We have demonstrated that porous PGSA could support the growth and differentiation of encapsulated human ESCs [89], and our in vivo experiments have also shown that porous PGSA, unlike non-porous PGSA, promotes tissue ingrowth and vascularization. PLLA- or PLGA-based polymers are the foremost scaffolds investigated for tissue engineering. A highly porous PLLA-PLGA scaffold was constructed and implanted into mice to study the therapeutic properties of ECs derived from hESCs [22]. Subcutaneous transplantation of PLLA-PLGA scaffolds seeded with hESC-derived ECs revealed that functional blood-carrying microvessels had formed and anastomosed with the host vasculature amidst fibrous connective tissue after 7 days. Yang et al. [90] recently synthesized poly(β-amino esters) (PBAEs), and formed nanoparticles with DNA. In their study, they modified human stem cells to express an angiogenic gene encoding VEGF and found that the transplantation of PBAE-/VEGF-modified stem cells significantly enhanced angiogenesis in a mouse subcutaneous model and in a hind limb ischemia model.

3.3 Scaffold Types

As discussed above, polymeric scaffolds are mostly fabricated into hydrogels [17, 63], or either 2D mesh or 3D tubular scaffolds by electrospinning [52, 87]. Additionally, other than traditional vascular scaffolds, self-assembling peptide scaffolds [96], or the hybrid scaffold fabricated from electrospun fibers and hydrogels [97] have also been synthesized and tested.

3.3.1 Hydrogels

Hydrogels are 3D crosslinked hydrophilic polymeric networks that can imbibe a large amount of media and keep their structures. Generally, hydrogels have excellent compatibility and can protect cells or growth factors from being attacked by the hostile environment. Due to their structural and functional similarities to natural tissues, hydrogels have numerous pharmaceutical and biomedical applications [1, 60, 98, 99]. Cell encapsulation within biodegradable hydrogels for vascular engineering offers many attractive features [100]. Hydrogels provide structural support for stem or progenitor cells, enabling them to differentiate and form different vascular structures according to the local microenvironment. Upon implantation, biodegradable hydrogels can be degraded in vivo, facilitating the integration of the microvascular construct. Protein-based collagen and fibrin gels have long been known to support vascular and tissue regeneration [71, 101]. Furthermore, hydrogels are also extraordinary for being able to load and release different growth factors needed in vascular regeneration [56, 75, 77, 102]. We demonstrated [62] that VEGF-loaded hydrogels promoted tissue ingrowth and facilitated hydrogel degradation. Other than release of growth factors, Kraehenbuehl et al. [103] reported that the release of thymosin β4 (Tβ4) triggered matrix metalloproteinase (MMP) secretion by encapsulated human umbilical vein endothelial cells (HUVECs) and induced vascular-like network formation on the PEG-hydrogels.

3.3.2 Electrospun Fibers

The electrospinning of nanofibers can be dated back to the early 1930s, but it received too little attention until 30 years ago [104]. Electrospun fibers, nonwoven fibers that can be fabricated into 2D mesh or 3D tubular scaffolds, have been intensively studied for tissue engineering [87, 92, 94, 105, 106]. Electrospun fibers have been made from a wide range of materials, such as silk [72–74], collagen [69, 70], PLGA [92], PCL [94, 95], etc. Recently, the electrospinning of tubular structures as scaffolds for vascular regeneration has drawn more attention [49, 51, 73, 107, 108]. Unlike hydrogel scaffolds, in which the stem cells differentiate and form different vascular structures according to the local microenvironment, cell differentiation usually occurs on the tubule surfaces, and vessel formation is thus based on the scaffold structures. Therefore, the development of blood vessels from electrospun fibers depends largely on the scaffold dimensions. To achieve desirable characteristics, polymer blends are used to improve the mechanical properties of the tubular fibers [109]. Multilayered structures are also often investigated as means to change the properties of these tubular structures [110, 111]. However, the application of tubular electrospun fibers is still restricted due to their size limitations as they cannot be fabricated with diameters below the micron scale. Zhang et al. developed a unique static collecting method with combinatorial electric fields that could fabricate nanofibrous tubes with different microscopic architectures and macroscopic 3D tubular structures [112].

3.3.3 Other Scaffolds

In addition to the above scaffolds, some polymeric scaffolds that can respond to external stimuli, such as pH [113], temperature [114], and glucose [115], are known as intelligent or smart scaffolds [116]. Those responses can induce changes in release and swelling behaviors, greatly improving the formation of new vessels. Under some circumstances, injectable hydrogels, soft scaffolds, membranes, solid-load-bearing scaffolds, or immunoprotective microencapsulation are more desirable [117]. For example, novel thermally-sensitive neutral solutions based on chitosan/polyol salt combinations were reported to encapsulate living cells and therapeutic proteins at room temperatures but formed monolithic gels at body temperature [74]. When injected in vivo, the liquid formulations turn into gel implants in situ, and this system was used successfully to deliver biologically active growth factors in vivo, as well as serving as an encapsulating matrix for living chondrocytes for tissue engineering applications. The development of smart scaffolds, because of their intelligent properties and easy, less invasive handling, could expedite the transition of vascular engineering scaffolds from bench to clinic.

3.4 Vascular Engineering Scaffold Properties

3.4.1 Degradation Properties

Degradation is critical in designing scaffolds for vascular engineering. Cell encapsulation may prevent their spreading and differentiation within the scaffolds, while degradation could break this limitation and facilitate cell spreading within the scaffolds. Kloxin et al. [118] demonstrated that the photodegradation of a hydrogel promoted cell spreading. Burdick et al. [119] indirectly supported the importance of structure breakdown by the sequential crosslinking of HA-based hydrogel. In their research, further crosslinking could retard cell spreading in the hydrogel. Furthermore, degradation is also one of the major release mechanisms, i.e., diffusion and degradation. For vascular engineering, scaffolds provide more than support for cell growth and differentiation—they also act as carriers for various growth factors. An appropriate degradation rate is essential for continuous growth factor release, which is indispensable for angiogenesis. Thus, the degradation property of scaffolds is crucial in vascular engineering.

3.4.2 Substrate Topography

Surface topography significantly affects cell attachment and orientation [120]. ECM proteins, both in vitro and in vivo, provide mammalian cells with biophysical cues, including specific surface chemistry and rich 3D surface topography [121] with features on the nanometer length scale [122]. ECM substrates provide chemical and physical external cues that dictate a variety of cell responses. Therefore, not only the milieu of soluble, diffusible factors, but also the adhesive, mechanical interactions with scaffolding materials, both natural and synthetic, control select cell functions, including cell attachment, migration, proliferation, differentiation, and regulation of genes [123, 124]. Xu et al. [91] showed that SMCs attached and migrated along the axis of aligned nanofibers, and the adhesion and proliferation rate of SMCs on the aligned nanofibrous scaffold were significantly improved than on the plane polymer films. Cellular responses to synthetic micro- and nanofabricated substrates via contact guidance have been shown in both ECs and SMCs, including morphology, motility, and proliferation [125–127]. We have also shown that nanotopography enhanced capillary tube formation and also facilitated the creation of organized vascular structures [128].

3.4.3 Mechanical Stimulation

Body movement can cause significant deformation, such as stretch, compression, or torsion, and these deformations will cause a wide change in mechanical properties. When applied to scaffolds, these mechanical forces can affect the release profile from the scaffold and further improve tissue development. Kim et al. [129] demonstrated that short-term application of strain increased the proliferation of SMCs adherent to scaffolds and the expression of collagen and elastin, while long-term application of cyclic strain upregulated elastin and collagen gene expression and led to increased organization in tissues. After vascular structures form, blood flow can induce a series of mechanical changes. These mechanical forces can further affect the cell phenotype, proliferation, and differentiation; mechanical factors (e.g., shear stress and cyclic strain) thus play a vital role in vascular regeneration. The role of shear stress on vascular remodeling has been extensively investigated [130–134]. During embryonic development, hydrodynamic shear was found to induce differentiation of ESCs into ECs or vascular wall cells, and even to induce cardiovascular commitment through epigenetic histone modification [135–138]. Recently, ECs derived from human ESCs were shown to respond functionally to changes in fluid shear stress by modulating gene expression and cell morphology [139]. Ye et al. [140] demonstrated that shear stress, as well as vascular SMCs, promotes endothelial differentiation of EPCs via activation of Akt, which may provide new insight into vascular regeneration. Not only can mechanical cues regulate cell differentiation, they can also modulate the response of cells to their microenvironment; a recent study indicates that mechanical stimulation of MSCs creates an angiogenesis-promoting environment [141]. In vitro engineering of vessel walls has been achieved by cyclic strain, which induced MSCs to differentiate toward SMCs; promoted secretion of ECM proteins, such as collagen I and fibronectin; and enabled the engineered production of a complete vessel from hMSCs and luminal ECs [142].

4 Inclusion of Vascular Stem and Somatic Cells into Biomaterials

Though the derivation of individual vascular cell types and the development of compatible biomaterials have played pivotal roles in vascular regeneration, the picture is not complete without the integration of ECs and SMCs into host vasculatures. To this end, functional biomaterials have been exploited to harness cells' abilities to repair damaged vasculatures or to vascularize engineered tissue. These biomaterials serve as scaffolds to guide cells into a specific arrangement or to provide environments conducive to directed differentiation. The creation of these functional composite systems allows for the localized presence of regulatory factors and physical constraints to promote differentiation or to serve as a scaffold in a controlled manner. The impact of biomaterials on vascular stem cells and ESC-derived vascular cells has already received attention, as indicated by the incorporation of several biomaterials into standard phenotype-analysis assays, such as Matrigel and collagen, reflecting the importance of the ECM in cell fate and differentiation. Confining these cells within biomaterials provides precise control over their 3D environment, and thus control over their fate. The studies described here exemplify the advances in vascular tissue engineering that have arisen from integration of cells within biomaterials.

4.1 Biomaterials to Engineer Blood Vessels

One of the earliest studies to seed cells into biomaterials to model the blood vessel was conducted by Weinberg and Bell [50]; they incorporated bovine SMCs, fibroblasts, and ECs sequentially into a hydrated collagen gel supported by a Dacron mesh. The researchers found that their construct had a functional endothelium and could support physiologic pressures without bursting. Drawbacks of their construct rested on the longitudinal, rather than radial, orientation of SMCs and collagen and the thinner linings of SMCs and collagen than is observed in the human body. Despite these drawbacks, Weinberg and Bell's construct still offered important information about the interaction between cell types and set the stage for future blood vessel engineering endeavors.

Wu et al. [143] also demonstrated the formation of microvessels within scaffolds. In this study, Wu et al. aimed to engineer microvessels by seeding EPC-derived ECs and SMCs onto PGA-PLLA-based scaffolds. They isolated EPCs from adult peripheral blood using CD34+ CD133+ cells and expanded and differentiated them into ECs. These EPC-derived ECs were then seeded with SMCs onto biodegradable scaffolds and appeared to form layered tissue embedded with microvessels. Histological sections of the scaffolds seeded with ECs revealed microvessels lined with CD31+; though α-SMA was noticed throughout the scaffold, it was not evident that microvessels were lined with an inner lining of ECs surrounded by a layer of SMCs.

4.2 Biomaterials to Deliver Cells to Host Vasculature

Kraehenbuehl et al. [103] developed a novel hydrogel that responds to cellular behavior. Making use of the fact that ECs secrete MMPs to degrade basement membrane and allow angiogenesis, the researchers crosslinked MMP-sensitive peptides to form their PEG-based hydrogel. Thus, hydrogel degradation was a direct result of cell-secreted MMP and its interaction with the hydrogel. The researchers demonstrated that incorporation of Tb4, a peptide known to enhance vascular cell survival, promoted HUVEC survival and MMP-2 and MMP-9 secretion. Therefore, it could support cell ingrowth such that this hydrogel could promote incorporation of ECs and Tβ4 into the host's vasculature. Though the study focuses on in vitro characterization, application of this hydrogel in vivo could deliver Tβ4 and HUVECs, which have been shown to upregulate MMP [144], to ischemic tissue; thus, the encapsulated growth factors and cells would be released in a controlled manner.

Cao et al. [145] looked into promoting angiogenesis by delivering VEGF and DAPT, a gamma secretase inhibitor, via an injectable alginate hydrogel. This composite system restored blood flow in ischemic hind limbs faster than either factor alone. Interestingly, direct injection of DAPT without the hydrogel did not improve blood flow, highlighting the fact that the delivery vehicle indeed makes a difference in the incorporation of this molecule.

4.3 Biomaterials to Induce Differentiation

A recent study by Ferreira et al. [17]. explored the possibility of utilizing dextran-based hydrogels as a means of differentiating hESCs into vascular cells. They created a novel bioactive scaffold by tethering the RGD peptide and incorporating

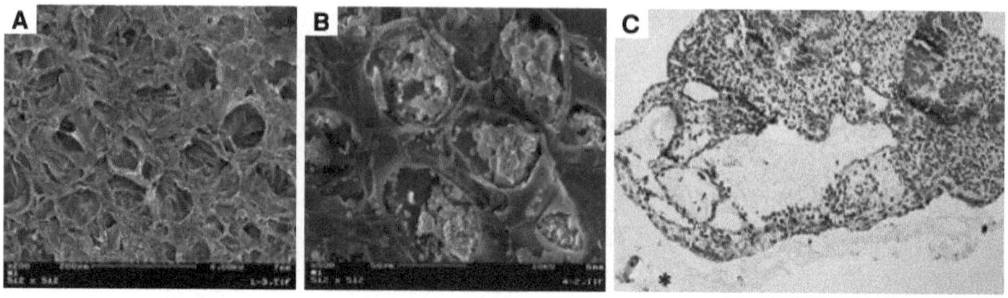

Fig. 2 SEM images of alginate scafford porous structure (**a**), alginate scaffold with EBs formed within the pores (**b**), and histological section of CD34+ cells lining the voids (**c**). Adapted from [58] with permission from Elsevier

VEGF to allow in vivo transplantation of the cell and gel composite for direct implantation. Fibronectin, abundant during embryonic vascular development, is the glycoprotein to which cells adhere via the RGD peptide. Encapsulation of hESCs in this scaffold actually decreased differentiation into the endoderm and ectoderm lineages. Incorporating these soluble factors into the structure of the hydrogel permitted directed differentiation into cells of mesoderm origin—specifically, vascular cells.

In an earlier study, we identified a novel porous alginate scaffold capable of supporting efficient formation of EBs from hESCs and of inducing vasculogenesis within its confined environment [58]. The hydrophilic alginate scaffold, which is made of naturally occurring polysaccharides found in brown sea algae, can impose physical and chemical constraints in the immediate vicinity of the undifferentiated hESCs, guiding their differentiation and the formation of EBs. SEM images of cell-seeded scaffolds after 30 days showed that EBs were formed mainly within the pores (Fig. 2).

Interestingly, the scaffolds appeared to induce vessel formation, indicated by the presence of CD34+ cells lining voids (Fig. 2). Taken together, the findings of this study highlight the potency of the microenvironment in directing differentiation of hESCs and implicate alginate scaffolds as effective platforms for engineering complex vascular arrangements.

5 Future Perspectives

Vascular regeneration offers permanent solutions for vascular diseases and reduces the need for invasive surgeries; however, many challenges must be overcome before it becomes useful in clinical practice. It will require multidisciplinary collaboration among biologists, materials scientists, engineers and physicians. The reconstruction of vascular microenvironments that present an instructive 3D environment for vascular formation is highly desirable, but many issues remain to be solved. The newly developed technologies can always facilitate the progress of vascular engineering, but in vivo studies and clinical practice will determine whether current technologies can actually treat vascular diseases.

Future technologies must focus on further elucidating the differentiation pathway from ESCs into vascular cells, as this can help to develop highly efficient derivation schemes from pluripotent cells. Furthermore, the translation from these cells to induced pluripotent stem cells must also be explored and verified in order to bring patient-specific therapies to the clinic, eliminating much of the risk of immune system rejection by the host. Another important issue that EC derivation faces is that the cells of the arteries, veins, and capillaries differ from one another. Characterizing the ECs derived from hECS must be thoroughly investigated to ensure that the proper EC population is transplanted into the correct portion of the vasculature.

From the point of view of clinical practice, the dream of a functional and transplantable organ remains out of reach and cannot be realized without the inclusion of cells into biomimetic scaffolds. The successful understanding and manipulation of vascular cells within the scaffolds will definitely expedite this transition to eliminating vascular diseases—or at least making them as treatable as other routine diseases.

6 Conclusion

Vascular diseases are responsible for more than half of all deaths in western countries. Most traditional approaches to treating vascular diseases are either invasive or have effects that are just temporary. However, many challenges remain prior to bringing the therapy to the clinic, as the formation of blood vessels is a complicated process. Recent progress has been made in engineering microenvironments comprised of biodegradable scaffolds that include proper signaling cues to guide the development of stem cells or mature cells into vessels suitable for cell-based vascular therapy. These functional biomaterials—which can be used as environments to stimulate the generation of blood vessels, to deliver cells to angiogenic areas of the vasculature, or to promote differentiation of cells—are critical, while the successful integration of biomaterial scaffolds with cells promises a bright future for patients afflicted with vascular diseases.

References

1. Sun, G., Gerecht, S.: Vascular regeneration: engineering the stem cell microenvironment. Regener. Med. **4**(3), 435–447 (2009)
2. Sata, M., Nagai, R.: Vascular regeneration and remodeling by circulating progenitor cells. In: Cardiovascular Regeneration Therapies Using Tissue Engineering Approaches, pp. 117–127. Springer, Tokyo (2005)
3. Asahara, T., Murohara, T., Sullivan, A., Silver, M., van der Zee, R., Li, T., Witzenbichler, B., Schatteman, G., Isner, J.M.: Isolation of putative progenitor endothelial cells for angiogenesis. Science **275**(5302), 964–966 (1997)

4. Ferreira, L.S., Gerecht, S., Shieh, H.F., Watson, N., Rupnick, M.A., Dallabrida, S.M., Vunjak-Novakovic, G., Langer, R.: Vascular progenitor cells isolated from human embryonic stem cells give rise to endothelial and smooth muscle like cells and form vascular networks in vivo. Circ. Res. **101**(3), 286–294 (2007)
5. Wagenseil, J.E., Mecham, R.P.: Vascular extracellular matrix and arterial mechanics. Physiol. Rev. **89**(3), 957–989 (2009)
6. Galis, Z.S., Khatri, J.J.: Matrix metalloproteinases in vascular remodeling and atherogenesis: the good, the bad, and the ugly. Circ. Res. **90**(3), 251–262 (2002)
7. Baluk, P., Hashizume, H., McDonald, D.M.: Cellular abnormalities of blood vessels as targets in cancer. Curr. Opin. Genet. Dev. **15**(1), 102–111 (2005)
8. Davis, G.E., Senger, D.R.: Endothelial extracellular matrix: biosynthesis, remodeling, and functions during vascular morphogenesis and neovessel stabilization. Circ. Res. **97**(11), 1093–1107 (2005)
9. Li, J., Zhang, Y.-P., Kirsner, R.S.: Angiogenesis in wound repair: angiogenic growth factors and the extracellular matrix. Microsc. Res. Tech. **60**(1), 107–114 (2003)
10. Stupack, D.G., Cheresh, D.A., Gerald, P.S.: Integrins and angiogenesis. Curr. Top. Dev. Biol. **64**, 207–238 (2004)
11. Wijelath, E.S., Rahman, S., Murray, J., Patel, Y., Savidge, G., Sobel, M.: Fibronectin promotes VEGF-induced CD34+ cell differentiation into endothelial cells. J. Vasc. Surg. **39**(3), 655–660 (2004)
12. Stupack, D.G., Cheresh, D.A.: ECM remodeling regulates angiogenesis: endothelial integrins look for new ligands. Sci. STKE **2002**(119), pe7 (2002)
13. Silva, E.A., Kim, E.-S., Kong, H.J., Mooney, D.J.: Material-based deployment enhances efficacy of endothelial progenitor cells. Proc. Natl. Acad. Sci. USA **105**(38), 14347–14352 (2008)
14. Risau, W., Flamme, I.: Vasculogenesis. Ann. Rev. Cell Dev. Biol. **11**(1), 73–91 (1995)
15. Carmeliet, P.: Mechanisms of angiogenesis and arteriogenesis. Nat. Med. **6**(4), 389–395 (2000)
16. Thomson, J.A., Itskovitz-Eldor, J., Shapiro, S.S., Waknitz, M.A., Swiergiel, J.J., Marshall, V.S., Jones, J.M.: Embryonic stem cell lines derived from human blastocysts. Science **282**(5391), 1145–1147 (1998)
17. Ferreira, L.S., Gerecht, S., Fuller, J., Shieh, H.F., Vunjak-Novakovic, G., Langer, R.: Bioactive hydrogel scaffolds for controllable vascular differentiation of human embryonic stem cells. Biomaterials **28**(17), 2706–2717 (2007)
18. Hanjaya-Putra, D., Gerecht, S.: Vascular engineering using human embryonic stem cells. Biotechnol. Prog. **25**(1), 2–9 (2009)
19. Levenberg, S., Zoldan, J., Basevitch, Y., Langer, R.: Endothelial potential of human embryonic stem cells. Blood **110**(3), 806–814 (2007)
20. Shalaby, F., Rossant, J., Yamaguchi, T.P., Gertsenstein, M., Wu, X.F., Breitman, M.L., Schuh, A.C.: Failure of blood-island formation and vasculogenesis in FLK-1-deficient mice. Nature **376**(6535), 62–66 (1995)
21. Ferrara, N., Carver-Moore, K., Chen, H., Dowd, M., Lu, L., O'Shea, K.S., Powell-Braxton, L., Hillan, K.J., Moore, M.W.: Heterozygous embryonic lethality induced by targeted inactivation of the VEGF gene. Nature **380**(6573), 439–442 (1996)
22. Levenberg, S., Golub, J.S., Amit, M., Itskovitz-Eldor, J., Langer, R.: Endothelial cells derived from human embryonic stem cells. Proc. Natl. Acad. Sci. USA **99**(7), 4391–4396 (2002)
23. Marchetti, S., Gimond, C., Iljin, K., Bourcier, C., Alitalo, K., Pouyssegur, J., Pages, G.: Endothelial cells genetically selected from differentiating mouse embryonic stem cells incorporate at sites of neovascularization in vivo. J. Cell Sci. **115**(10), 2075–2085 (2002)
24. Cao, F., Lin, S., Xie, X., Ray, P., Patel, M., Zhang, X., Drukker, M., Dylla, S.J., Connolly, A.J., Chen, X., Weissman, I.L., Gambhir, S.S., Wu, J C.: In vivo visualization of embryonic stem cell survival, proliferation, and migration after cardiac delivery. Circulation **113**(7), 1005–1014 (2006)

25. Gerecht-Nir, S., Ziskind, A., Cohen, S., Itskovitz-Eldor, J.: Human embryonic stem cells as an in vitro model for human vascular development and the induction of vascular differentiation. Lab. Invest. **83**(12), 1811–1820 (2003)
26. Kaufman, D.S., Lewis, R.L., Hanson, E.T., Auerbach, R., Plendl, J., Thomson, J.A.: Functional endothelial cells derived from rhesus monkey embryonic stem cells. Blood **103**(4), 1325–1332 (2004)
27. Vodyanik, M.A., Slukvin, I.I.: Hematoendothelial differentiation of human embryonic stem cells. In: Juan, S.B et al. (eds.) Current Protocols in Cell Biology, Chap. 23. Wiley, New York (2007)
28. Yamashita, J., Itoh, H., Hirashima, M., Kubo, H., Yurugi, T., Ogawa, M., Naito, M., Nakao, K., Nishikawa, S.: Identification of vascular progenitor cells in ES cells and vascular regeneration. J. Hypertens. **18**, S189 (2000)
29. Narazaki, G., Uosaki, H., Teranishi, M., Okita, K., Kim, B., Matsuoka, S., Yamanaka, S., Yamashita, J.K.: Directed and systematic differentiation of cardiovascular cells from mouse induced pluripotent stem cells. Circulation **118**(5), 498–506 (2008)
30. Sone, M., Itoh, H., Yamahara, K., Yamashita, J.K., Yurugi-Kobayashi, T., Nonoguchi, A., Suzuki, Y., Chao, T.H., Sawada, N., Fukunaga, Y., Miyashita, K., Park, K., Oyamada, N., Sawada, N., Taura, D., Tamura, N., Kondo, Y., Nito, S., Suemori, H., Nakatsuji, N., Nishikawa, S.I., Nakao, K.: Pathway for differentiation of human embryonic stem cells to vascular cell components and their potential for vascular regeneration. Arterioscler. Thromb. Vasc. Biol. **27**(10), 2127–2134 (2007)
31. Taura, D., Sone, M., Homma, K., Oyamada, N., Takahashi, K., Tamura, N., Yamanaka, S., Nakao, K.: Induction and isolation of vascular cells from human induced pluripotent stem cells—brief report. Arterioscler. Thromb. Vasc. Biol. **29**(7), 1100–1103 (2009)
32. Huang, H., Zhao, X., Chen, L., Xu, C., Yao, X., Lu, Y., Dai, L., Zhang, M.: Differentiation of human embryonic stem cells into smooth muscle cells in adherent monolayer culture. Biochem. Biophys. Res. Commun. **351**(2), 321–327 (2006)
33. Xie, C.-Q., Huang, H., Wei, S., Song, L.-S., Zhang, J., Ritchie, R.P., Chen, L., Zhang, M., Chen, Y.E.: A comparison of murine smooth muscle cells generated from embryonic versus induced pluripotent stem cells. Stem Cells Dev. **18**(5), 741–748 (2009)
34. Xiao, Q., Luo, Z., Pepe, A.E., Margariti, A., Zeng, L., Xu, Q.: Embryonic stem cell differentiation into smooth muscle cells is mediated by Nox4-produced H_2O_2. Am. J. Phys. Cell Physiol. **296**(4) (2009)
35. Sinha, S., Wamhoff, B.R., Hoofnagle, M.H., Thomas, J., Neppl, R.L., Deering, T., Helmke, B.P., Bowles, D.K., Somlyo, A.V., Owens, G.K.: Assessment of contractility of purified smooth muscle cells derived from embryonic stem cells. Stem Cells **24**(7), 1678–1688 (2006)
36. Riha, G.M., Wang, X., Wang, H., Chai, H., Mu, H., Lin, P.H., Lumsden, A.B., Yao, Q., Chen, C.: Cyclic strain induces vascular smooth muscle cell differentiation from murine embryonic mesenchymal progenitor cells. Surgery **141**(3), 394–402 (2007)
37. Shintani, S., Murohara, T., Ikeda, H., Ueno, T., Honma, T., Katoh, A., Sasaki, K.-I., Shimada, T., Oike, Y., Imaizumi, T.: Mobilization of endothelial progenitor cells in patients with acute myocardial infarction. Circulation **103**(23), 2776–2779 (2001)
38. Takahashi, T., Kalka, C., Masuda, H., Chen, D., Silver, M., Kearney, M., Magner, M., Isner, J.M., Asahara, T.: Ischemia- and cytokine-induced mobilization of bone marrow-derived endothelial progenitor cells for neovascularization. Nat. Med. **5**(4), 434–438 (1999)
39. Lee, S.T., Chu, K., Jung, K.H., Park, H.K., Kim, D.H., Bahn, J.J., Kim, J.H., Oh, M.J., Lee, S.K., Kim, M., Roh, J.K.: Reduced circulating angiogenic cells in Alzheimer disease. Neurology **72**(21), 1858–1863 (2009)
40. Tepper, O.M., Galiano, R.D., Capla, J.M., Kalka, C., Gagne, P.J., Jacobowitz, G.R., Levine, J.P., Gurtner, G.C.: Human endothelial progenitor cells from type II diabetics exhibit impaired proliferation, adhesion, and incorporation into vascular structures. Circulation **106** (22), 2781–2786 (2002)

41. Dzau, V.J., Gnecchi, M., Pachori, A.S., Morello, F., Melo, L.G.: Therapeutic potential of endothelial progenitor cells in cardiovascular diseases. Hypertension **46**(1), 7–18 (2005)
42. Hill, J.M., Zalos, G., Halcox, J.P.J., Schenke, W.H., Waclawiw, M.A., Quyyumi, A.A., Finkel, T.: Circulating endothelial progenitor cells, vascular function, and cardiovascular risk. N. Engl. J. Med. **348**(7), 593–600 (2003)
43. Mead, L.E., Prater, D., Yoder, M.C., Ingram, D.A.: Isolation and characterization of endothelial progenitor cells from human blood. Curr. Protoc. Stem Cell Biol. **6**, 2C.1.1.–2C.1.27 (2008)
44. Broxmeyer, H.E., Srour, E., Orschell, C., Ingram, D.A., Cooper, S., Plett, P.A., Mead, L.E., Yoder, M.C.: Cord blood-derived stem and progenitor cells. In: Klimanskaya, I., Lanza, R. (eds.) Methods in Enzymology, vol. 419, pp. 439–473. Academic Press/Elsevier Science, San Diego (2006)
45. Vittet, D., Prandini, M.H., Berthier, R., Schweitzer, A., Martin-Sisteron, H., Uzan, G., Dejana, E.: Embryonic stem cells differentiate in vitro to endothelial cells through successive maturation steps. Blood **88**(9), 3424–3431 (1996)
46. Choi, K.-D., Junying, Y., Kim, S.-O., Giorgia, S., William, R., Maxim, V., James, T., Igor, S.: Hematopoietic and endothelial differentiation of human induced pluripotent stem cells. Stem Cells **27**(3), 559–567 (2009)
47. Ross, R.: Atherosclerosis—an inflammatory disease. N. Engl. J. Med. **340**(2), 115–126 (1999)
48. Blank, R.S., Swartz, E.A., Thompson, M.M., Olson, E.N., Owens, G.K.: A retinoic acid-induced clonal cell line derived from multipotential P19 embryonal carcinoma cells expresses smooth muscle characteristics. Circ. Res. **76**(5), 742–749 (1995)
49. Van Tomme, S.R., Hennink, W.E.: Biodegradable dextran hydrogels for protein delivery applications. Expert Rev. Med. Dev. **4**(2), 147–164 (2007)
50. Weinberg, C.B., Bell, E.: A blood vessel model constructed from collagen and cultured vascular cells. Science **231**(4736), 397–400 (1986)
51. Thomas, V., Zhang, X., Catledge, S.A., Vohra, Y.K.: Functionally graded electrospun scaffolds with tunable mechanical properties for vascular tissue regeneration. Biomed. Mater. **2**(4), 224–232 (2007)
52. He, W., Yong, T., Ma, Z.W., Inai, R., Teo, W.E., Ramakrishna, S.: Biodegradable polymer nanofiber mesh to maintain functions of endothelial cells. Tissue Eng. **12**(9), 2457–2466 (2006)
53. Swartz, D.D., Russell, J.A., Andreadis, S.T.: Engineering of fibrin-based functional and implantable small-diameter blood vessels. Am. J. Physiol. Heart Circ. Physiol. **288**(3), H1451–H1460 (2005)
54. Beamish, J.A., Fu, A.Y., Choi, A.-J., Haq, N.A., Kottke-Marchant, K., Marchant, R.E.: The influence of RGD-bearing hydrogels on the re-expression of contractile vascular smooth muscle cell phenotype. Biomaterials **30**(25), 4127–4135 (2009)
55. Freeman, I., Cohen, S.: The influence of the sequential delivery of angiogenic factors from affinity-binding alginate scaffolds on vascularization. Biomaterials **30**(11), 2122–2131 (2009)
56. Kedem, A., Perets, A., Gamlieli-Bonshtein, I., Dvir-Ginzberg, M., Mizrahi, S., Cohen, S.: Vascular endothelial growth factor-releasing scaffolds enhance vascularization and engraftment of hepatocytes transplanted on liver lobes. Tissue Eng. **11**(5–6), 715–722 (2005)
57. Loebsack, A., Greene, K., Wyatt, S., Culberson, C., Austin, C., Beiler, R., Roland, W., Eiselt, P., Rowley, J., Burg, K., Mooney, D., Holder, W., Halberstadt, C.: In vivo characterization of a porous hydrogel material for use as a tissue bulking agent. J. Biomed. Mater. Res. **57**(4), 575–581 (2001)
58. Gerecht-Nir, S., Cohen, S., Ziskind, A., Itskovitz-Eldor, J.: Three-dimensional porous alginate scaffolds provide a conducive environment for generation of well-vascularized embryoid bodies from human embryonic stem cells. Biotechnol. Bioeng. **88**(3), 313–320 (2004)

59. Chupa, J.M., Foster, A.M., Sumner, S.R., Madihally, S.V., Matthew, H.W.T.: Vascular cell responses to polysaccharide materials: in vitro and in vivo evaluations. Biomaterials **21**(22), 2315–2322 (2000)
60. Sun, G., Chu, C.-C.: Synthesis, characterization of biodegradable dextran-allyl isocyanate-ethylamine/polyethylene glycol-diacrylate hydrogels and their in vitro release of albumin. Carbohydr. Polym. **65**(3), 273–287 (2006)
61. Stephen, P., Massia, J.S.: Immobilized RGD peptides on surface-grafted dextran promote biospecific cell attachment. J. Biomed. Mater. Res. **56**(3), 390–399 (2001)
62. Sun, G., Shen, Y.I., Ho, C.C., Kusuma, S., Gerecht, S.: Functional groups affect physical and biological properties of dextran-based hydrogels. J. Biomed. Mater. Res. Part A **9999** (9999), NA (2009)
63. Peattie, R.A., Nayate, A.P., Firpo, M.A., Shelby, J., Fisher, R.J., Prestwich, G.D.: Stimulation of in vivo angiogenesis by cytokine-loaded hyaluronic acid hydrogel implants. Biomaterials **25**(14), 2789–2798 (2004)
64. Giavaresi, G., Torricelli, P., Fornasari, P.M., Giardino, R., Barbucci, R., Leone, G.: Blood vessel formation after soft-tissue implantation of hyaluronanbased hydrogel supplemented with copper ions. Biomaterials **26**(16), 3001–3008 (2005)
65. Slevin, M., Krupinski, J., Kumar, S., Gaffney, J.: Angiogenic oligosaccharides of hyaluronan induce protein tyrosine kinase activity in endothelial cells and activate a cytoplasmic signal transduction pathway resulting in proliferation. Lab. Invest. **78**(8), 987–1003 (1998)
66. Gerecht, S., Burdick, J.A., Ferreira, L.S., Townsend, S.A., Langer, R., Vunjak-Novakovic, G.: Hyaluronic acid hydrogel for controlled self-renewal and differentiation of human embryonic stem cells. Proc. Natl. Acad. Sci. USA **104**(27), 11298–11303 (2007)
67. Autissier, A., Letourneur, D., Visage, C.L.: Pullulan-based hydrogel for smooth muscle cell culture. J. Biomed. Mater. Res. Part A **82A**(2), 336–342 (2007)
68. Koch, S., Yao, C., Grieb, G., Prevel, P., Noah, E.M., Steffens, G.C.M.: Enhancing angiogenesis in collagen matrices by covalent incorporation of VEGF. J. Mater. Sci. Mater. Med. **17**(8), 735–741 (2006)
69. Matthews, J.A., Wnek, G.E., Simpson, D.G., Bowlin, G.L.: Electrospinning of collagen nanofibers. Biomacromolecules **3**(2), 232–238 (2002)
70. Matthews, J.A., Boland, E.D., Wnek, G.E., Simpson, D.G., Bowlin, G.L.: Electrospinning of collagen type II: a feasibility study. J. Bioact. Compat. Polym. **18**(2), 125–134 (2003).
71. Shaikh, F.M., Callanan, A., Kavanagh, E.G., Burke, P.E., Grace, P.A., McGloughlin, T.M.: Fibrin: a natural biodegradable scaffold in vascular tissue engineering. Cells Tissues Organs **188**(4), 333–346 (2008)
72. Jin, H.J., Chen, J.S., Karageorgiou, V., Altman, G.H., Kaplan, D.L.: Human bone marrow stromal cell responses on electrospun silk fibroin mats. Biomaterials **25**(6), 1039–1047 (2004)
73. Soffer, L., Wang, X.Y., Mang, X.H., Kluge, J., Dorfmann, L., Kaplan, D.L., Leisk, G.: Silk-based electrospun tubular scaffolds for tissue-engineered vascular grafts. J. Biomater. Sci. Polym. Ed. **19**(5), 653–664 (2008)
74. Jin, H.J., Fridrikh, S.V., Rutledge, G.C., Kaplan, D.L.: Electrospinning *Bombyx mori* silk with poly(ethylene oxide). Biomacromolecules **3**(6), 1233–1239 (2002)
75. Silva, E.A., Mooney, D.J.: Spatiotemporal control of vascular endothelial growth factor delivery from injectable hydrogels enhances angiogenesis. J. Thromb. Haem. **5**(3), 590–598 (2007)
76. Chenite, A., Chaput, C., Wang, D., Combes, C., Buschmann, M.D., Hoemann, C.D., Leroux, J.C., Atkinson, B.L., Binette, F., Selmani, A.: Novel injectable neutral solutions of chitosan form biodegradable gels in situ. Biomaterials **21**(21), 2155–2161 (2000)
77. Ishihara, M., Obara, K., Ishizuka, T., Fujita, M., Sato, M., Masuoka, K., Saitō, Y., Yura, H., Matsui, T., Hattori, H., Kikuchi, M., Kurita, A.: Controlled release of fibroblast growth factors and heparin from photocrosslinked chitosan hydrogels and subsequent effect on in vivo vascularization. J. Biomed. Mater. Res. Part A **64A**(3), 551–559 (2003)

78. Bettinger, C.J., Bruggeman, J.P., Borenstein, J.T., Langer, R.S.: Amino alcohol-based degradable poly(ester amide) elastomers. Biomaterials **29**(15), 2315–2325 (2008)
79. Yamanouchi, D., Wu, J., Lazar, A.N., Craig Kent, K., Chu, C.-C., Liu, B.: Biodegradable arginine-based poly(ester-amide)s as non-viral gene delivery reagents. Biomaterials **29**(22), 3269–3277 (2008)
80. Levenberg, S., Huang, N.F., Lavik, E., Rogers, A.B., Itskovitz-Eldor, J., Langer, R.: Differentiation of human embryonic stem cells on three-dimensional polymer scaffolds. Proc. Natl. Acad. Sci. USA **100**(22), 12741–12746 (2003)
81. Wang, G.J., Hsueh, C.C., Hsu, S.H., Hung, H.S.: Fabrication of PLGA microvessel scaffolds with circular microchannels using soft lithography. J. Micromech. Microeng. **17**(10), 2000–2005 (2007)
82. Ma, Z.W., He, W., Yong, T., Ramakrishna, S.: Grafting of gelatin on electrospun poly(caprolactone) nanofibers to improve endothelial cell spreading and proliferation and to control cell orientation. Tissue Eng. **28**(3), 1149–1158 (2005)
83. Williamson, M.R., Woollard, K.J., Griffiths, H.R., Coombes, A.G.A.: Gravity spun polycaprolactone fibers for applications in vascular tissue engineering: proliferation and function of human vascular endothelial cells. Tissue Eng. **12**(1), 45–51 (2006)
84. Pankajakshan, D., Kalliyana, K., Krishnan, L.K.: Vascular tissue generation in response to signaling molecules integrated with a novel poly(epsilon-caprolactone)-fibrin hybrid scaffold. J. Tissue Eng. Regen. Med. **1**(5), 389–397 (2007)
85. Sarkar, S., Lee, G.Y., Wong, J.Y., Desai, T.A.: Development and characterization of a porous micro-patterned scaffold for vascular tissue engineering applications. Biomaterials **27**(27), 4775–4782 (2006)
86. Dong, Y.X., Yong, T., Liao, S., Chan, C.K., Ramakrishna, S.: Long-term viability of coronary artery smooth muscle cells on poly(L-lactide-co-epsilon-caprolactone) nanofibrous scaffold indicates its potential for blood vessel tissue engineering. J. R. Soc. Interface **5**(26), 1109–1118 (2008)
87. Hemmrich, K., Salber, J., Meersch, M., Wiesemann, U., Gries, T., Pallua, N., Klee, D.: Three-dimensional nonwoven scaffolds from a novel biodegradable poly(ester amide) for tissue engineering applications. J. Mater. Sci. Mater. Med. **19**(1), 257–267 (2008)
88. Nijst, C.L.E., Bruggeman, J.P., Karp, J.M., Ferreira, L., Zumbuehl, A., Bettinger, C.J., Langer, R.: Synthesis and characterization of photocurable elastomers from poly(glycerol-co-sebacate). Biomacromolecules **8**(10), 3067–3073 (2007)
89. Gerecht, S., Townsend, S.A., Pressler, H., Zhu, H., Nijst, C.L.E., Bruggeman, J.P., Nichol, J.W., Langer, R.: A porous photocurable elastomer for cell encapsulation and culture. Biomaterials **28**(32), 4826–4835 (2007)
90. Yang, F., Cho, S.-W.C., Son, S.M., Bogatyrev, S.R., Singh, D., Green, J.J., Mei, Y., Park, S., Bhang, S.H., Kim, B.-S., Langer, R., Anderson, D.G.: Genetic engineering of human stem cells for enhanced angiogenesis using biodegradable polymeric nanoparticles. Proc. Natl. Acad. Sci. USA **107**(8) 3317–3322 (2009)
91. Richardson, T.P., Peters, M.C., Ennett, A.B., Mooney, D.J.: Polymeric system for dual growth factor delivery. Nat. Biotech. **19**(11), 1029–1034 (2001)
92. Zong, X., Bien, H., Chung, C.-Y., Yin, L., Fang, D., Hsiao, B.S., Chu, B., Entcheva, E.: Electrospun fine-textured scaffolds for heart tissue constructs. Biomaterials **26**(26), 5330–5338 (2005)
93. Xu, C.Y., Inai, R., Kotaki, M., Ramakrishna, S.: Aligned biodegradable nanofibrous structure: a potential scaffold for blood vessel engineering. Biomaterials **25**(5), 877–886 (2004)
94. Yoshimoto, H., Shin, Y.M., Terai, H., Vacanti, J.P.: A biodegradable nanofiber scaffold by electrospinning and its potential for bone tissue engineering. Biomaterials **24**(12), 2077–2082 (2003)
95. Shin, M., Yoshimoto, H., Vacanti, J.P.: In vivo bone tissue engineering using mesenchymal stem cells on a novel electrospun nanofibrous scaffold. Tissue Eng. **10**(1–2), 33–41 (2004)

96. Davis, M.E., Motion, J.P.M., Narmoneva, D.A., Takahashi, T., Hakuno, D., Kamm, R.D., Zhang, S., Lee, R.T.: Injectable self-assembling peptide nanofibers create intramyocardial microenvironments for endothelial cells. Circulation **111**(4), 442–450 (2005)
97. Ekaputra, A.K., Prestwich, G.D., Cool, S.M., Hutmacher, D.W.: Combining electrospun scaffolds with electrosprayed hydrogels leads to three-dimensional cellularization of hybrid constructs. Biomacromolecules **9**(8), 2097–2103 (2008)
98. Kashyap, N., Kumar, N., Kumar, M.N.V.R.: Hydrogels for pharmaceutical and biomedical applications. **22**(2), 107–150 (2005)
99. Jagur-Grodzinski, J.: Polymeric gels and hydrogels for biomedical and pharmaceutical applications. Polym. Adv. Technol. **9999**(9999), n/a (2009)
100. Nicodemus, G.D., Bryant, S.J.: Cell encapsulation in biodegradable hydrogels for tissue engineering applications. Tissue Eng. Part B Rev. **14**(2), 149–165 (2008)
101. Seliktar, D., Black, R.A., Vito, R.P., Nerem, R.M.: Dynamic mechanical conditioning of collagen-gel blood vessel constructs induces remodeling in vitro. Ann. Biomed. Eng. **28**(4), 351–362 (2000)
102. Yamamoto, M., Ikada, Y., Tabata, Y.: Controlled release of growth factors based on biodegradation of gelatin hydrogel. J. Biomater. Sci. Polym. Ed. **12**(1), 77–88 (2001)
103. Kraehenbuehl, T.P., Ferreira, L.S., Zammaretti, P., Hubbell, J.A., Langer, R.: Cell-responsive hydrogel for encapsulation of vascular cells. Biomaterials **30**(26), 4318–4324 (2009)
104. Subbiah, T., Bhat, G.S., Tock, R.W., Parameswaran, S., Ramkumar, S.S.: Electrospinning of nanofibers. J. Appl. Polym. Sci. **96**(2), 557–569 (2005)
105. Kidoaki, S., Kwon, I.K., Matsuda, T.: Mesoscopic spatial designs of nano- and microfiber meshes for tissue-engineering matrix and scaffold based on newly devised multilayering and mixing electrospinning techniques. Biomaterials **26**(1), 37–46 (2005)
106. Yang, F., Murugan, R., Wang, S., Ramakrishna, S.: Electrospinning of nano/micro scale poly(L-lactic acid) aligned fibers and their potential in neural tissue engineering. Biomaterials **26**(15), 2603–2610 (2005)
107. Stankus, J.J., Soletti, L., Fujimoto, K., Hong, Y., Vorp, D.A., Wagner, W.R.: Fabrication of cell microintegrated blood vessel constructs through electrohydrodynamic atomization. Biomaterials **28**(17), 2738–2746 (2007)
108. Jeong, S.I., Kim, S.Y., Cho, S.K., Chong, M.S., Kim, K.S., Kim, H., Lee, S.B., Lee, Y.M.: Tissue-engineered vascular grafts composed of marine collagen and PLGA fibers using pulsatile perfusion bioreactors. Biomaterials **28**(6), 1115–1122 (2007)
109. Lee, S.J., Yoo, J.J., Lim, G.J., Atala, A., Stitze, J.: In vitro evaluation of electrospun nanofiber scaffolds for vascular graft application. J. Biomed. Mater. Res. A **83A**(4), 999–1008 (2007)
110. Qi, H.X., Kong, S.L., Hu, P., Zhang, J., Lin, W.J., Song, F.: In: Kim, H.S., Li,Y.B., Lee, S.W. (eds.) Biomimic Tubulose Scaffolds with Multilayer Prepared by Electrospinning for Tissue Engineering, pp. 882–885. Trans Tech, Zurich (2006)
111. Pham, Q.P., Sharma, U., Mikos, A.G.: Electrospun poly(ε-caprolactone) microfiber and multilayer nanofiber/microfiber scaffolds: characterization of scaffolds and measurement of cellular infiltration. Biomacromolecules **7**(10), 2796–2805 (2006)
112. Zhang, D., Chang, J.: Electrospinning of three-dimensional nanofibrous tubes with controllable architectures. Nano Lett. **8**(10), 3283–3287 (2008)
113. Park, T.G., Hoffman, A.S.: Synthesis and characterization of pH- and or temperature-sensitive hydrogels. J. Appl. Polym. Sci. **46**(4), 659–671 (1992)
114. Sun, G.M., Zhang, X.Z., Chu, C.C.: Formulation and characterization of chitosan-based hydrogel films having both temperature and pH sensitivity. J. Mater. Sci. Mater. Med. **18**(8), 1563–1577 (2007)
115. Klumb, L.A., Horbett, T.A.: Design of insulin delivery devices based on glucose sensitive membranes. J. Contr. Rel. **18**(1), 59–80 (1992)
116. Moroni, L., De Wijn, J.R., Van Blitterswijk, C.A.: Integrating novel technologies to fabricate smart scaffolds. J. Biomater. Sci. Polym. Ed. **19**(5), 543–572 (2008)

117. Sittinger, M., Hutmacher, D.W., Risbud, M.V.: Current strategies for cell delivery in cartilage and bone regeneration. Curr. Opin. Biotechnol. **15**(5), 411–418 (2004)
118. Kloxin, A.M., Kasko, A.M., Salinas, C.N., Anseth, K.S.: Photodegradable hydrogels for dynamic tuning of physical and chemical properties. Science **324**(5923), 59–63 (2009)
119. Khetan, S., Katz, J.S., Burdick, J.A.: Sequential crosslinking to control cellular spreading in 3-dimensional hydrogels. Soft Matter **5**(8), 1601–1606 (2009)
120. Deutsch, J., Motlagh, D., Russell, B., Desai, T.A.: Fabrication of microtextured membranes for cardiac myocyte attachment and orientation. J. Biomed. Mater. Res. **53**(3), 267–275 (2000)
121. Timpl, R.: Macromolecular organization of basement membranes. Curr. Opin. Cell Biol. **8**(5), 618–624 (1996)
122. Abrams, G.A., Schaus, S.S., Goodman, S.L., Nealey, P.F., Murphy, C.J.: Nanoscale topography of the corneal epithelial basement membrane and Descemet's membrane of the human. Cornea **19**(1), 57–64 (2000)
123. Watt, F.M., Jordan, P.W., O'Neill, C.H.: Cell shape controls terminal differentiation of human epidermal keratinocytes. Proc. Natl. Acad. Sci. USA **85**(15), 5576–5580 (1988)
124. McBeath, R., Pirone, D.M., Nelson, C.M., Bhadriraju, K., Chen, C.S.: Cell shape, cytoskeletal tension, and RhoA regulate stem cell lineage commitment. Dev. Cell **6**(4), 483–495 (2004)
125. Yim, E.K., Reano, R.M., Pang, S.W., Yee, A.F., Chen, C.S., Leong, K.W.: Nanopattern-induced changes in morphology and motility of smooth muscle cells. Biomaterials **26**(26), 5405–5413 (2005)
126. Pasqui, D., Rossi, A., Barbucci, R., Lamponi, S., Gerli, R., Weber, E.: Hyaluronan and sulphated hyaluronan micropatterns: effect of chemical and topographic cues on lymphatic endothelial cell alignment and proliferation. Lymphology **38**(2), 50–65 (2005)
127. Rossi, A., Pasqui, D., Barbucci, R., Gerli, R., Weber, E.: The topography of microstructured surfaces differently affects fibrillin deposition by blood and lymphatic endothelial cells in culture. Tissue Eng. Part A **15**(3), 525–533 (2009)
128. Bettinger, C.J., Zhang, Z., Gerecht, S., Borenstein, J.T., Langer, R.: Enhancement of in vitro capillary tube formation by substrate nanotopography. Adv. Mater. **20**(1), 99–103 (2008)
129. Kim, B.S., Nikolovski, J., Bonadio, J., Mooney, D.J.: Cyclic mechanical strain regulates the development of engineered smooth muscle tissue. Nat. Biotechnol. **17**(10), 979–983 (1999)
130. Nerem, R.M.: Vascular fluid mechanics, the arterial wall, and atherosclerosis. J. Biomech. Eng. **114**(3), 274–282 (1992)
131. Mo, M., Eskin, S.G., Schilling, W.P.: Flow-induced changes in Ca^{2+} signaling of vascular endothelial cells: effect of shear–stress and ATP. Am. J. Physiol. **260**(5), H1698-H1707 (1991)
132. De Keulenaer, G.W., Chappell, D.C., Ishizaka, N., Nerem, R.M., Alexander, R.W., Griendling, K.K.: Oscillatory and steady laminar shear stress differentially affect human endothelial redox state—role of a superoxide-producing NADH oxidase. Circ. Res. **82**(10), 1094–1101 (1998)
133. Shyy, J.Y., Chien, S.: Role of integrins in endothelial mechanosensing of shear stress. Circ. Res. **91**(9), 769–775 (2002)
134. Stone, P.H., Coskun, A.U., Kinlay, S., Clark, M.E., Sonka, M., Wahle, A., Ilegbusi, O.J., Yeghiazarians, Y., Popma, J.J., Orav, J., Kuntz, R.E., Feldman, C.L.: Effect of endothelial shear stress on the progression of coronary artery disease, vascular remodeling, and in-stent restenosis in humans—in vivo 6-month follow-up study. Circulation **108**(4), 438–444 (2003)
135. Huang, H., Nakayama, Y., Qin, K., Yamamoto, K., Ando, J., Yamashita, J., Itoh, H., Kanda, K., Yaku, H., Okamoto, Y., Nemoto, Y.: Differentiation from embryonic stem cells to vascular wall cells under in vitro pulsatile flow loading. J. Art. Org. **8**(2), 110–118 (2005)
136. Wang, H., Riha, G.M., Yan, S., Li, M., Chai, H., Yang, H., Yao, Q., Chen, C.: Shear stress induces endothelial differentiation from a murine embryonic mesenchymal progenitor cell line. Arterioscler. Thromb. Vasc. Biol. **25**(9), 1817–1823 (2005)

137. Illi, B., Scopece, A., Nanni, S., Farsetti, A., Morgante, L., Biglioli, P., Capogrossi, M.C., Gaetano, C.: Epigenetic histone modification and cardiovascular lineage programming in mouse embryonic stem cells exposed to laminar shear stress. Circ. Res. **96**(5), 501–508 (2005)
138. Zeng, L.F., Xiao, Q.Z., Margariti, A., Zhang, Z.Y., Zampetaki, A., Patel, S., Capogrossi, M.C., Hu, Y.H., Xu, Q.B.: HDAC3 is crucial in shear- and VEGF-induced stem cell differentiation toward endothelial cells. J. Cell Biol. **174**(7), 1059–1069 (2006)
139. Metallo, C.M., Vodyanik, M.A., Pablo, J.J.D., Slukvin, I.I., Palecek, S.P.: The response of human embryonic stem cell-derived endothelial cells to shear stress. Biotechnol. Bioeng. **100**(4), 830–837 (2008)
140. Ye, C., Bai, L., Yan, Z.Q., Wang, Y.H., Jiang, Z.L.: Shear stress and vascular smooth muscle cells promote endothelial differentiation of endothelial progenitor cells via activation of Akt. Clin. Biomech.**23**, S118-S124 (2008)
141. Kasper, G., Dankert, N., Tuischer, J., Hoeft, M., Gaber, T., Glaeser, J.D., Zander, D., Tschirschmann, M., Thompson, M., Matziolis, G., Duda, G.N.: Mesenchymal stem cells regulate angiogenesis according to their mechanical environment. Stem Cells **25**(4), 903–910 (2007)
142. Gong, Z.D., Niklason, L.E.: Small-diameter human vessel wall engineered from bone marrow-derived mesenchymal stem cells (hMSCs). FASEB J. **22**(6), 1635–1648 (2008)
143. Wu, X., Rabkin-Aikawa, E., Guleserian, K.J., Perry, T.E., Masuda, Y., Sutherland, F.W.H., Schoen, F.J., Mayer, J.E., Jr, Bischoff, J.: Tissue-engineered microvessels on three-dimensional biodegradable scaffolds using human endothelial progenitor cells. Am. J. Physiol. Heart Circ. Physiol. **287**(2), H480–H487 (2004)
144. Hiraoka, N., Allen, E., Apel, I.J., Gyetko, M.R., Weiss, S.J.: Matrix metalloproteinases regulate neovascularization by acting as pericellular fibrinolysins. **95**(3), 365–377 (1998)
145. Cao, L., Arany, P.R., Wang, Y.S., Mooney, D.J.: Promoting angiogenesis via manipulation of VEGF responsiveness with notch signaling. Biomaterials **30**(25), 4085–4093 (2009)

Synthetic Niches for Stem Cell Differentiation into T cells

Ankur Singh and Krishnendu Roy

Abstract T cell development from hematopoietic stem cells takes place in the thymus under precisely controlled intercellular signaling between the stem cells and thymic stromal and epithelial cells. In vitro or ex vivo development of mature T cells from stem cells faces two primary hurdles; one being the inability of culture conditions to provide a three dimensional thymic niche with lineage-specific signaling like notch, and the second being the need for efficient positive or negative selection processes to achieve antigen-specific, functional T cells. Recent research has focused on development of 2D and 3D niches mimicking the thymic microenvironment by first identifying soluble or immobilized factors essential for T cell differentiation followed by presenting them efficiently to stem and progenitor cells. This chapter discusses how T cells are generated in the thymus and current approaches, including biomaterial-based strategies, towards ex vivo or in vitro differentiation of stem cells into T cells using signaling molecules.

Abbreviations

BMP	Bone morphogenic protein
CCL21	Chemokine (C–C motif) ligand 21
CCR	CC chemokine receptors
CMV	Cytomegalovirus
CTL	Cytotoxic T lymphocytes
CXCL	Chemokine (C–X–C motif) ligand
DL1, DLL1, Delta1	Delta-like ligand 1
DL4, DLL4, Delta4	Delta-like ligand 4
DN	Double negative

A. Singh and K. Roy (✉)
Department of Biomedical Engineering, The University of Texas at Austin, C0800, 1 University Station, Austin TX, 78712, USA
e-mail: kroy@mail.utexas.edu

Stud Mechanobiol Tissue Eng Biomater (2010) 2: 225–245
DOI: 10.1007/8415_2010_11

Published online: 12 June 2010

DP	Double positive
FTOC	Fetal thymic organ cultures
GMCSF	Granulocyte–macrophage colony-stimulating factor
HSC	Hematopoietic stem cell
HPC	Hematopoietic progenitor cell
HES	Hairy enhancer of split
ICAM	Intercellular adhesion molecule
ICC	3D inverted colloidal crystal
I-Delta	Immobilized Delta ligand
IL-7	Interleukin-7
LCMV	Lymphocytic choriomeningitis virus
MART	Melanoma antigen recognized by T-cells
MHC	Major histocompatibility complex
NI-Delta	Non-immobilized Delta ligand
NY-ESO-1	Cancer-testis antigen
RAG	Recombination activating genes
RTOC	Reaggregate thymic organ cultures
SCA	Stem cell antigen
SCF	Stem cell factor
SHH	Sonic hedgehog proteins
TCR	T cell receptor
VCAM	Vascular cell adhesion molecule

1 Introduction

Generation, maintenance or replacement of a healthy, functional T cell population remains a critical aspect of immunocompetence in patients with AIDS, bone marrow transplants, immunodeficiency disorders like Di George's Syndrome, as well as cancer patients suffering T cell depletion with repetitive exposure to tumor suppressive therapies [1–3]. Transfusion of T lymphocytes, referred to as adoptive T cell therapy, has been extensively studied for immunotherapy against cancer and infectious diseases. However, even after several decades of research, T lineage development and functional priming has remained one of the most obscure and difficult tasks to accomplish. High throughput generation of functional T cells outside human or animal body has not been achieved yet. This is primarily because we have not been able to completely disentangle the complex network of molecular signals, cues and niche requirements for stepwise T cell regeneration from stem and progenitor cells. Progenitors in the thymus are inefficient in maintaining long term T cell development. They need replacements from the bone marrow and several preclinical studies have showed that intrathymic or intravenous transfusion of thymocytes do not result in desired long term self renewal of

T cells. It is only when bone marrow is transferred intravenously (and not intra-thymically) that long term self renewal is observed [4]. Understanding the complete T cell reconstitution and the biomolecular phenomenon directly in the body is intricate and can only be accomplished through killing the embryo [5], thus designing comparatively simpler in vitro system mimicking thymic microenvironments may help in understanding the underlying biology and identifying factors necessary for controlled, on-demand, therapeutic T cell regeneration.

Advances in cell transplantation strategies have enabled ex vivo generation and expansion of tumor antigen specific cytotoxic T lymphocytes (CTL) from patient-derived blood cells [6]. Although manipulation of ex vivo conditions to completely mimic in vivo environment may not be practical, ex vivo manipulation can offer superior specificity, magnitude, and phenotypic alterations than conventional in vivo targeted therapy. Such ex vivo generation of patient-specific T lymphocytes followed by transplantation, a process known as adoptive T cell therapy, has its own constrains like limited availability of donor cells suitable for collection, efficiency and time lag involved with cell expansion, manipulation and transfer, etc. [7]. Thus, there exists a need to develop in vitro systems that can efficiently generate functional T cells, starting from a potentially unlimited cell source, stem cells.

In recent years, considerable efforts have been made to understand and direct the differentiation of T cells from stem and progenitor cells, using in vitro or ex vivo systems. The self renewal ability of stem cells and their controlled differentiation into a specific lineage, in vivo, is governed by the tissue specific stem cell niche [8]. Explicit understanding of how the progenitors expand and differentiate in T cell niches can greatly facilitate the design and development of strategies to generate T cells with desired characteristics, in vitro. Recent studies have identified a series of complex molecular interactions taking place in the three dimensional thymic microarchitecture during T cell development. This chapter focuses on discussing the T cell development niche and use of signaling biomolecules, with or without biomaterials for generation of T Lymphocytes.

2 The T Cell Niche

Mammalian T cell development primarily happens in the thymus through differentiation of bone marrow hematopoietic stem cell (HSC)-derived progenitors. Relatively few molecular cues have been identified that facilitates rapid proliferation and sequential differentiation of these HSC-derived T cell progenitors. Although there could be several progenitor cell types that have the ability to enter T cell lymphopoiesis, classically T cell precursors are distinguished by their expression of thymic chemokine receptors, thymic growth factor receptors, the stem cell markers c-Kit, and stem-cell antigen 1 (Sca1) [9]. c-Kit binds to stem-cell factor (SCF), a cytokine that contributes to the self renewal and maintenance of HSCs in vivo, and also promote adhesion of progenitors with stromal cells to keep the differentiating cells in the niche [10].

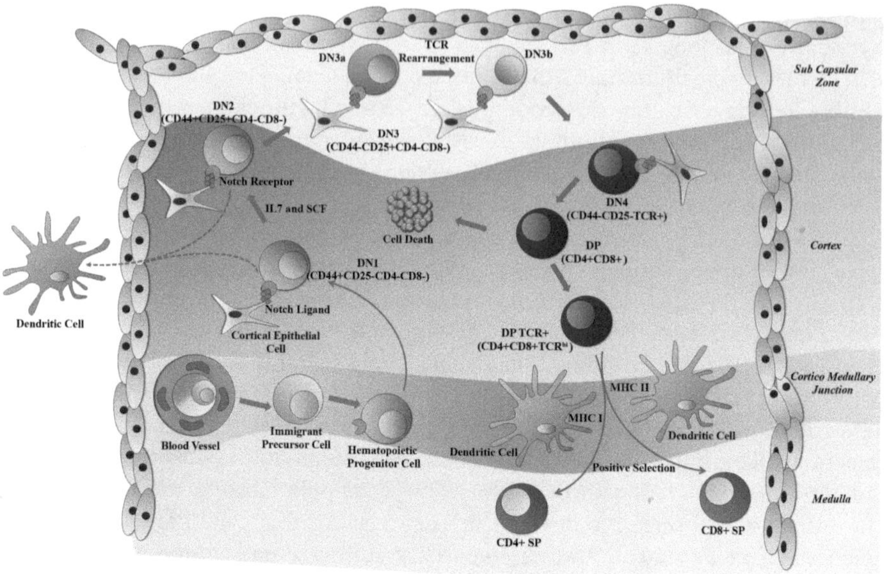

Fig. 1 Thymic microacrchitecture and developmental stages for T cells (modified with permission from [9, 11, 12])

The thymic microarchitecture is compartmentalized into medullary and cortical sections (Fig. 1), each having stromal cells that secrete specific cytokines, growth factors and provide inter-cellular signaling. The most notable among these stromal cell-stem cell signal is the Notch–Delta-like ligand interaction [11]. Lymphoid precursors migrate from the bone marrow, enter the cortico-medullary junction of the thymus through blood vessels and then migrate into the cortex and other distinct subdivisions of the thymic architecture. These cells then undergo T-cell lineage commitment and finally get converted into $CD4^+$ single positive (SP) and $CD8^+$ SP cells in the thymic medulla [9, 11]. Rearrangement at the T cell receptor (TCR) gene locus is a hallmark of T cell lineage commitment however the progression through "pro T cell stages" i.e. $CD4^-CD8^-$ double negative (DN) stages (1 through 3) is independent of TCR rearrangement (Fig. 1). The different DN stages can further be marked by expression of CD25 and CD44 surface markers, as summarized in Table 1 and Fig. 1. The proliferation of T cell progenitors occur rapidly through DN1 to DN3 stages setting the T lineage identity, thereafter pausing between DN3a and DN3b for extensive TCR rearrangement. The proliferation continues thereafter till double positive ($CD4^+$ $CD8^+$, DP) cells are generated. It should however be noted that at the early "pro-T cell stages", DN1 and DN2 stage cells can also differentiate into dendritic cells, macrophages and natural killer cells under appropriate environment [9] (Fig. 1).

At the DN3 stage, complete V(D)J rearrangement takes place followed by β- or γδ-TCR selection thus generating DN3b stage cells. Cells at DN3 stage prior to TCR selection are referred to as DN3a cells. The $TCR\beta^+$ cells, if successful,

Table 1 Phenotypic markers of murine early T cell lineage precursors (Adapted with permission from [9])

	ETP	DN2	DN3a	DN3b	DN4
c-Kit	High	High-mid	Low	Low	Low
CD44	High	High	Low	Low	Low
CD25	Neg	High	High	High-mid	Low
Proliferation	+	+	−	+	+

express TCRα chain and the pre TCR complex followed by expression of CD4 and CD8 surface markers, as the progenitors essentially progress from DN thymocytes to the DP stage [9, 11, 12].

At the next stage, the DP thymocytes undergo positive selection through their interactions with self-peptide loaded major histocompatibility complex (MHC) molecules on cortical epithelial cells [9]. Autoreactive thymocytes expressing TCR with high affinity for self antigens undergo negative selection and are destroyed at this stage while the surviving cells undergo positive selection based on their recognition ability towards MHC I and MHC II expressed on dendritic or medullary epithelial cells thereby differentiating into $CD8^+$ and $CD4^+$ mature SP T cells, respectively (Fig. 1).

2.1 T Cell Receptor Gene Rearrangement

The antigen specificity of T cells is determined solely by the phenotypic characteristics of the heterodimeric TCR $\alpha\beta$ complex. Both α and β chains of the TCR are comprised of a variable (V) and a constant (C) region. The T cell receptor α chain consists of variable (V_α) and junctional gene (J_α), while β chain is composed of V_β, J_β, and diversity (D) gene segments (Fig. 2) [13]. The β chain rearrangement takes place first with rearrangement of D gene segments into J_β, and subsequent rearrangement of V_β to DJ_β. Thymocytes that fail to develop β chains successfully, can either die or make rearrangements at γ and δ loci. After successful rearrangement of β chain, it combines with a surrogate α chain, pTα, and forms a β:pTα heterodimer followed by a complexation with CD3 molecules [14, 15].

The β gene rearrangement stops at this stage by deactivation of recombination activating gene (RAG)-2 proteins present at the end of VDJ genes [14, 15]. The cells then differentiate into DN4 stage and resume rapid proliferation that was halted at the DN3 stage. Consequently, a large numbers CD4 and CD8 expressing cells are generated and rearrangement of α genes occurs for successful development of α chains leading to DP TCR^+ cells.

2.2 T Cell Microenvironment

The thymic microenvironment is a complex milieu of progenitors, supportive stromal cells, signaling interactions like Delta and Notch, cells undergoing clonal

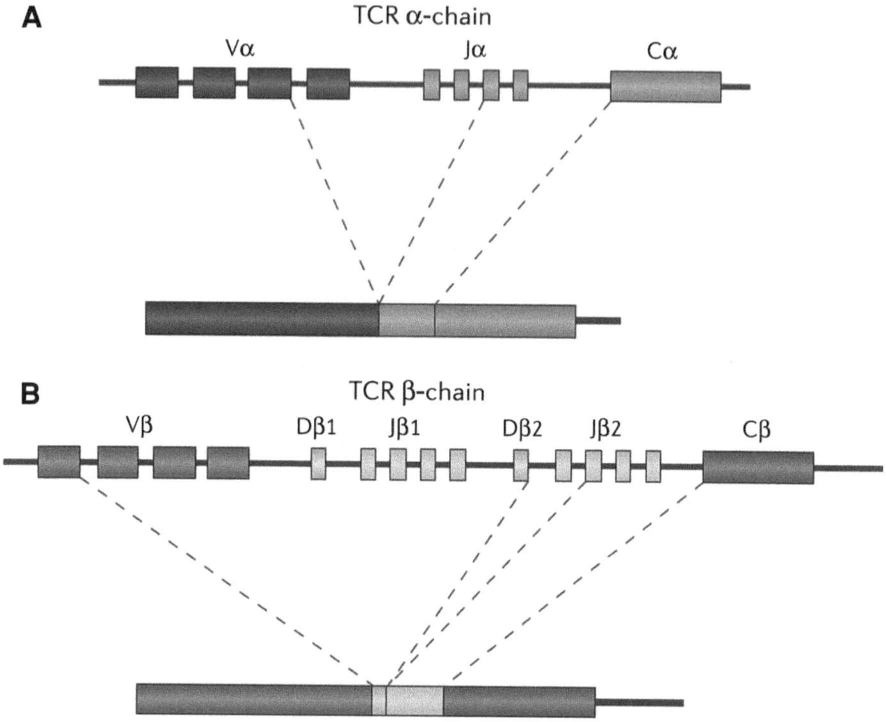

Fig. 2 T-cell receptor (*TCR*) gene rearrangement. Functional TCRs are heterodimers consisting of an α-chain and a β-chain that are generated by somatic gene recombination of variable (V), diversity (D) and junctional (J) gene segments for the β-chain, and V and J gene segments for the α-chain. In the human TCR loci, there are 42Vβ, 2Dβ, 12Jβ, 43Vα and 58Jα functional gene segments. In the mouse TCR loci, there are 35Vβ, 2Dβ, 12Jβ, 71Vα and 51Jα functional gene segments. During T-cell development, gene segments recombine and are spliced together with the constant region (**c**) to form the functional αβTCR, with each T cell expressing only one type of recombined receptor complex (adapted with permission from [13])

selection through MHC–TCR interactions, as well as abundant adhesion, expansion and homing molecules present as extracellular matrix. The entry of lymphoid precursors into the cortico-medullary junction of the thymus is facilitated by chemokines likes Chemokine (C–C motif) ligand-21 (referred as CCL21) and adhesion molecules like laminin and P-selectin. During the developmental stages DN1 and 2, cytokines like IL-7 promote survival and rapid proliferation, while the *RAG* gene helps in TCR rearrangements. Important signaling components include Notch (through the stromal cell ligands Delta-1 or 4) [16–18], bone morphogenic protein family (BMP2/4) [19, 20], Kit ligand [21] and the hedgehog protein family [22]. The sonic hedgehog proteins (SHH) produced by thymic epithelial cells is an essential morphogen and signals the proliferation and maturation through all DNs and DP stages as well as plays role in TCR-repertoire selection at the DP to SP stage [22]. Hedgehog signaling is still in its exploratory stage and not much work has been performed in using these for in vitro generation of T cells. Notch signaling on the other hand, has been extensively studied and is discussed in the following sections of this chapter.

Chemotactic migration of early, immature T cells during DN2 and DN3 stages is provided with the help of chemokines like CXCL12, CCL25 and adhesion molecules like Vascular cell adhesion molecule (VCAM) and laminin. Following the DN to DP differentiation, DP thymocytes undergo positive selection by MHCs [23] and self peptide recognition through their interaction with cortical epithelial cells. This is further mediated by intercellular adhesion molecule (ICAM-1) and IL-7 and T cell co-stimulatory molecules. Finally the migration from cortex to the medullary region of thymus is facilitated by chemotactic signals, CCL19 and CCL21, which bind to the chemokine receptor CCR7.

The notch pathway has attracted great research interest over the past decade and is one of the most extensively characterized signaling pathways for understanding T cell development as well as for in vitro differentiation of progenitors into early T cells. Active notch is involved in several developmental lineage decisions, e.g. during hematopoiesis and lymphopoiesis, in self renewal of stem cells as well as in fetal and postnatal development [24–26]. Notch signaling consists of an interplay of four Notch receptors, namely Notch 1–4 and their ligands Delta-1, 2, 3, 4 and Jagged 1, 2 [24]. Notch signaling results in activation of several genes including Hairy enhancer of split (Hes) family (*Hes1* and *Hes5*) [24, 27, 28]. As discussed later in the chapter, exposure to notch ligands Jagged and Delta can promote HSC maintenance, differentiation and self renewal. However, a comparative study with stromal cells transfected to express Jagged and delta showed that T cell development and maturation was dependant on Delta ligands and not Jagged [29].

Notch receptors and ligands are present both in bone marrow and thymus. However cells committed to the B cell lineage have notch signaling inactivated by the B lineage commitment factor Pax5. On the other hand, for T cell development it is essential to have notch in the "ON" state. Of all the Notch ligands, Delta 1 has been characterized extensively and thymic stromal cells expressing these ligands have been able to induce ex vivo and in vitro T cell differentiation [24].

3 T Cell Differentiation Through Co-culture

Early studies for ex vivo generation of T cells heavily relied on using fetal thymic organ cultures (FTOCs) and re-aggregated thymic organ cultures (RTOCs) [23, 30, 31]. FTOCs, initially developed by Owen and Jenkinson, were essentially fetal thymic lobes devoid of thymocytes thus serving as a native microarchitecture and niche for T cell precursor differentiation [30]. These systems have been extensively studied for generating CD4 and CD8 SP cells from human bone marrow and cord blood cells [32]. These studies provided useful insight into T cell differentiation mechanisms and capabilities of various progenitors to undergo differentiation under the influence of cytokines and other factors [33, 34]. Despite initial success these thymic systems suffer from disadvantages including inefficient differentiation, complicated isolation process and maintenance of distinct signals [30]. Nevertheless, FTOC and RTOC are still in use for direct cell differentiation or for adoptive transfer of ex vivo/in vitro differentiated T cell progenitors (discussed later in the chapter). Recent advancements have focused on obtaining T cells or progenitors from differentiation of embryonic stem cells (ESC or ES), hematopoietic Stem cells (HSCs), and cord blood cells using stromal cell co-culture; cytokine cocktails of stem cell factors, interleukins (IL-6, IL-7, IL-11), c-kit ligand, flt3 ligand and direct influence of thymic niche components like Notch ligands.

In absence of molecular cues, embryonic stem cells (ESC or ES) differentiate in vitro into clusters of cells called embryoid bodies [35, 36]. These cystic structures are composed of germ layer-derived hematopoietic progenitors that have the potential to differentiate into erythromyeloid and blood cell lineages in vitro, but have failed in in vivo translation [37–39]. Further, it has been extremely difficult to generate functional T cells from ES cells in vitro. There is emerging evidence that hematopoietic stem cell precursors derived from bone marrow can only commit to T cells lineage once they are in the thymus under the influence of Notch signals [36, 40]. One of the earliest reports of Notch signaling mediated HSC differentiation was through retroviral-mediated expression of intracellular domain of Notch-1 (N1Ic) in $lin^{-}sca\text{-}1^{+}c\text{-}kit^{+}$ mouse progenitors was from the work of Varnum-Finney et al. [41], as discussed later in the chapter.

Honjo et al. [42] employed an in vitro ES cell differentiation system that utilizes the macrophage colony-stimulating factor (M-CSF)-deficient stromal cell line, OP9, to specifically induce lymphopoiesis and early erythro-myeloid lineages. Adult hematopoiesis occurs within the bone marrow and marrow stromal cells secrete soluble factors necessary for hematopoiesis. However, the OP9 coculture system is only able to generate B cells, and not T cells [37, 42]. Other researchers demonstrated in vitro T-cell differentiation from ES cells but only from early stages of ES cell differentiation (developmental prehematopoietic precursors, $Flk1^{+}CD45^{-}$ subset) and required transfer of this prehematopoietic subset into reaggregate thymic organ cultures (RTOCs) [43]. A three-dimensional microarchitecture of the thymus is essential for such lineage differentiation since thymic stromal cell monolayer cultures (TSMC) have failed to support differentiation of T cells [44]. However this complete thymus niche may not be an absolute necessity for differentiation of ES cells into T cell lineage and there is a need for a simpler system to overcome the cumbersome arrangements involving fresh thymus isolation and mixed coculture methods.

Recent advancements have shown the ability to induce T cell differentiation from HSC progenitors and ESCs using the Notch-OP9 based system, where OP9 cells are retrovirally transfected with Notch ligand Delta-1 at the 5′ of the internal ribosomal entry site [36, 45]. Zúñiga-Pflücker et al. showed that co-culture of ESCs on OP9-DL1 cells efficiently direct the differentiation of ESCs towards T cell lineage, in vitro. ESC line R1, which are expanded on murine embryonic fibroblasts (MEFs), when co-cultured with OP9-DL1 cells (Fig. 3a), engage the

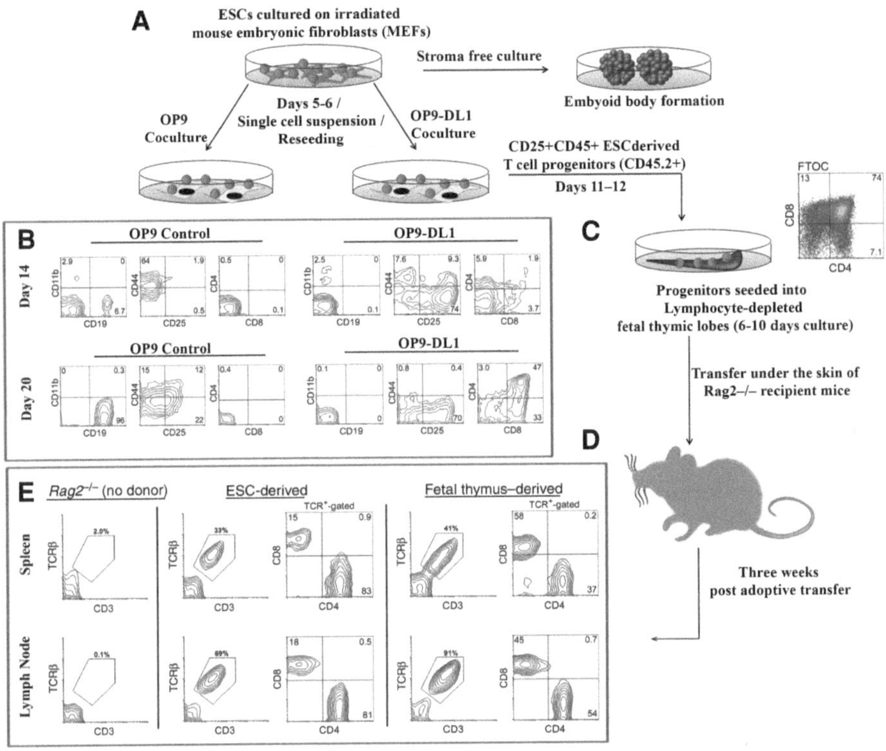

Fig. 3 T cell development in vitro and in vivo: **a** ESCs expanded on MEFs were induced to differentiate by culture on OP9-control or OP9-DL1 cells. Nonadherent hematopoietic cells were collected on days 8, 14 and 20 of culture and analyzed by flow cytometry (Day 8 data not included in this figure, see Schmitt et al. [36]). **b** Myeloid progenitors were characterized by CD11b expression; B cell progenitors were identified by CD19 expression (from OP9-control cultures); early CD4 CD8 DN T cell progenitors were distinguished by high CD25 (IL-2Rα) expression (from OP9-DL1 cultures) and at later time points, CD4 CD8 DP T cells were evident. Data are representative of three independent experiments. Numbers in quadrants indicate percentages of each population. **c** ESC-derived T cell progenitors can reconstitute FTOC. $CD45.2^{+}$ $CD25^{+}$ ESC-derived T cell progenitors (1 × 1E5) were seeded into CD45.1 congenic, deoxyguanosine-treated fetal thymic lobes and were cultured for 10 days. Only T cells of donor origin (CD45.2) are found. *Numbers* in quadrants indicate percentages of each population. **d** ESC-derived T cell progenitors can reconstitute an immune-deficient host. **e** ESC-derived T cell progenitors were seeded into fetal thymic lobes from day 14 from CD45 congenic B6 mice and were cultured in FTOC. After 5 days, the reconstituted fetal thymic lobes or nondepleted fetal thymic lobes were implanted under the skin of RAG2-deficient mice. After 3 weeks, the spleens and lymph nodes of recipient mice (*center* and *right*) as well as those of an unmanipulated $Rag2^{-/-}$ mouse (*left*) were analyzed by flow cytometry for TCR+ donor-derived T cells. *Numbers* in quadrants and near outlined areas indicate percentages of each population (quantitative data in **b**, **c**, **e** adapted with permission from Schmitt et al. [36], pictorial schematics reconstructed from description from Schmitt et al. [36])

Notch ligand Delta-1 on OP9-DL1 cells through notch receptor and start differentiating into early T cell precursors marked with CD25 and/or CD44 antigens as early as day 14 (Fig. 3b). Control cells (cultured on OP9 cells without notch ligands) failed to show CD25 and/or CD44 antigens, markers of conventional T cell (Fig. 3b). Interestingly, the ESC-OP9-DL1 coculture resulted in high $CD4^+$ $CD8^+$ (double positive, DP) cells indicating the efficacy of OP9 presented notch signaling for T cell differentiation (Fig. 3b, c). Further, TCR rearrangement was observed in ESC derived T cells. These ESC derived T cell precursors were capable of differentiating into mature T cells when seeded into FTOCs as $CD25^+$ DN T cell precursors reconstituted into $CD4^+$ $CD8^+$ DP cells (Fig. 3c). Finally when implanted under the skin of *Rag2*$^{-/-}$ mice, these DP cell reconstituted thymic lobes resulted in TCRβ+ donor-derived T cells as indicated by TCRβ population in ESC derived T cells at levels comparable to fetal thymus derived T cells (Fig. 3d, e).

Significant developments have been reported in differentiation of T cells from human embryonic stem cells (hESCs). Genetically manipulated hESCs were used to generate T cells by first coculturing on OP9 cells, in vitro, and then transferring these cells into human thymic tissues of immunodeficient mice, in vivo, thus resulting in functional T cells [46]. Earlier reports have also shown generation of DP αβ T cells within 35 days from human bone marrow $CD34^+$ progenitor cells cocultured with OP9-DL1 [28]. This study did not include thymic organ culture. Other studies include retroviral transduction of hHSCs with TCR gene and co-culturing hHSCs with OP9-DL1 for extrathymic differentiation into T cells with tumor antigen-specific TCR [47].

Despite these advancements the far reaching impact of OP9-DL1 system has been limited by (a) inability of OP9 cells to express MHCII on the surface and thus participate in positive selection of CD4 SP T cells and (b) non-specific genetic manipulation and proto-oncogene activation while retrovirally transfecting stromal cells for notch expression cannot be ruled out. In addition, the effectiveness of such stromal cell co-culture using human stroma and stem cells are yet to be validated.

4 T Cell Differentiation Through Immobilization of Notch Ligands

The rate of T cell reconstitution in vivo, following stem cell transplantation, is directly dependent on the number of infused stem cells and their degree of pre-differentiation. T cell committed progenitors result in faster reconstitution as compared to non-committed lymphoid progenitors [48–50]. Therefore, methods to differentiate stem cells into early T-lineage committed cells or matured, functional T cells prior to transplantation could significantly improve patient care, both for stem cell therapy as well as for adoptive immunotherapy.

Notch signaling plays a significant physiological role in T cell versus B cell fate during lymphoid differentiation [51]. While presentation of notch through co-culture with transduced stromal cells result in differentiation of HSCs and ESCs into T cell lineage, the possibility of interference of factors secreted by notch presenting cells and contamination of retrovirally transfected cells in the final population, cannot be ruled out. Thus cell-free notch ligand presentation has been explored as an alternative strategy. One of the earliest attempts to develop a stromal cell-free system for Notch signaling mediated cellular differentiation was by Varnum-Finney et al. [52] Monomeric and dimeric forms of Notch ligand Delta-1 resulted in inhibition of C2C12 myoblast differentiation and HES-1 transactivation in U20S cells only when the Delta-1 was immobilized on plastic surface as compared to soluble forms.

4.1 T Cell Differentiation Through Plate Immobilization

Cocultures of hematopoietic precursor cells with immobilized, engineered Notch ligands have been shown to generate a multilog increase in T-cell precursors as compared to modest improvement with Notch ligand presented in solution or on cell surfaces [50, 51]. Further, density of Notch ligand plays important role too as low densities of Delta-1 can result in differentiation of $lin^{-}Sca\text{-}1^{+}c\text{-}kit^{+}$ (LSK) hematopoietic progenitors into $Sca\text{-}1^{+}c\text{-}kit^{+}$ cells, positive for both early T cell progenitors ($Thy1^{+}CD25^{+}$) and B cell precursors ($B220^{+}CD43^{-/lo}$). On the other hand, high densities of Delta-1 primarily promote differentiation into T lymphoid precursors [53]. Varnum-Finney et al. showed that by using an immobilized Notch ligand $Delta1^{ext\text{-}IgG}$, where the Delta1 extracellular domain was fused to the Fc domain of human immunoglobulin G1, a multi-log increase in the number of T cell progenitors can be obtained [51] (Fig. 4a, b). These precursors were efficient at inducing short-term and long-term T-cell reconstitution in nonobese diabetic/severe combined immunodeficiency (NOD/SCID) mice [51].

In these studies, immobilization of Notch ligand was necessary for T cell generation. Interestingly, inclusion of interleukin-7 (IL-7) in regular growth factor cocktail (SCF, IL-6, IL-11, and Flt-3l; collectively referred as 4GF) used with immobilized $Delta1^{ext\text{-}IgG}$ resulted in higher number of cells committed to the lymphoid lineage [51].

As indicated in Fig. 4c, after 14 days of culture, $Lin^{-}Sca\text{-}1^{+}c\text{-}kit^{+}$ hematopoietic cells incubated with immobilized ligands (I-$Delta1^{ext\text{-}IgG}$) appeared as blast cells with high expression levels of $Sca\text{-}1^{+}c\text{-}kit^{+}$, and low levels of the granulocyte-associated antigen GR-1, a myeloid lineage marker (Fig. 4c). Early T cell differentiation was confirmed after 28 days of culture with up regulation of CD25 antigen in cells exposed to I-$Delta1^{ext\text{-}IgG}$ [51], 4GF and IL-7 as compared to negative controls. I-$Delta1^{ext\text{-}IgG}$ with only 4GF resulted in lower levels of CD25. Immobilized ligands offer the ability to quantitatively control the amount of Notch signaling, an advantage the OP9-DL1 system does not provide. Similarly using exogenous cytokines such as stem cell factor, several interleukins, flt-3 ligand, and thrombopoietin and immobilization of Delta ligands at different concentrations, Deanley et al. [54] demonstrated the ex vivo differentiation and in vivo reconstitution of T cells from cord blood cells in a dose dependent manner.

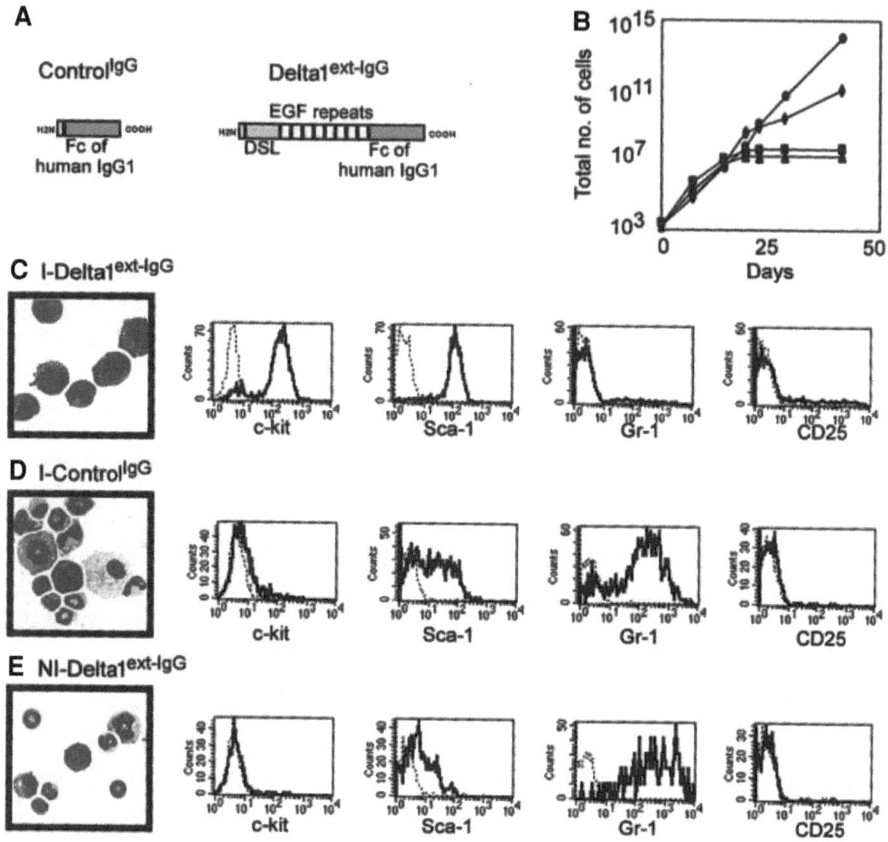

Fig. 4 Culture of Lin$^-$Sca-1$^+$c-kit$^+$ hematopoietic cells with immobilized Notch ligand, Delta1ext-IgG, and hematopoietic growth factors. **a** Schematic diagram of engineered Notch ligand consisting of the extracellular domain of Delta1 fused to the Fc portion of human IgG1 (Delta1ext-IgG). Delta1ext-IgG includes the Delta-Serrate-Lag1 (*DSL*) domain and epidermal growth factor (*EGF*) repeats. The control molecule consists of a signal peptide fused to the Fc portion of human IgG1 (ControlIgG). **b** Total number of cells generated over time was determined during culture with 4GF and I-Delta1ext-IgG (*filled diamond*); 4GF, IL-7, and I-Delta1ext-IgG (*filled circle*); 4GF and I-ControlIgG (*filled square*); 4GF and NI-Delta1ext-IgG (*filled triangle*). **c–e**) Morphology and phenotype of cultured cells. Wright–Giemsa-stained cytospin preparations (far *left* and fluorescence histograms *right panels*) of cells cultured for 14 days with 4GF and I-Delta1ext-IgG (**c**), I-ControlIgG (**d**), or NI-Delta1ext-IgG (**e**). Cells were stained with monoclonal antibodies that recognize Sca-1, c-Kit, GR-1, or CD25 epitopes. Original magnification, ×400. *X*-axis, log fluorescence intensity; *y*-axis, cell number; *solid line* staining with test antibody; *dotted line* staining with isotype-matched control antibody (adapted with permission from [51]). *I* Immobilized; *NI* non immobilized; *GR-1* Myeloid lineage marker

Although such systems have demonstrated potential in T cell commitment ex vivo in a cell-free manner, the expense and stability of the plate-adsorbed ligand are legitimate concerns for long-term culture. Ligand immobilization and OP9-DL1 systems are essentially a 2D, plate-based culture method with significant limitations to scale up processing. There is a need to evaluate bioreactor or 3D scaffold-based systems that might enhance the T-lineage differentiation of non-adherent cells like HSCs and lymphocytes and would allow easy scale up.

4.2 T Cell Differentiation Through Notch–Ligand Presenting Microbeads

Presentation of signaling molecules like Notch through microbeads can overcome drawbacks associated with OP9-DL1 and plate coating approaches. First, this could allow large-scale, bioreactor-based suspension culture of ESCs and HSCs as well as the differentiating T cells. Second, it enables the analysis of ligand density along with exposure time (since the beads could be added or removed during culture) on T cell differentiation. Taqvi et al. [55] demonstrated that by conjugating the notch ligand DLL4 on the surface of magnetic polystyrene microbeads acting as artificial stromal cells, it is possible to signal bone marrow derived HSCs (BMHSCs) into T cell lineage.

In this work, biotinylated anti-Histidine antibodies against recombinant DLL4 (tagged with Histiding, HIS) was attached to streptavidin functionalized microbeads. DLL4 notch signaling using the bead system was first demonstrated by inhibition of myotube formation in C2C12 myoblast cells. $Lin^{-}cKit^{+}Sca1^{+}$ BMHSCs were studied for differentiation into early T cell progenitors ($Thy1.2^{+}$) by incubation with DLL4 functionalized microbeads, co-cultured with OP9 cells directly or in-directly using Transwell^{TM} inserts. The study demonstrated that HSC-stromal cell contact is not essential for T cell development and soluble growth factors secreted by stromal cells that are not in direct contact might support differentiation. While co-incubation with non-functionalized microbeads resulted in generation of mostly $CD19^{+}$ B cells, those incubated with DLL4 functionalized beads differentiated into both $Thy1.2^{+}$ and $CD19^{+}$ cells from BMHSCs [55].

5 Generation of Antigen-specific T Cells from Stem Cells

One of the major limitations of adoptive immunotherapy is that patients with malignant tumors already have pre-existing tumor specific T cells thus making the isolation difficult [47]. Other complications include, morbidity associated with patient cells isolation, long culture-time and inefficiency of in vitro "training" of autologous T cells, limited availability of Ag-specific T cells and the ultimate risk of graft-versus-host disease. While antigen specific T cells can be effectively generated using the OP9-DL1 coculture system followed by culture in fetal thymic lobes and adoptive transfer into mouse, the system however requires implantation to reconstitute the T cell compartment, in vivo, and cannot be used as a pharmaceutical entity. Direct generation of antigen-specific cytotoxic T cells or effector T cells in vitro, that can then be used as therapeutic, transplantable cells, could offer a more efficient system for disease specific, on-demand adoptive immunotherapy.

5.1 Retroviral Transduction of T Cell Receptors

It has been long argued that a clinically relevant T cell immunotherapy would preferably require the adoptive transfer of autologous T cells expressing antigen-specific T-cell receptor (TCR). The antigen specificity of T cells is dependent on expression of TCR αβ complexes that binds to MHCs on stromal cells or dendritic cells during positive selection process in the thymus. Although it is possible to introduce TCR α and β genes in T cells through retroviral transduction, the off-shelf availability and risk of expressing additional endogenous TCRs remain potential drawbacks [56]. An alternative could be retroviral TCR gene transfer into mouse or human stem cells (ESCs and HSCs) followed by T cell differentiation [56]. In this strategy, HSCs from bone marrow can be transfected with retroviral carriers of TCR α and β chain encoding genes [57]. This process prevents expression of any endogenous TCR thus improving the specificity by reducing TCR mismatching as well as less competition for thymic expression under influence from CD3. It should be noted that such transfection of stem cells does not result in TCR expression unless transferred into a thymic niche from bone marrow environment (Fig. 5). This can be solely attributed to expression of CD3 by progenitors in thymus and not in bone marrow [56]. CD3 expression in early T cell progenitors leads to surface expression of TCR complexes.

Yang and Baltimore performed one of the initial studies to pre-program HSCs to differentiate into T cells with desired antigen specificity in a mouse tumor model for

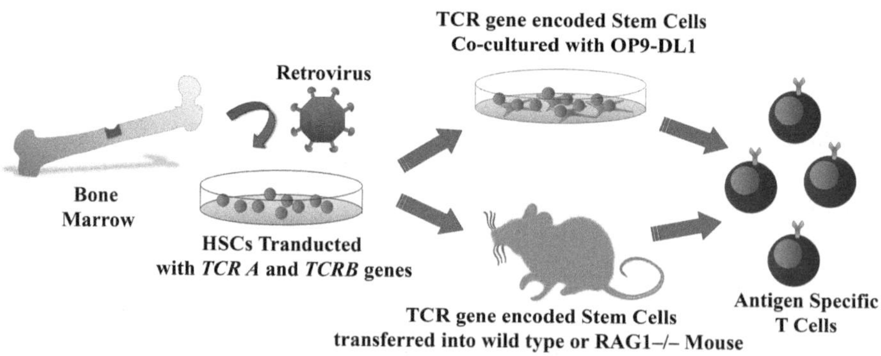

Fig. 5 T-cell receptor (*TCR*) gene transfer into bone marrow derived hematopoietic stem cells (*HSC*). The TCR remains intracellular post-infection until it is transferred into a thymic microenvironment; co-culture with OP9-DL1 or transfer into mouse. In the thymic microenvironment, HSCs express CD3 that promotes surface expression of TCR chains. Thymic differentiation of TCR gene transducted HSC results into antigen specific T cells

chicken ovalbumin [58]. HSCs isolated from wild type B6 and RAG1$^{-/-}$ mouse were transduced with MOT1 or MOT2 viruses (vectors expressing OT1 or OT2 TCR). The retrovirally infected HSCs when transferred into wild type or RAG1$^{-/-}$ resulted in development of Sca-1^{+}c-kit^{+}, CD8^{+} CD4^{+} DP cells, as well as later conversion into either CD8 or CD4 single-positive T cells. The system further exhibited significant tumor suppression in E.G7 mouse tumor model. Legrand et al. [59] performed retroviral transduction of TCRαβ genes into human CD34^{+}CD1a^{-} HSCs followed by coculture with OP9-DL1 cells which resulted in in vitro production of large populations of antigen specific T cells against melanoma (MART-1), viral (CMV), and minor histocompatibility (HA-2).

Similarly, Morgan et al. [47] generated human p53-TCR or ESO expressing T cells from p53 to TCR or NY-ESO-1 transduced human umbilical cord blood derived HSCs. NY-ESO-1 is a germ cell antigen abnormally expressed by various tumor cells. T cell differentiation was achieved by coculture with OP9-DL1. In this study co-culture resulted in differentiation of HSCs into T cells expressing T cell lineage markers CD1a, CD7, CD25, CD27, CD44, CD3, CD4, and CD4^{+} CD8^{+} DP cells, irrespective of transduction (Fig. 6a). TCR mediated antigen specificity of the in vitro generated T cells was confirmed by recognizing specific tumor antigen epitopes on p53 or ESO (control) peptide pulsed antigen presenting cells (T2), with production of GMCSF and IL-2 (Fig. 6b). Similarly, p53–TCR transduced HSCs derived T cells resulted in specific lysis of p53–peptide pulsed target cells (Fig. 6c).

In tumor immunotherapy there is growing evidence that CD4^{+} T cells play helping role in further stimulating CD8^{+} T cells and thus in vitro generation of antigen specific CD4^{+} T cells might be beneficial towards adoptive immunotherapy. Dai et al. [60] recently reported in vitro generation of such CD4 T cells by coculturing OT2 CD4 TCR expressing retrovirally transduced adult bone marrow cells with OP9-DL1 cells. Lack of expression of MHC II molecules on OP9 cells makes it challenging to generate antigen specific CD4 T cells, Dai et al. ectopically expressed I-A^{b} in OP9-DL1 cells. Polyclonal anti-CD3/CD28 stimulation of CD4 T cells resulted in proliferated and expression of activated T cell markers. These cells also secreted CD4 T cell specific cytokines IFN-γ and IL-2 further establishing the functional maturity. In comparison to OP9-DL1 coculture, MHCII expressing OP9-DL1-IA^{b} resulted in markedly higher production of IFN-γ and IL-2 suggesting that ectopic expression of MHCII can further enhance the antigen specific generation of CD4 T cells.

Despite the numerous advantages discussed with the TRC gene transfer there exist several disadvantages. While this system has high specificity for viral antigens, it may impose several problems with tumor self-antigens. The self antigen-specific T cells can either get deleted in the thymus itself or can result in tolerance if these T cells reach the peripheral tissues [56]. Further, retroviral transduction may result in viral mutagenesis in the stem cells making it a risky system in long term clinical studies [57]. Thus it may be desirable to design a system that can be as efficient as viral transduction but does not involve use of such potentially pathogenic carriers.

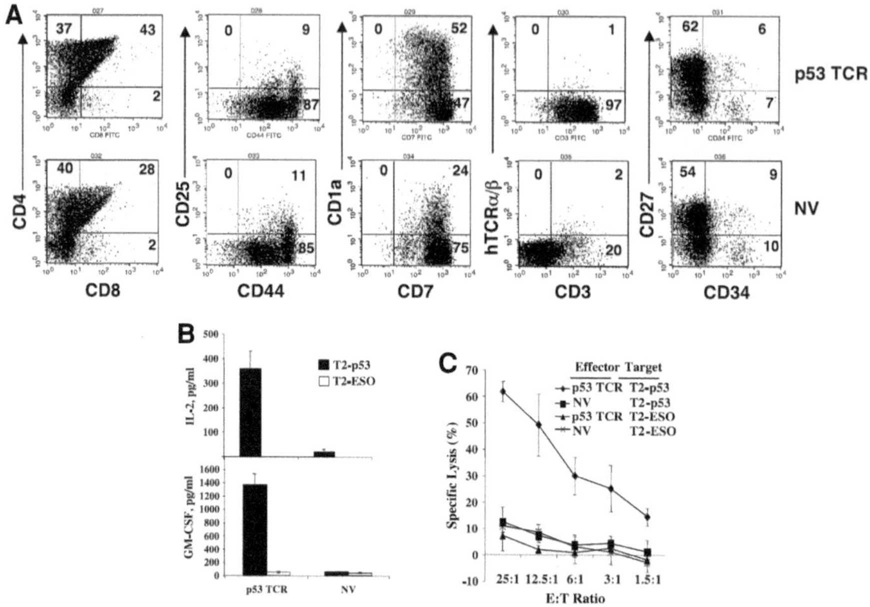

Fig. 6 T cells committed from genetically engendered HSCs and antigen-specific functional responses of the T cells in vitro developed from tumor antigen TCR-transduced HSCs. **a** Flow cytometry analysis for T-cell markers for MSGV-p53-AIB—transduced HSCs (*p53 TCR*) or nontransduced HSCs (*NV*) cocultured with OP9-DL1 cell line for 29 days. Representative of four experiments. **b** T cells generated by coculturing MSGV-p53-AIB (p53 TCR)—transduced HSCs with OP9-DL1 for 42 days [nontransduced HSCs (NV) were used as control] were cocultured with p53 peptide—pulsed T2 cell line (T2-p53); NY-ESO-1 p156-165V peptide—pulsed T2 cells (T2-ESO) were used as a control. Production of IL-2 and GM-CSF was determined by ELISA. Representative of three experiments. **d** T cells generated by coculturing MSGV-p53-AIB—transduced HSCs (p53 TCR) or nontransduced HSCs (NV) with OP9-DL1 for 44 days were cocultured with 51Cr-labeled, p53 peptide—pulsed T2 (T2-p53) or NY-ESO-1 peptide—pulsed T2 (T2-ESO) for 4 h, and 51Cr release was measured. Representative of two experiments. Adapted with permission from [47]]

5.2 T Cell Differentiation in a Three-dimensional Matrix

Biomaterials based 3D scaffolds promote cultures through cell–cell and cell–material interactions, maintenance of physiological cellular morphologies and provide mechanical support [61]. 3D scaffolds for ES cell differentiation have been developed from natural polymers such as fibrin [62] and hyaluronic acid [63], as well as synthetic materials like poly (lactic-co-glycolic acid) and poly(ethylene glycol) (PEG) [64]. However, fabrication of 3D scaffolds has focused mainly on chondrogenesis, osteogenesis, and vascularization [63]. Bone marrow and thymic co-culture on 3D scaffolds have resulted in significant T and B cell differentiation of stem cells as compared to monolayered co-cultures [65, 66]. Direct hematopoietic differentiation of embryonic stem cells in 3D scaffold-based cultures and their subsequent differentiation into antigen-specific, functional T cells, have only been recently reported [67]. Kotov et al. have generated B cells by culturing human HSCs in a porous polyacrylamide scaffold of a similar architecture [65, 66], and have also shown possibility of early, immature T cell differentiation from human HSCs on DLL1-coated 3D inverted colloidal crystal (ICC) scaffolds [67].

Current design rationale behind these 3D scaffolds for T cell differentiation from stem cells is based on the mechanically soft nature of the thymic microenvironment and the presence of key cell signaling molecules, e.g. notch ligands and MHCs. Scaffolds, whether natural or polymeric, should provide 3D mechanical support for stem cell differentiation and specific ligands to create an artificial thymic niche that supports differentiation of ES cells or HSCs into mature, therapeutic T cells. The scaffolds should also provide cell–cell contacts and it would be advantageous to have a porous network with-in the scaffolds to provide nutrient and waste transport, as well as predict the diffusion and migration characteristics of the cells cultured on these scaffolds [68, 69]. A schematic representation of an artificial 3D microenvironment mimicking thymic micro-environment is shown in Fig. 7a. The inner surface of the scaffolds can be modified with chemical moieties like clay or with specific biomolecules like ligands and receptors [67]. The polymer architecture for high-throughput generation of T cells should be easily modifiable to incorporate factors that need to be immobilized or sequestered for cell signaling.

Kotov et al. designed notch functionalized 3D inverted colloidal crystal (ICC) scaffolds for ex vivo T-cell development of human HSCs [67]. The microarchitecture of this artificial thymic microenvironment is a highly organized spherical pores that are interconnected with adjacent pores (Fig. 7b, i). These polyacrylamide hydrogel ICC scaffolds are formed using layer-by-layer (LBL) molecular assembly technique and have 110 μm uniform pore size with ~30 μm large interconnecting channels [67]. The 3D scaffold addresses most of the ideal design characteristics mentioned earlier for mimicking thymic architecture [69, 70] except that it lacks any T cell positive selection mechanism. The ICC hydrogels have LBL coating of clay and poly(diallyl dimethylammonium chloride) (PDDA) that promotes stromal cell adhesion and immobilization of Notch ligands through electrostatic interactions [67]. Figure 7b, ii represents DLL1 coated well on the surface of the ICC scaffold [67] and the intricate design resulted in dynamic culture with pore surfaces densely covered with cells (Fig. 7b, iii). The repeated proliferation and notch induction resulted in differentiation of HSCs into T-cell lineage as these cells started expressing CD3, TCRα, and CD117 (Fig. 7b, iv). Finally after 28 days of continuous culture, these HSCs differentiated into $CD4^+$ T cells (Fig. 7b, v, vi).

The study shows the possibility of generating a 3D system for high throughput regeneration of T cells from stem cells, ex vivo, however several aspects of the T cell development like positive selection, antigen specificity, graft versus host (GVH) rejection upon transplantation needs to be addressed.

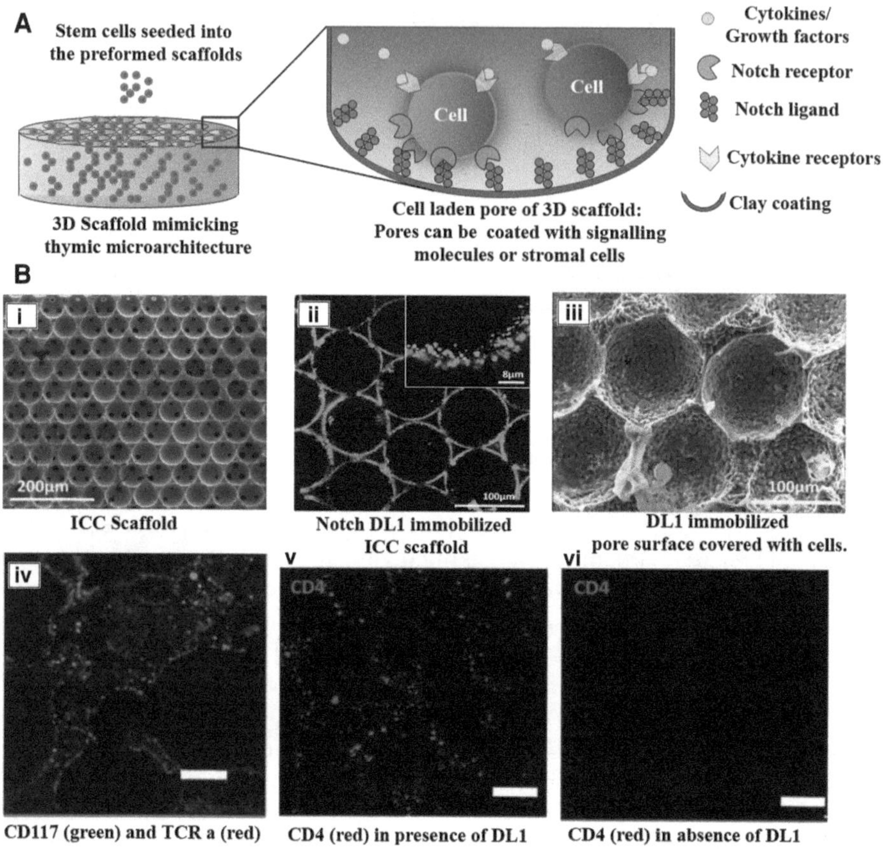

Fig. 7 **a** General schematic representing an artificial 3D scaffold mimicking thymic microenvironment and architecture. The pre-fabricated scaffolds can be seeded with stem cells or thymic progenitors and under the influence of signaling molecules like Notch and growthfactors like IL-7 and SCF, precursor cells can differentiate into matured T cells. **b** Representative 3D inverted colloidal crystal (*ICC*) hydrogel scaffolds. *i* Scanning electron micrograph (SEM) of hydrogel ICC scaffold with interconnecting pores, *ii* confocal image of a DL-1 notch–ligand-coated hydrogel ICC scaffold. For imaging purposes, DL-1 notch ligands were bound to antibodies conjugated to FITC., *iii* the Delta 1 notch–ligand immobilized pore surface is densely covered with $CD34^+$ HSCs. Restricted cellular population mimicked the thymic microenvironment, *iv* CD117 and TCRα after 22 days of culture. (*Scale bar* 40 mm, see online version [67]) *v*, *vi* CD4 surface marker analysis after 28 days of culture in presence of notch and in absence of notch (*Scale bar* 80 mm) (Figure B adapted with permission from [67])

Acknowledgments We thank Myung Hee Kim, Jian Lin, Eileen Dawson, and Dr. Irina Fernandez at The University of Texas at Austin for comments and discussion on the chapter. We thank Dr. Sabia Taqvi for providing an insight into the details of notch and microbead system which is documented in her doctoral thesis at The University of Texas at Austin.

References

1. Mackall, C.L., et al.: T-cell regeneration after bone marrow transplantation: differential CD45 isoform expression on thymic-derived versus thymic-independent progeny. Blood **82**(8), 2585–2594 (1993)
2. Mackall, C.L., Hakim, F.T., Gress, R.E.: T-cell regeneration: all repertoires are not created equal. Immunol. Today **18**(5), 245–251 (1997)
3. Yee, C.: Adoptive T cell therapy: addressing challenges in cancer immunotherapy. J. Transl. Med. **3**(1), 17 (2005)
4. Bhandoola, A., Sambandam, A.: From stem cell to T cell: one route or many? Nat. Rev. Immunol. **6**(2), 117–126 (2006)
5. Rothenberg, E.V.: From totipotency to T in a dish. Nat. Immunol. **5**(4), 359–360 (2004)
6. Meidenbauer, N., et al.: Survival and tumor localization of adoptively transferred Melan-A-specific T cells in melanoma patients. J. Immunol. **170**(4), 2161–2169 (2003)
7. Daley, G.Q.: From embryos to embryoid bodies: generating blood from embryonic stem cells. Ann. N Y Acad. Sci. **996**, 122–131 (2003)
8. Dellatore, S.M., Garcia, A.S., Miller, W.M.: Mimicking stem cell niches to increase stem cell expansion. Curr. Opin. Biotechnol. **19**(5), 534–540 (2008)
9. Rothenberg, E.V., Moore, J.E., Yui, M.A.: Launching the T-cell-lineage developmental programme. Nat. Rev. Immunol. **8**(1), 9–21 (2008)
10. Broudy, V.C.: Stem cell factor and hematopoiesis. Blood **90**(4), 1345–1364 (1997)
11. Zuniga-Pflucker, J.C.: T-cell development made simple. Nat. Rev. Immunol. **4**(1), 67–72 (2004)
12. Taghon, T., et al.: Developmental and molecular characterization of emerging beta- and gamma delta-selected pre-T cells in the adult mouse thymus. Immunity **24**(1), 53–64 (2006)
13. Turner, S.J., et al.: Structural determinants of T-cell receptor bias in immunity. Nat. Rev. Immunol. **6**(12): 883–894 (2006)
14. Janeway, C.: Immunobiology : the immune system in health and disease, vol. xxiii, pp. 823, 6th edn. Garland Science, New York (2005)
15. Murphy, K.P., et al.: Janeway's immunobiology, vol. xxi, pp. 887, 7th edn. Garland Science, New York (2008)
16. Besseyrias, V., et al.: Hierarchy of Notch–Delta interactions promoting T cell lineage commitment and maturation. J. Exp. Med. **204**(2), 331–343 (2007)
17. de La Coste, A., et al.: In vivo and in absence of a thymus, the enforced expression of the Notch ligands delta-1 or delta-4 promotes T cell development with specific unique effects. J. Immunol. **174**(5), 2730–2737 (2005)
18. Lefort, N., et al.: Short exposure to Notch ligand Delta-4 is sufficient to induce T-cell differentiation program and to increase the T cell potential of primary human CD34+ cells. Exp. Hematol. **34**(12), 1720–1729 (2006)
19. Hager-Theodorides, A.L., et al.: Bone morphogenetic protein 2/4 signaling regulates early thymocyte differentiation. J. Immunol. **169**(10), 5496–5504 (2002)
20. Tsai, P.T., Lee, R.A., Wu, H.: BMP4 acts upstream of FGF in modulating thymic stroma and regulating thymopoiesis. Blood **102**(12), 3947–3953 (2003)
21. Massa, S., et al.: Critical role for c-kit (CD117) in T cell lineage commitment and early thymocyte development in vitro. Eur. J. Immunol. **36**(3), 526–532 (2006)
22. Crompton, T., Outram, S.V., Hager-Theodorides, A.L.: Sonic hedgehog signalling in T-cell development and activation. Nat. Rev. Immunol. **7**(9), 726–735 (2007)
23. Anderson, G., et al.: MHC class II-positive epithelium and mesenchyme cells are both required for T-cell development in the thymus. Nature **362**(6415), 70–73 (1993)
24. Radtke, F., et al.: Notch regulation of lymphocyte development and function. Nat. Immunol. **5**(3), 247–253 (2004)
25. Hozumi, K., et al.: Active form of Notch members can enforce T lymphopoiesis on lymphoid progenitors in the monolayer culture specific for B cell development. J. Immunol. **170**(10), 4973–4979 (2003)

26. De Smedt, M., et al.: Active form of Notch imposes T cell fate in human progenitor cells. J. Immunol. **169**(6), 3021–3029 (2002)
27. Davis, R.L., Turner, D.L.: Vertebrate hairy and enhancer of split related proteins: transcriptional repressors regulating cellular differentiation and embryonic patterning. Oncogene **20**(58), 8342–8357 (2001)
28. De Smedt, M., Hoebeke, I., Plum, J.: Human bone marrow CD34+ progenitor cells mature to T cells on OP9-DL1 stromal cell line without thymus microenvironment. Blood Cells Mol. Dis. **33**(3), 227–232 (2004)
29. Lehar, S.M., et al.: Notch ligands Delta 1 and Jagged1 transmit distinct signals to T-cell precursors. Blood **105**(4), 1440–1447 (2005)
30. de Pooter, R., Zuniga-Pflucker, J.C.: T-cell potential and development in vitro: the OP9-DL1 approach. Curr. Opin. Immunol. **19**(2), 163–168 (2007)
31. Lehar, S.M., Bevan, M.J.: T cell development in culture. Immunity **17**(6), 689–692 (2002)
32. Yeoman, H., et al.: Human bone marrow and umbilical cord blood cells generate CD4+ and CD8+ single-positive T cells in murine fetal thymus organ culture. Proc. Natl Acad. Sci. USA **90**(22), 10778–107782 (1993)
33. Weekx, S.F., et al.: Generation of T cells from adult human hematopoietic stem cells and progenitors in a fetal thymic organ culture system: stimulation by tumor necrosis factor-alpha. Blood **95**(9), 2806–2812 (2000)
34. Res, P., et al.: CD34+ CD38dim cells in the human thymus can differentiate into T, natural killer, and dendritic cells but are distinct from pluripotent stem cells. Blood **87**(12), 5196–5206 (1996)
35. Doetschman, T.C., et al.: The in vitro development of blastocyst-derived embryonic stem cell lines: formation of visceral yolk sac, blood islands and myocardium. J. Embryol. Exp. Morphol. **87**, 27–45 (1985)
36. Schmitt, T.M., et al.: Induction of T cell development and establishment of T cell competence from embryonic stem cells differentiated in vitro. Nat. Immunol. **5**(4), 410–417 (2004)
37. Cho, S.K., et al.: Functional characterization of B lymphocytes generated in vitro from embryonic stem cells. Proc. Natl Acad. Sci. USA **96**(17), 9797–9802 (1999)
38. Potocnik, A.J., Nielsen, P.J., Eichmann, K.: In vitro generation of lymphoid precursors from embryonic stem cells. Embo J. **13**(22), 5274–5283 (1994)
39. Gutierrez-Ramos, J.C., Palacios, R.: In vitro differentiation of embryonic stem cells into lymphocyte precursors able to generate T and B lymphocytes in vivo. Proc. Natl Acad. Sci. USA **89**(19), 9171–9175 (1992)
40. Heinzel, K., et al.: Bone marrow-derived hemopoietic precursors commit to the T cell lineage only after arrival in the thymic microenvironment. J. Immunol. **178**(2), 858–868 (2007)
41. Varnum-Finney, B., et al.: Pluripotent, cytokine-dependent, hematopoietic stem cells are immortalized by constitutive Notch1 signaling. Nat. Med. **6**(11), 1278–1281 (2000)
42. Nakano, T., Kodama, H., Honjo, T.: Generation of lymphohematopoietic cells from embryonic stem cells in culture. Science, **265**(5175), 1098–1101 (1994)
43. de Pooter, R.F., et al.: In vitro generation of T lymphocytes from embryonic stem cell-derived prehematopoietic progenitors. Blood **102**(5), 1649–1653 (2003)
44. Mohtashami, M., Zuniga-Pflucker, J.C.: Three-dimensional architecture of the thymus is required to maintain delta-like expression necessary for inducing T cell development. J. Immunol. **176**(2), 730–734 (2006)
45. Schmitt, T.M., Zuniga-Pflucker, J.C.: Induction of T cell development from hematopoietic progenitor cells by delta-like-1 in vitro. Immunity **17**(6), 749–756 (2002)
46. Galic, Z., et al.: T lineage differentiation from human embryonic stem cells. Proc. Natl Acad. Sci. USA **103**(31), 11742–11747 (2006)
47. Zhao, Y., et al.: Extrathymic generation of tumor-specific T cells from genetically engineered human hematopoietic stem cells via Notch signaling. Cancer Res. **67**(6), 2425–2429 (2007)
48. Uchida, N., et al.: High doses of purified stem cells cause early hematopoietic recovery in syngeneic and allogeneic hosts. J. Clin. Invest. **101**(5), 961–966 (1998)

49. Bhattacharya, D., et al.: Rapid lymphocyte reconstitution of unconditioned immunodeficient mice with non-self-renewing multipotent hematopoietic progenitors. Cell Cycle **5**(11), 1135–1139 (2006)
50. Dallas, M.H., et al.: Enhanced T-cell reconstitution by hematopoietic progenitors expanded ex vivo using the Notch ligand Delta1. Blood **109**(8), 3579–3587 (2007)
51. Varnum-Finney, B., Brashem-Stein, C., Bernstein, I.D.: Combined effects of Notch signaling and cytokines induce a multiple log increase in precursors with lymphoid and myeloid reconstituting ability. Blood **101**(5), 1784–1789 (2003)
52. Varnum-Finney, B., et al.: Immobilization of Notch ligand, Delta-1, is required for induction of notch signaling. J. Cell Sci. **113** (Pt 23), 4313–4318 (2000)
53. Dallas, M.H., et al.: Density of the Notch ligand Delta1 determines generation of B and T cell precursors from hematopoietic stem cells. J. Exp. Med. **201**(9), 1361–1366 (2005)
54. Delaney, C., et al.: Dose-dependent effects of the Notch ligand Delta1 on ex vivo differentiation and in vivo marrow repopulating ability of cord blood cells. Blood **106**(8), 2693–2699 (2005)
55. Taqvi, S., Dixit, L., Roy, K.: Biomaterial-based notch signaling for the differentiation of hematopoietic stem cells into T cells. J. Biomed. Mater. Res. A **79**(3), 689–697 (2006)
56. Stauss, H.J., et al.: Monoclonal T-cell receptors: new reagents for cancer therapy. Mol. Ther. **15**(10), 1744–1750 (2007)
57. Kessels, H.W., et al.: Immunotherapy through TCR gene transfer. Nat. Immunol. **2**(10), 957–961 (2001)
58. Yang, L., Baltimore, D.: Long-term in vivo provision of antigen-specific T cell immunity by programming hematopoietic stem cells. Proc. Natl Acad. Sci. USA **102**(12), 4518–4523 (2005).
59. van Lent, A.U., et al.: Functional human antigen-specific T cells produced in vitro using retroviral T cell receptor transfer into hematopoietic progenitors. J. Immunol. **179**(8), 4959–4968 (2007)
60. Dai, B., Wang, P.: In vitro differentiation of adult bone marrow progenitors into antigen-specific CD4 helper T cells using engineered stromal cells expressing a notch ligand and a major histocompatibility complex class II protein. Stem Cells Dev. **18**(2), 235–245 (2009)
61. Hwang, N.S., Varghese, S., Elisseeff, J.: Controlled differentiation of stem cells. Adv. Drug Deliv. Rev. **60**(2), 199–214 (2008)
62. Willerth, S.M., et al.: The effects of soluble growth factors on embryonic stem cell differentiation inside of fibrin scaffolds. Stem Cells **25**(9), 2235–2244 (2007)
63. Williams, C.G., et al.: In vitro chondrogenesis of bone marrow-derived mesenchymal stem cells in a photopolymerizing hydrogel. Tissue Eng. **9**(4), 679–688 (2003)
64. Levenberg, S., et al.: Differentiation of human embryonic stem cells on three-dimensional polymer scaffolds. Proc. Natl Acad. Sci. USA **100**(22), 12741–12746 (2003)
65. Poznansky, M.C., et al.: Efficient generation of human T cells from a tissue-engineered thymic organoid. Nat. Biotechnol. **18**(7), 729–734 (2000)
66. Nichols, J.E., et al.: In vitro analog of human bone marrow from 3D scaffolds with biomimetic inverted colloidal crystal geometry. Biomaterials **30**(6), 1071–1079 (2009)
67. Lee, J., Kotov, N.A.: Notch ligand presenting acellular 3D microenvironments for ex vivo human hematopoietic stem-cell culture made by layer-by-layer assembly. Small **5**(9), 1008–10013 (2009)
68. Inanc, B., Elcin, A.E., Elcin, Y.M.: Human embryonic stem cell differentiation on tissue engineering scaffolds: effects of NGF and retinoic acid induction. Tissue Eng. Part A **14**(6), 955–964 (2008)
69. Shanbhag, S., Woo Lee, J., Kotov, N.: Diffusion in three-dimensionally ordered scaffolds with inverted colloidal crystal geometry. Biomaterials **26**(27), 5581–5585 (2005)
70. Shanbhag, S., Wang, S., Kotov, N.A.: Cell distribution profiles in three-dimensional scaffolds with inverted-colloidal-crystal geometry: modeling and experimental investigations. Small **1**(12), 1208–1214 (2005)

Understanding Hypoxic Environments: Biomaterials Approaches to Neural Stabilization and Regeneration after Ischemia

Jennie B. Leach and Elizabeth M. Powell

Abstract Hypoxic/ischemic brain damage results in permanent neurological dysfunction. Though currently there are no effective treatments for this condition, exciting advances in neural stem/progenitor cell (NSPC) biology promise cellular based therapeutics. Major strategies seek to transplant exogenous NSPCs or recruit endogenous NSPCs in order to protect injured neurons or replace the function of lost neurons, but current methods are hindered by poor regulation of NSPC survival, proliferation, migration and integration into the existing environment. This chapter provides a review of the biology including response to hypoxia and current methods to manipulate NSPCs in vitro. We highlight recent applications of utilizing biomaterials to control NSPC response. Despite advanced technologies to synthesize and probe biomaterial systems, many efforts provide incremental improvement compared to transplants of simply NSPCs alone. We challenge the biomaterials community to view the NSPC as a component of the biomaterial and define the biological response of the NSPC within the environment, whether natural, man-made, or a hybrid of each. Accordingly, interactive collaboration among engineers, neurobiologists and clinical neurologists will lead to breakthroughs in basic science and advance biomaterials technology to achieve commercial and therapeutic solutions for human disease.

J. B. Leach (✉)
Department of Chemical & Biochemical Engineering, UMBC, 1000 Hilltop Circle, Baltimore, MD 21250, USA
e-mail: jleach@umbc.edu

E. M. Powell
Department of Anatomy & Neurobiology, University of Maryland School of Medicine, HSF II S251, 20 Penn Street, Baltimore, MD 21201, USA
e-mail: epowe001@umaryland.edu

E. M. Powell
Fischell Department of Bioengineering, University of Maryland, Baltimore, USA

Stud Mechanobiol Tissue Eng Biomater (2010) 2: 247–274
DOI: 10.1007/8415_2010_9

Published online: 12 June 2010

1 Ischemic Brain Damage in Adult and Neonatal Humans

Several types of injury can cause hypoxic/ischemic damage in the central nervous system [1]. Events such as heart attack and asphyxia (e.g., carbon monoxide inhalation and drowning) can result in near complete cessation of oxygenated blood flow to the brain. This type of injury predominately results in the death of particularly susceptible neurons, such as the pyramidal neurons of the hippocampus. Stroke, which occurs due to rupture or occlusion of the cerebral artery, is more common and causes focal ischemia. This injury causes neuronal death in a focused area that is surrounded by a region of damage that is partially reversible depending on the severity and location of injury. In infants, perinatal asphyxia during birth causes hypoxic/ischemic brain injury and is the most common cause of neurological disability in children [2]. The major outcome of the cessation of blood flow in the brain is cell death due to reduced supplies of oxygen and metabolites. Cell death can also be caused during reperfusion injury, wherein blood flow resumes to the damaged tissue. Due to vessel damage and a disrupted blood–brain barrier, the already compromised tissue is flooded with ions, hemoglobin, inflammatory factors and blood cells. The long-term outcome of the most extreme degree of ischemic brain damage is the formation of a permanent cystic cavity surrounded by impenetrable glial scar tissue; neurological function of this tissue is thus permanently lost.

Despite recent advances in critical care of adult and infant patients with hypoxic/ischemic brain damage, effective treatments that prevent permanent disability are not available. Fortunately, neurological studies in the past several decades have culminated to a very exciting point in which multidisciplinary work combining advances in neurobiology, stem cell biology, clinical medicine and biomaterials are poised to make significant advances towards treating hypoxic/ischemic brain injuries. For example, recent Phase I and II clinical trials of new stroke therapies indicate that human cell transplants remain postmitotic up to 2 years after transplantation, supporting safety and feasibility [3–6]. Unfortunately, functional improvement resulting from cell transplants was not significant. This review focuses on the effect of hypoxic/ischemic injury on the brain and its populations of neural stem/progenitor cells (NSPCs), endeavors by biomaterials scientists to improve the success of NSPC transplants, and suggestions for promising avenues of future research to design biomaterials for NSPC implants in patients afflicted by stroke and asphyxia.

2 Response of NSPCs to Ischemic Brain Damage

Ischemia and hypoxia induces neurogenesis in the brain. Interestingly, this response occurs even in the older populations (ages 60–78), wherein the incidence of stroke is higher [7–10]. After fetal development, neurogenesis in the human brain is most commonly observed in two regions: in the subventricular zone of the cerebral cortex and the hippocampal dentate gyrus. In the resting, uninjured brain, the neuroglial precursors simultaneously divide and migrate along the rostral migratory stream on the way to the olfactory bulb, where they will become interneurons [11–13]. Normal hippocampal neurogenesis is dependent on multiple on environmental influences, including exercise, learning cognitive tasks and anti-depressant use [14–18]. Current stem cell therapies have focused on harnessing the potential of these endogenous NSPCs.

In the injured brain, the niches of NSPCs may be activated. The cells are induced to proliferate, migrate towards the injury site, differentiate into neurons, and function to support the survival of injured cells or re-establish functional connectivity [19–28] (Fig. 1). Despite the non-permissive environment generated during stroke and reperfusion, NSPCs are relatively resistant to hypoxia/ischemia [29] (discussed in "Physiological Hypoxia" and Hypoxic/Ischemic Injury section in more detail) and potentially damaging inflammatory molecules and metabolites, such as reactive oxygen species [30]. Moreover, hypoxia promotes NSPC proliferation directly [25, 31–36] and indirectly via chemotactic gradients of growth factors such as vascular endothelial growth factor (VEGF), fibroblast growth factor 2 (FGF2) and epithelial growth factor (EGF) that are established by neurons and glia during hypoxia [13, 37, 38]. However, the details of the neural stem cell response are not completely clear and may impact the various cell types or neuronal phenotypes differently [32]. The underlying signaling processes are difficult to elucidate due to the complexity, variability and dynamics inherent to in vivo studies of hypoxic brain injury as well as the fact that in vitro studies typically investigate individual factors that are added exogenously to a culture carried out on a non-physiological and two-dimensional (2D) substrate.

Another key factor involved in hypoxic/ischemic injury that influences NSPC response is angiogenesis. The growth of new blood vessels may be a key factor

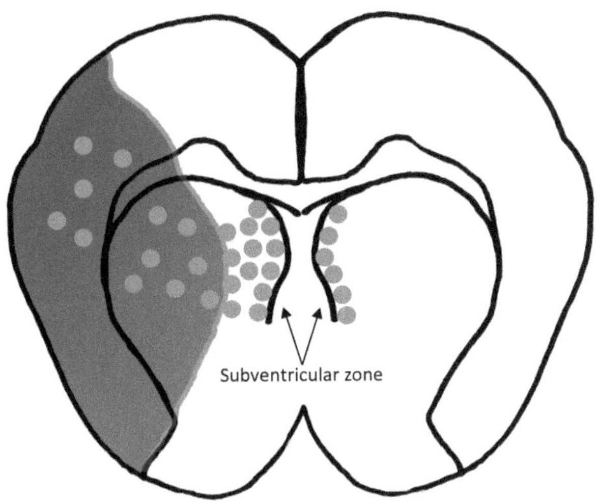

Fig. 1 Schematic of post-stroke neurogenesis. The *shaded* area indicates global ischemia in the cortex and striatum (as is induced in rodent models by middle cerebral artery occlusion) wherein NSPCs (*circles*) migrate from the subventricular zone towards the injury area. Figure adapted from [19, 43]

towards the success of stroke therapies for direct effects on re-establishing perfusion and the blood–brain barrier as well as indirect effects on NSPC response. Development and stroke studies have elucidated remarkable synergy between the processes of angiogenesis and neurogenesis [39–44]. For example, after hypoxic/ ischemic injury, angiogenesis-related genes encoding VEGF, FGF2 and EGF are upregulated [45] and produce factors which also promote neurogenesis [46–49]. Moreover, the vasculature is a key structural component of the NSPC niche [48, 50, 51] and provides a pathway to guide NSPC migration during development and in adults [52].

Despite the promise suggested by these endogenous repair mechanisms in the brain, most NSPCs that migrate into sites of damage die before they differentiate and provide functional benefit. Indeed, NSPCs only successfully replace as few as 0.2% of the original neuronal population [23], which does not provide for adequate replacement of function. Increased plasticity of neurons in the tissue local to the injury has been hypothesized to provide some benefit, but the loss of neuronal connectivity and architecture in the cystic cavity provides a significant barrier towards re-establishment of the original tissue functionality [53]. Thus, a more complete understanding of how to influence NSPC response will likely result in improved therapeutic strategies for treating hypoxic/ischemic injury.

3 NSPC Implants to Treat Ischemic Brain Damage

Mammalian NSPCs can differentiate into neurons [54–56], astrocytes [57–59], oligodendrocytes [60, 61] and possibly even endothelial cells [62]. Therefore, it is plausible that repair by NSPCs could result in fully restored tissue following hypoxic/ischemic brain injury. As mentioned above, the full potential of NSPC transplants has not yet been achieved in humans, and numerous studies in animal models have not succeeded in providing reproducible functional restoration (for a recent comprehensive review, see [63]). Nonetheless, promising results have been realized: NSPCs are capable of surviving transplantation, can migrate into the area of injury, differentiate into mature neurons, and form functional connections with surrounding cells [54, 63–65]. Also, transplanted NSPCs may help promote the survival and function of cells in the injury site [60, 66–68], which may alone provide significant benefit [53]. Thus, two major outcomes have been targeted by the therapeutic application of NSPC implants: (1) treat neuronal death and possible loss of circuitry by achieving long-term NSPC survival and functional integration or (2) treat neuronal injury by more short-term NSPC survival that allows for neuroprotection and recruitment of endogenous NSPCs (Fig. 2).

Unfortunately, the majority of transplanted NSPCs often dies [69–71] or differentiates into glia [57–59]. Furthermore, the age of the NSPC source may play an underappreciated role in their ability to augment repair: a recent study by Takahashi et al. indicates that NSPCs derived from embryonic animals respond very differently from those derived from adults when transplanted into stroke

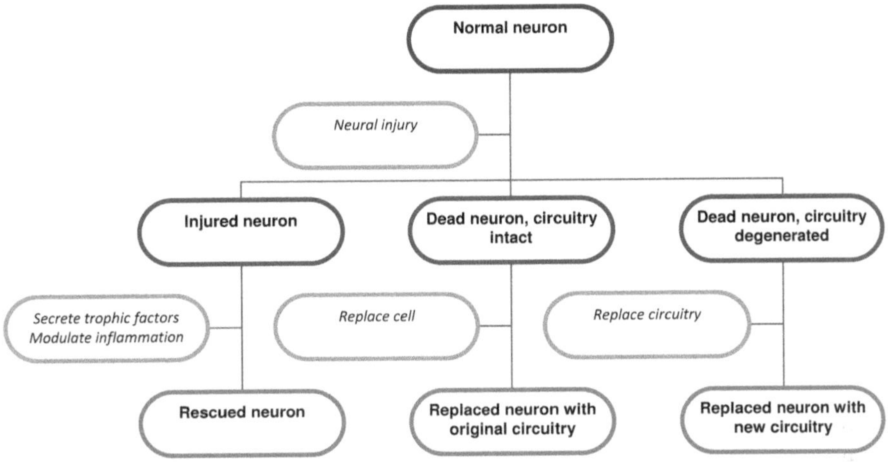

Fig. 2 Roles of NSPCs to treat neural injury. Transplanted or endogenous NSPCs may serve to promote survival of injured neurons by secreting trophic factors or modulating inflammation. If the neuron has already died but suitable scaffolding cues remain in the injury site, NSPCs may have the potential to replace lost circuitry. In the presence of a cystic cavity or excessive glial scarring, NSPCs may even aid in the formation of new circuitry. Schematic adapted from [53]

models in rats [72]. Whereas both populations of NSPCs were associated with reduced infarct volumes, embryonic NSPCs had better survival and lower immune response in the injury site than adult NSPCs. Yet, in animal studies, embryonic NSPCs more often yield teratomas or other tumors at the site of implantation [73–75]. Despite their few numbers, however, adult NSPCs were associated with greater therapeutic benefit, perhaps due to the production of more environmentally appropriate trophic factors that support the survival of impaired host cells. Thus, significant questions remain to be answered regarding the most effective means to promote NSPC survival, direct fate and identify the most appropriate source and age of cells to implant.

4 NSPC Isolation and Culture: State-of-the-Art

NSPCs can be isolated from mammals of all ages, from embryo to adult. NSPC is a general term that includes progenitor cells with varying expansion and differentiation potential. Developmental biologists are defining the multiple types of progenitors, including initial neuroepithelium of the neural tube, radial glial neuroprogenitors of the embryonic brain, transiently amplifying cells of the rostral migratory stream and the adult stem cells of the dentate gyrus of the hippocampus. The molecular mechanisms of cell proliferation and fate determination are largely unknown. Most in vitro studies favor embryonic or fetal rodents and use standard procedures for isolating particular regions of brain tissue. NSPCs are typically propagated as neurospheres (non-adherent cell aggregates) in serum-free medium containing supplements such as N2 or B27 as well as FGF2 and EGF [76, 77]. Upon manipulation of growth factor content, addition of serum, or plating on an adherent substrate, the cells differentiate within days in the in vitro environment into neurons, astrocytes and oligodendrocytes [76, 78]. NSPCs can also be propagated on adherent substrates as neurospheres or dissociated cells [79, 80].

The details of these methods can significantly influence NSPC survival and differentiation [81, 82]. Perhaps for this reason then, methods have favored one of the two formats for some cell types: NSPCs isolated from the hippocampus are usually grown as dissociated cells on laminin-coated substrates [83], whereas those derived from the subventricular zone destined for the rostral migratory stream are usually cultured as neurospheres [84]. Cells from these sources also vary in terms of growth factor supplementation during culture: hippocampal NSPCs are typically expanded in medium containing EGF and FGF2, and differentiated in medium without growth factors, but brain-derived growth factor (BDNF) can also be used to promote neuronal survival and differentiation [66]. Subventricular zone NSPCs may be differentiated into glial cells in the presence of EGF and specifically astrocytes with bone morphogenic proteins [85–87]. These culture conditions reflect the current knowledge of mechanisms of stem cell biology. The biological response to a single growth factor is dependent upon origin and age of the NSPCs and the composition of the environment: i.e., extracellular matrix molecules and configuration [88]. Adherent substrates coated with laminin have also proven to promote NSPC survival, proliferation and differentiation into neurons from human embryonic stem cells [89, 90] as well as postnatal humans [91] and embryonic mouse [92]. In vivo, multiple growth factors act simultaneously. A few studies have addressed responses to multiple growth factors and show synergistic and sometimes opposing effects. For hippocampal stem cells, EGF in culture promotes proliferation, whereas in the presence of BDNF, cell survival and differentiation are increased, but some degree of proliferation is maintained [93]. The sequence of the addition of either EGF, BDNF or NGF altered proliferation, differentiation, neurosphere size and individual cell morphology of embryonic heterogeneous striatal NSPCs [94]. These culture techniques consequently reflect a combination of established cell and tissue culture techniques and limited knowledge of molecular mechanisms that regulate cell proliferation, fate determination and survival.

Other non-neural sources of NSPCs are being explored, including the possibility of deriving neurons from skin and blood. Bone marrow mesenchymal cells may have the potential to differentiate into a variety of cell types, including NSPCs, but the results have been unclear as to whether this is a promising source of neural progenitors for therapeutic use [66, 95–98]. Human umbilical cord derived NSPCs also show promise in delaying the progression of neurological disease and degeneration [99]. Lastly, induced progenitor cells (iPCs) provide mechanisms to possibly elicit the neural cell phenotype from nearly any adult cell [100–102].

Despite their widespread use, culture of NSPCs as floating neurospheres or on adherent substrates share the general disadvantage of being highly artificial environments. While neurospheres offer the advantage of potentially providing a sphere of many cells that were propagated clonally from a single cell [103], most NSPC protocols describe the use of dissociated cells that coalesce into chimeric spheres. Regardless, the internal microenvironment in these systems is heterogeneous across a population of neurospheres as well as within a single neurosphere. Therefore, cell fate is greatly influenced by factors that are not controllable nor easily measured or interpreted [104–106]. For example, gradients of soluble molecules, differential cell–cell contacts and extracellular matrix deposition form heterogeneities or cellular subdomains. Moreover, cell viability may be compromised in neurosphere culture [66], particularly when the core of larger neurospheres becomes hypoxic and necrotic [107]. Finally, neurosphere cell populations change over time in culture and become enriched in glial cells [59, 108], thus providing an additional source of experimental variability. One may argue that adherent culture of dissociated cells provides a more uniform and controllable microenvironment [84], however, the stiff and two-dimensional nature of commonly used culture substrates forces the cells into yet a different non-physiological environment that may produce misleading results [109–112]. Indeed, the effects of suspension versus adherent culture on cell proliferation and fate are not well understood and few studies have been undertaken to elucidate specific differences [113, 114].

5 Biomaterials Use in NSPC Applications: State-of-the-Art

Biomaterials have a recognized role in their ability to mimic the physiological milieu in terms of structure and biochemistry and present the additional advantage of providing a reproducible scaffold to support in vitro culture and transplants. With advances in biomaterials synthesis, processing, and analysis in the past few decades, new technologies for controlling biomaterial physical properties [115–118], processing into relevant three-dimensional architectures [119, 120], as well as sophisticated methods to pattern cells [121–123] and present and release bioactive compounds [124–132], including soluble and immobilized gradients [125, 133–136] are now possible.

Given these recent advances, there have been a number of studies investigating the interactions between NSPCs and biomaterials (Tables 1, 2, 3). In general, work in this area has focused on the effect of the biomaterial microenvironment on NSPC differentiation. A variety of biomaterials have been investigated, including naturally-derived biopolymers [129, 131, 137–148], self-assembled peptides [149–151], and synthetic polymers [115–117, 125, 152–158]. Each of these classes of biomaterials include specific examples that are degradable [116, 117, 137, 154–156] and non-degradable [118, 125, 137].

Nearly all biomaterials studies evaluate cell viability, either with known standard cell lines, such as mouse fibroblasts, or with primary NSPCs and neuronal cell

Table 1 NSPC responses investigated using 2D biomaterial substrates

Cell response	Biomaterial	Source of NSPC	References
Viability			
Greater survival on stiff versus soft substrates			
	Crosslinked RGD- heparin-PEG gels (G' 0.2–5 kPa) that release FGF2	Embryonic rodent mesencephalon	[115]
Greater cell numbers on substrates with adhesive/bioactive molecules versus without			
	Chimeric protein of keratin and adhesive domain of laminin	Embryonic rodent striatum	[211]
	Poly(sialic acid) hydrogel modified with poly-ornithine/laminin, PLL, collagen	Fetal rodent mesencephalon	[140]
	Substrates with immobilized EGF	Embryonic rodent striatum	[135]
Proliferation			
Greater proliferation on substrates with adhesive peptides versus without			
	Self-assembled peptides containing RGD	Adult rodent hippocampus	[149]
	Interpenetrating network of various polymers modified with RGD or IKVAV	Adult rodent hippocampus	[158]
Neuronal fate			
Greater neuronal marker expression and neurite growth on soft versus stiff substrates			
	PDMS coated with poly-ornithine (YM < 12–750 kPa)	Embryonic rodent cerebral cortex	[118]
	PEG-PLL-laminin gels (EM 1–15 kPa) with varying chemical properties	Perinatal rodent brain	[116]
Greater neuronal differentiation on substrates modified with specific biological molecules or chemical groups			
	Glass modified with chemical groups (e.g. amino, carboxyl)	Embryonic rodent cerebral cortex	[145]
	Crosslinked chitosan modified with IFN-γ, BDNF, EP	Adult rodent SVZ/SEZ	[212]
	Poly(sialic acid) hydrogel modified with poly-ornithine/laminin, PLL, collagen	Fetal rodent mesencephalon	[145]
	Interpenetrating network of various polymers modified with RGD or IKVAV	Adult rodent hippocampus	[158]
Greater neurite growth on micropatterned substrates and co-culture with astrocytes			
	Micropatterned polymer for non-contact co-culture	Adult rodent hippocampus	[123, 213]
Migration			
Greater migration on substrates modified with specific biological molecules or chemical groups			
	PEG-PLL gels coated with laminin with varying chemical, mechanical properties	Perinatal rodent whole brain	[116]
	Glass modified with chemical groups (e.g., amino, carboxyl)	Embryonic rodent cerebral cortex	[145]
Function			
Greater expression of synaptic markers on soft versus stiff substrates			
	PDMS coated with poly-ornithine (YM < 12–750 kPa)	Embryonic rodent cerebral cortex	[118]

BDNF brain derived neurotrophic factor, *EM* elastic modulus, *EP* erythropoietin, *FGF2* fibroblast growth factor-2, *IFN-γ* interferon-γ, *IKVAV* Ile-Lys-Val-Ala-Val laminin-derived adhesive peptide, *PDMS* poly (dimethyl siloxane), *PEG* poly(ethylene glycol), *PLL* poly(L-lysine), *RGD* Arg-Gly-Asp fibronectin-derived adhesive peptide, *SEZ* subependymal zone, *SVZ* subventricular zone, *YM* Young's modulus

Table 2 NSPC responses investigated using 3D biomaterial scaffolds

Cell response	Biomaterial	Source of NSPC	References
Viability			
Maximal viability at moderate cell density			
	Type I collagen seeded with 1e4–1e8 cells/ml	Embryonic rodent striatum	[147]
Maximal viability at low gel density			
	Type I collagen (0.3–3 mg/ml)	Embryonic rodent striatum	[147]
Greater viability in scaffolds containing specific biological molecules			
	Type I collagen, agarose	Embryonic rodent cortex	[143, 144]
	PEG-PLA gel containing FGF2 and collagen	Embryonic rodent forebrain	[117, 156]
	Porous PLGA scaffold coated with Matrigel or fibronectin	Human embryonic cell line	[154, 155]
Greater viability in degradable versus non-degradable gels			
	Alginate	Adult rodent hippocampus	[137]
Proliferation			
Greater proliferation on stiff versus soft scaffolds			
	Alginate gels (EM 0.2–20 kPa)	Adult rodent hippocampus	[138]
Greater proliferation in scaffolds containing specific biological molecules			
	Porous PLGA scaffold coated with Matrigel or fibronectin	Human embryonic cell line	[154, 155]
Neuronal fate			
Greater neuronal differentiation in scaffolds containing specific biological molecules			
	HA/collagen sponge	Adult rodent SVZ/SEZ	[146]
	Type 1 collagen, agarose	Embryonic rodent cortex	[143, 144]
	Self-assembled peptides containing IKVAV	Embryonic rodent cortex	[150]
Greater neuronal differentiation on soft versus stiff scaffolds			
	Alginate gels (EM 0.2–20 kPa)	Adult rodent hippocampus	[138]
Migration			
Maximal migration at low gel density			
	Type 1 collagen (0.3–3 mg/ml)	Embryonic rodent striatum	[147]
Function			
Greater polarization and/or excitability in adhesive gels			
	Type I collagen, agarose	Embryonic rodent cortex	[141, 143, 144]
	PEG-PLA gel containing FGF2 and collagen	Embryonic rodent forebrain	[117, 156]

EM elastic modulus, *FGF2* fibroblast growth factor-2, *HA* hyaluronic acid, *IKVAV* Ile-Lys-Val-Ala-Val laminin-derived adhesive peptide, *PEG* poly(ethylene glycol), *PLA* poly(lactic acid), *PLGA* poly(lactic-co-glycolic acid), *SEZ* subependymal zone, *SVZ* subventricular zone

lines. Further characterizations of the materials include expression analysis of neuronal and glial markers and general description of morphology, namely neurite outgrowth. In some advanced stages, neuronal function is measured using

Table 3 NSPC responses investigated using controlled release of bioactive factors

Cell response	Biomaterial	Source of NSPC	References
Viability			
Greater cell numbers with controlled release versus soluble factor			
	Retinoic acid-releasing PLGA microspheres coated with PLL	Embryonic rodent cell line	[132]
Neuronal fate			
Greater neuronal differentiation with controlled release versus soluble factor			
	Retinoic acid-releasing PLGA microspheres coated with PLL	Embryonic rodent cell line	[132]
	NT-3-releasing chitosan carriers	Perinatal rodent spinal cord	[131]
Greater neurite growth with controlled release versus soluble factor			
	NT-3-releasing chitosan carriers	Perinatal rodent spinal cord	[131]

PLGA poly(lactic-co-glycolic acid), *PLL* poly(L-lysine), *NT-3* neurotrophin-3

electrophysiology techniques or photosensing dyes (sensitive to voltage or the presence of calcium).

6 Current Challenges in Biomaterials for NSPC Applications

Whereas some success has been noted in the use of biomaterial implants to reduce inflammation and glial scar in the brain [65, 115, 159], most investigations of biomaterials for NSPC applications have been implemented in vitro and the few transplant studies undertaken have not improved upon the level of success described above for NSPC transplants alone. We and others [114, 160] view the primary challenge towards the successful application of biomaterials for NSPC culture and transplantation as not a lack of engineering tools, but rather an insufficiency of biological inspiration. This result could stem from NSPC biology being a rather recent and currently unfolding field in combination with an inherent communication barrier between the fields of neurobiology and biomaterials engineering.

Studies of NSPC–biomaterials interactions seem to be largely driven by biomaterials innovations and result in the characterization of induced differentiation by the biomaterials environment or controlled release of factors well-recognized to influence NSPC proliferation and differentiation. In other words, biomaterials design in general is not being driven by the need to answer challenging biological questions, which the field is certainly well-poised to tackle in collaboration with neurobiologists and clinicians. For example, it is not clear why many studies focus on the characterization NSPC differentiation at the exclusion of promoting cell survival and expansion. Indeed, it is likely that successful therapeutic strategies will require large numbers of cells and in general original sources of NSPCs only provide limited numbers of viable cells. Perhaps the focus on promoting cell adhesion, a popular strategy for terminally differentiated cells, has distracted researchers away from exploring the possibility that adhesion promotes cell differentiation and thus studies have overlooked strategies to promote earlier stages in neurogenesis. A final but significant challenge towards implementing biomaterials for NSPC therapies is that there is a lack of side-by-side comparisons between multiple materials (exceptions being [116, 161, 162]) and standardization of cell sources, culture methods, target outcomes and assessment tools.

7 Potential of Biomaterials for Reverse-engineering NSPC Microenvironments

Despite the great recent advances in our understanding of NSPC biology, this exciting field has the possibility of stagnating because of the limitations of current culture platforms and transplantation methods. Given the technological sophistication of biomaterials science, it is highly plausible that in collaboration with neurobiologists and clinical neurologists, biomaterials engineers can provide the tools and analytical methods to make groundbreaking discoveries in NSPC biology and therapeutic application. We suggest here several potential avenues where biomaterials engineers may make specific contributions in the near future.

7.1 Neurosphere Culture

One area that biomaterials engineers could make an immediate impact is to consider neurospheres as a type of biomaterial. Several lines of reasoning support this proposition. First, neurosphere culture is one of the few methods of propagating stem cells in the undifferentiated state. What makes this possible? As mentioned above, side-by-side comparison of the signaling events that occur in the neurosphere versus dissociated cells may make this possible. However, due to the heterogeneity of the neurosphere microenvironment, this approach poses significant challenges to the application of current biological tools. Yet by reverse-engineering the neurosphere environment, biomaterials engineers may be able to identify specific and perhaps unique physical and biochemical cues presented within the neurosphere by mimicking this microenvironment within tunable three-dimensional synthetic scaffolds. Specific questions include:

1. What are the dominant factors that NSPCs use to communicate with other cells in neurosphere culture? Presumably, cell–cell contacts and soluble elements are important, but which specific pathways are involved? Is extracellular matrix secreted and does it initiate integrin-mediated signaling events? Does lack of substrate adhesion contribute to maintenance of the undifferentiated state or is it an artifact of culture in a compliant three-dimensional microenvironment? How does the undifferentiated cell, which is largely connected to its neighbors

through cell–cell contacts, remodel its environment during and after differentiation? Can this process be controlled with clever new materials that are hybrids of classical biopolymers and natural molecules?
2. Gradients of soluble molecules, including nutrients, waste, growth factors and oxygen, persist within neurosphere culture. At what critical size or cell number do these factors begin to present dominant effects? Could these gradients be beneficial? For example, NSPCs are relatively resistant to hypoxia/ischemia as compared to differentiated cells; could this effect be at play in neurosphere culture to promote the undifferentiated state? Under what conditions does the necrotic core form? How does it influence the state of the cells in the immediate environment of the core? Can the biochemical environment be mimicked in a different geometry that is more suitable for large scale production?
3. What specific properties of the neurosphere should be mimicked in biomaterial scaffolds? Is this the preferred environment or is it a 'stray path' to try to understand an already artificial environment [163]? Or, perhaps neurosphere culture provides a point of inspiration towards developing a new strategy to afford dynamic cell adhesion to biomaterials wherein the scaffold is initially non-adhesive to promote NSPC expansion and then adhesive towards cells to promote differentiation (analogous to the temperature sensitive polymer poly-*N*-isopropylacrylamide, which has typically been implemented to promote cell adhesion first and then detachment [164, 165]). Similar methods could be developed for dynamic growth factor delivery (as already demonstrated to be feasible for promoting angiogenesis [166]) or controllable hypoxia gradients [134, 167].

7.2 The Stem Cell Niche

After fetal development, a small fraction of NSPCs remain in the brain within the unique microenvironment of the neural stem cell niche [168, 169] (Fig. 3). Because neural stem cells are multipotent and have the potential for unlimited proliferation, strict control over their behavior is critical during development and maintenance of the stem cell population throughout the lifespan of the adult. However, stem cells are known to become activated and migrate towards sites of injury, necessitating regulation of signals that control stem cell quiescence as well as activation within the niche microenvironment. While the details behind these processes have not yet been made clear, it is likely that cell–cell signaling, growth factors and extracellular matrix contribute to this highly regulated process [170–172]. Specific mediators may include growth factors, such as FGF2, ciliary neurotrophic factor (CNTF) and transforming growth factor α (TGFα), neurotransmitters, hormones and integrin-mediated cell–matrix interactions. Comprehensive reviews of these factors are found in [168, 170, 173, 174]. Moreover, clinical studies have identified potential life experiences such as seizure [175], exercise [16], stress [176], injury [25] and medication use [15] to mediate neurogenesis in adults.

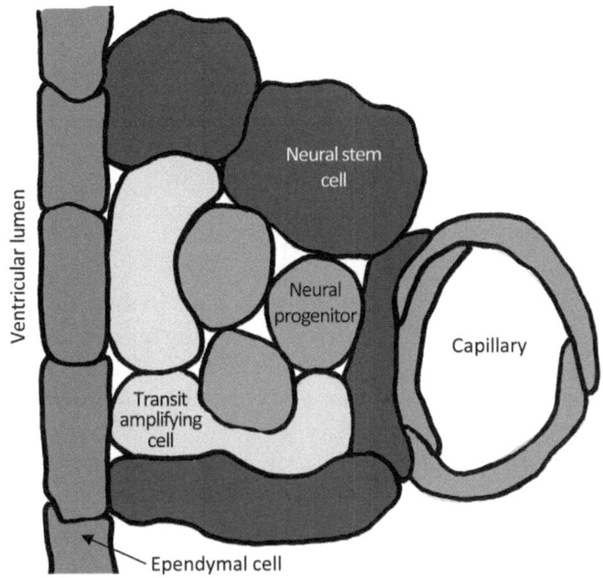

Fig. 3 The neural stem cell niche in the subventricular zone of the postnatal brain. Neural stem cells differentiate first into an intermediate cell type, transit amplifying cells, which rapidly proliferate, and then into neural progenitor cells. Stem cells remain in the niche in contact with ependymal cells as well as the endothelial cells and basal lamina of nearby capillaries. Figure adapted from [30]

Biomaterials scientists have begun to address these topics [91, 92, 177], yet even as a number of specific individual molecules have been suggested to regulate the neural stem cell niche, questions remain. For example, what dynamics and spatial gradients underplay stem cell quiescence and activation? Do synergistic interactions mediate this response? How does the stem cell niche balance cell survival versus differentiation? Are all niches the same? Can stem cells from one niche populate another?

7.3 "Physiological Hypoxia" and Hypoxic/Ischemic Injury

Oxygen is a key regulator of neural development, NSPC regulation and normal brain physiology [30]. Due to high levels of metabolic activity, oxygen transport is strictly regulated within the brain [178] and global oxygen concentrations are conserved among mammals, but vary by the specific region of the brain [30]. Reports have listed oxygen concentrations in the human brain to range from 0.6 to 8% in the midbrain (internal core) and pia (external surface), respectively [30], and 0.1–5% in various regions of the rat brain depending on proximity to major vasculature and metabolic load [179] (Table 4). In the developing mammalian embryo, when the NSPCs are expanding to form the brain, the oxygen concentrations are even lower. Air, however, contains 21% oxygen; this large discrepancy between the oxygen levels in the brain and in the air that we breath and expose cells during culture has inspired some to term normal low levels of oxygen in the brain as "physiological hypoxia" [180–182].

Table 4 Oxygen concentrations in the rat brain as measured by microelectrode [179]

Brain region	Oxygen concentration (%)
Gray cortex	2.5–5.3
White cortex	0.8–2.1
Hippocampus	2.6–3.9
Hypothalamus	1.4–2.1
Pons/fornix	0.1–0.4

Because cells within the brain vary in their differentiation state and the physiological levels of oxygen in their microenvironment, injuries such as stroke and asphyxia result in a non-homogeneous susceptibility to hypoxia in various subpopulations of the brain. For example, projection neurons in the neocortex and cerebellum are particularly sensitive to hypoxia and quickly die without sufficient oxygen. Interestingly, cells in neighboring regions of the brain can vary quite distinctly in their sensitivity to hypoxia: projection neurons in the CA2 and CA3 fields of the hippocampus are relatively resistant to hypoxia and survive through insults that cause nearby projection neurons in the CA1 area to die [183]. Focal areas of hypoxia are also linked to tumors in the brain and other tissues [184], suggesting that cancer cells are well adapted to tolerate hypoxic conditions. The mechanisms that give rise to this hierarchy in sensitivity to hypoxia are not well understood.

Culture studies have elucidated some key differences in how NSPCs respond to hypoxia (Fig. 4). The fundamental work in this area was initiated in the early 2000s, when several key studies demonstrated that low but physiologic levels of hypoxia promoted proliferation and inhibited differentiation of NSPCs derived from rat or human midbrain [39, 180–182]. However, as mentioned above, the relative susceptibility of NSPCs to hypoxia depends on the region of the brain. In culture carried out in air (20% oxygen), rat cortical and human postnatal NSPCs differentiate into neurons [10, 185], whereas mouse fetal cortical NSPCs undergo apoptosis [186]. Others have postulated that culture in air may also promote NSPC differentiation to astrocytes; an effect that is exacerbated at high cell densities [10]. Indeed, a consensus understanding about the effect of hypoxia on NSPCs has not been established [32, 36, 139, 187, 188], possibly due to varying cell type (region of brain, animal age) and culture methods (e.g., oxygen concentration, culture duration, presence or absence of growth factors, suspension or adherent culture) [36].

These relationships between oxygen concentration, culture models and NSPC biology have important implications for understanding hypoxic injury and developing more effective therapeutic strategies. For example, hypoxia is likely a trigger to quiescent NSPCs to break away from their niche and reactivate the low oxygen embryonic environment, inducing them to proliferate, migrate and differentiate [32]. Thus, it is plausible that a more detailed understanding of how hypoxia mediates NSPC biology could be implemented to produce large numbers of NSPCs for therapeutic use [10, 30, 36].

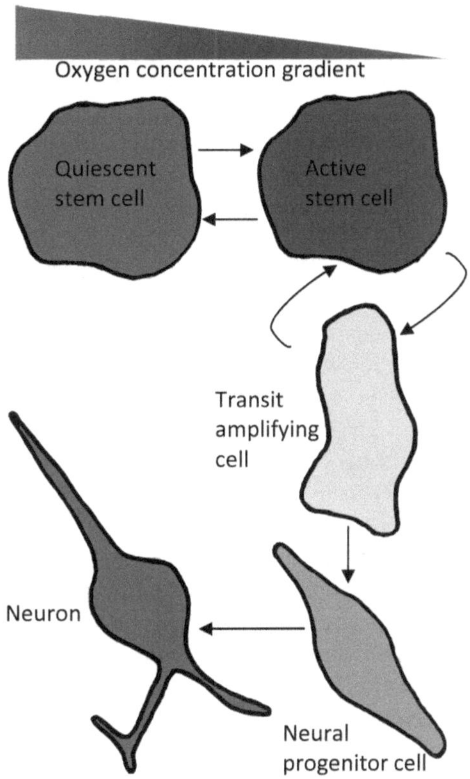

Fig. 4 Influence of oxygen concentration on NSPC activation and differentiation. Hypoxia can promote NSPC proliferation, differentiation and migration to areas where oxygen concentration is higher. Figure adapted from [30]

While several oxygen-sensitive signaling pathways have been identified, one clear mediator of cell response to hypoxia is the transcription factor hypoxia inducible factor 1 (HIF-1) [30]. When activated, HIF-1 binds to hypoxia response elements and initiates the transcription of genes that allow cells to survive in low levels of oxygen [189, 190]. At low oxygen concentrations, HIF-1 initiates signaling pathways that promote proliferation and generally inhibit differentiation in NSPCs; conversely, higher levels of oxygen promote NSPC differentiation and apoptosis. HIF-1 also mediates angiogenesis and initiates transcription of factors such as VEGF and erythropoietin that play roles in both angiogenesis and neurogenesis [39, 188, 191–193]. HIF-1 is upregulated after hypoxia/ischemia brain injury in adults [194] as well as neonatal humans [189].

Despite these advances, many mechanistic questions remain as to the details of how hypoxia directly and indirectly affects NSPC response. It is likely that biomaterials can play a significant role in elucidating this problem; for example:

1. In addition to hypoxia, other factors could also be altered in ischemic tissues that regulate NSPC response. For example, factors recognized to affect NSPC migration include architectures provided by astrocytes and blood vessels [30, 170], matrix molecules and their degradation by matrix metalloproteinases

[195–197] and pathfinding molecules such as slit/robo and ephrin B [168, 198, 199]. NSPC survival and proliferation are also key processes for healing hypoxic tissue; a large number of growth factors are known to be expressed in ischemic tissues [200–203] and may initiate signaling events directly or cause upregulated expression of cell-surface receptors [204, 205] resulting in increases sensitivity to cues from the cellular microenvironment. Thus, in vitro models of ischemic disease with tunable presentation of soluble and matrix cues could be useful to elucidate interactions in the neural stem cell niche.

2. Diatomic oxygen (as opposed to oxygen free radicals) is not produced in the body. It is difficult to disrupt oxygen-mediated pathways and there are few methods to directly quantify oxygen levels in three-dimensional systems with high resolution [184]. Nonetheless, recent advances in biomaterials engineering may provide adequate tools towards controlling oxygen gradients in culture [134, 167] and measuring oxygen concentration in complex in vitro culture scaffolds [206].
3. Reperfusion to hypoxic/ischemic injury returns vital oxygen and nutrients to weak tissue; however, the effects of reperfusion are not all positive. Oxidative stress is one major contributor towards tissue damage during reperfusion injury [207]. Moreover, rapid proliferation and migration of NSPCs are metabolically demanding and the high respiratory rates exacerbate local levels of reactive oxygen species in the injury site [208]. Interestingly, more oxidized states lead to NSPC differentiation whereas reduced states promote NSPC proliferation [208, 209]. Biomaterials technologies may provide new mechanisms to control concentrations of reactive oxygen species via quenching, adsorption or controlled release of factors to influence redox reactions in the extracellular environment (e.g., lactic acid; [210]). Similar approaches may target other aspects of reperfusion injury, such as hemoglobin and harmful inflammatory cues.

8 Conclusions

The stem cell field is rich with potential for therapies for previously intractable disorders, such as those related to hypoxic brain injury. Stem cells represent a unique biological material, one that is directly suitable for therapeutic use and can be expanded and tailored into specific cell types and therapies. Biologists have started to reveal the molecules and pathways necessary for supporting cell viability and driving cell behavior. However, many of the major scientific and clinical challenges in developing and implementing stem cell based therapies are engineering problems related to mass transfer of biological components and the dynamics of multi-component interactions. Biomedical and biochemical engineers are equipped with multiple tools for quantifying biological responses and provide better material environments to achieve specific cellular products. Clinicians have the expertise to define the health care problem and suggest tissue replacement strategies. Thus, collaborations between biologists, engineers and clinicians will define the goals, refine and advance the technology, and eventually provide commercial and therapeutic products for advancing human health.

Acknowledgments We gratefully acknowledge funding from NIH-NINDS (R01 NS065205; JBL), the Henry Luce Foundation (JBL), the Maryland Stem Cell Research Fund (EMP), March of Dimes Basil O'Connor Starter Scholar Award (EMP) and NIH-NIDA (R01 MH018826; EMP).

References

1. Bjorklund, A., Lindvall, O.: Cell replacement therapies for central nervous system disorders. Nat. Neurosci. **3**, 537–544 (2000)
2. Volpe, J.J.: Neurology of the Newborn. Saunders, Philadelphia (2001)
3. Bang, O.Y., Lee, J.S., Lee, P.H., Lee, G.: Autologous mesenchymal stem cell transplantation in stroke patients. Ann. Neurol. **57**, 874–882 (2005)
4. Kondziolka, D., Steinberg, G.K., Wechsler, L., Meltzer, C.C., Elder, E., Gebel, J., Decesare, S., Jovin, T., Zafonte, R., Lebowitz, J., Flickinger, J.C., Tong D., Marks, M.P., Jamieson, C., Luu, D., Bell-Stephens, T., Teraoka, J.: Neurotransplantation for patients with subcortical motor stroke: a phase 2 randomized trial. J. Neurosurg. **103**, 38–45 (2005)
5. Kondziolka, D., Wechsler, L., Goldstein, S., Meltzer, C., Thulborn, K.R., Gebel, J., Jannetta, P., DeCesare, S., Elder, E.M., McGrogan, M., Reitman, M.A., Bynum, L.: Transplantation of cultured human neuronal cells for patients with stroke. Neurology **55**, 565–569 (2000)
6. Nelson, P.T., Kondziolka, D., Wechsler, L., Goldstein, S., Gebel, J., DeCesare, S., Elder, E.M., Zhang, P.J., Jacobs, A., McGrogan, M., Lee, V.M., Trojanowski, J.Q.: Clonal human (hNT) neuron grafts for stroke therapy: neuropathology in a patient 27 months after implantation. Am. J. Pathol. **160**, 1201–1206 (2002)
7. Jin, K., Wang, X., Xie, L., Mao, X.O., Zhu, W., Wang, Y., Shen, J., Mao, Y., Banwait, S., Greenberg, D.A.: Evidence for stroke-induced neurogenesis in the human brain. Proc. Natl. Acad. Sci. USA **103**, 13198–13202 (2006)
8. Macas, J., Nern, C., Plate, K.H., Momma, S.: Increased generation of neuronal progenitors after ischemic injury in the aged adult human forebrain. J. Neurosci. **26**, 13114–13119 (2006)
9. Minger, S.L., Ekonomou, A., Carta, E,M., Chinoy, A., Perry, R.H., Ballard, C.G.: Endogenous neurogenesis in the human brain following cerebral infarction. Regen. Med. **2**, 69–74 (2007)
10. Pistollato, F., Chen, H.L., Schwartz, P.H., Basso, G., Panchision, D.M.: Oxygen tension controls the expansion of human CNS precursors and the generation of astrocytes and oligodendrocytes. Mol. Cell. Neurosci. **35**, 424–435 (2007)
11. Kornack, D.R., Rakic, P.: The generation, migration, and differentiation of olfactory neurons in the adult primate brain. Proc. Natl. Acad. Sci. USA **98**, 4752–4757 (2001)
12. Lledo, P.M., Merkle, F.T., Alvarez-Buylla, A.: Origin and function of olfactory bulb interneuron diversity. Trends Neurosci. **31**, 392–400 (2008)
13. Luskin, M.B.: Restricted proliferation and migration of postnatally generated neurons derived from the forebrain subventricular zone. Neuron **11**, 173–189 (1993)
14. Gould, E., Beylin, A., Tanapat, P., Reeves, A., Shors, T.J.: Learning enhances adult neurogenesis in the hippocampal formation. Nat. Neurosci. **2**, 260–265 (1999)
15. Santarelli, L., Saxe, M., Gross, C., Surget, A., Battaglia, F., Dulawa, S., Weisstaub, N., Lee, J., Duman, R., Arancio, O., Belzung, C., Hen, R.: Requirement of hippocampal neurogenesis for the behavioral effects of antidepressants. Science **301**, 805–809 (2003)

16. van Praag, H., Kempermann, G., Gage, F.H.: Running increases cell proliferation and neurogenesis in the adult mouse dentate gyrus. Nat. Neurosci. **2**, 266–270 (1999)
17. Will, B., Galani, R., Kelche, C., Rosenzweig, M.R.: Recovery from brain injury in animals: relative efficacy of environmental enrichment, physical exercise or formal training (1990–2002). Prog. Neurobiol. **72**, 167–182 (2004)
18. Wolf, S.A., Kronenberg, G., Lehmann, K., Blankenship, A., Overall, R., Staufenbiel, M., Kempermann, G.: Cognitive and physical activity differently modulate disease progression in the amyloid precursor protein (APP)-23 model of Alzheimer's disease. Biol. Psychiatry **60**, 1314–1323 (2006)
19. Kalluri, H.S., Dempsey, R.J.: Growth factors, stem cells, and stroke. Neurosurg. Focus **24**, E14 (2008)
20. Fagel, D.M., Ganat, Y., Silbereis, J., Ebbitt, T., Stewart, W., Zhang, H., Ment, L.R., Vaccarino, F.M.: Cortical neurogenesis enhanced by chronic perinatal hypoxia. Exp. Neurol. **199**, 77–91 (2006)
21. Hayashi, T., Iwai, M., Ikeda, T., Jin, G., Deguchi, K., Nagotani, S., Zhang, H., Sehara, Y., Nagano, I., Shoji, M., Ikenoue, T., Abe, K.: Neural precursor cells division and migration in neonatal rat brain after ischemic/hypoxic injury. Brain Res. **1038**, 41–49 (2005)
22. Sharp, F.R., Liu, J., Bernabeu, R.: Neurogenesis following brain ischemia. Brain Res. Dev. Brain Res. **134**, 23–30 (2002)
23. Arvidsson, A., Collin, T., Kirik, D., Kokaia, Z., Lindvall, O.: Neuronal replacement from endogenous precursors in the adult brain after stroke. Nat. Med. **8**, 963–970 (2002)
24. Kee, N.J., Preston, E., Wojtowicz, J.M.: Enhanced neurogenesis after transient global ischemia in the dentate gyrus of the rat. Exp. Brain Res. **136**, 313–320 (2001)
25. Jin, K., Minami, M., Lan, J.Q., Mao, X.O., Batteur, S., Simon, R.P., Greenberg, D.A.: Neurogenesis in dentate subgranular zone and rostral subventricular zone after focal cerebral ischemia in the rat. Proc. Natl. Acad. Sci. USA **98**, 4710–4715 (2001)
26. Yagita, Y., Kitagawa, K., Ohtsuki, T., Takasawa, K., Miyata, T., Okano, H., Hori, M., Matsumoto, M.: Neurogenesis by progenitor cells in the ischemic adult rat hippocampus. Stroke **32**, 1890–1896 (2001)
27. Takagi, Y., Nozaki, K., Takahashi, J., Yodoi, J., Ishikawa, M., Hashimoto, N.: Proliferation of neuronal precursor cells in the dentate gyrus is accelerated after transient forebrain ischemia in mice. Brain Res. **831**, 283–287 (1999)
28. Liu, J., Solway, K., Messing, R.O., Sharp, F.R.: Increased neurogenesis in the dentate gyrus after transient global ischemia in gerbils. J. Neurosci. **18**, 7768–7778 (1998)
29. Romanko, M.J., Rothstein, R.P., Levison, S.W.: Neural stem cells in the subventricular zone are resilient to hypoxia/ischemia whereas progenitors are vulnerable. J. Cereb. Blood Flow Metab. **24**, 814–825 (2004)
30. Panchision, D.M.: The role of oxygen in regulating neural stem cells in development and disease. J. Cell Physiol. **220**, 562–568 (2009)
31. Burgers, H.F., Schelshorn, D.W., Wagner, W., Kuschinsky, W., Maurer, M.H.: Acute anoxia stimulates proliferation in adult neural stem cells from the rat brain. Exp. Brain Res. **188**, 33–43 (2008)
32. Horie, N., So, K., Moriya, T., Kitagawa, N., Tsutsumi, K., Nagata, I., Shinohara, K.: Effects of oxygen concentration on the proliferation and differentiation of mouse neural stem cells in vitro. Cell. Mol. Neurobiol. **28**, 833–845 (2008)
33. Parent, J.M., Vexler, Z.S., Gong, C., Derugin, N., Ferriero, D.M.: Rat forebrain neurogenesis and striatal neuron replacement after focal stroke. Ann. Neurol. **52**, 802–813 (2002)
34. Zhang, C.P., Zhu, L.L., Zhao, T., Zhao, H., Huang, X., Ma, X., Wang, H., Fan, M.: Characteristics of neural stem cells expanded in lowered oxygen and the potential role of hypoxia-inducible factor-1alpha. Neurosignals **15**, 259–265 (2006)
35. Zhang, R.L., Zhang, Z.G., Zhang, L., Chopp, M.: Proliferation and differentiation of progenitor cells in the cortex and the subventricular zone in the adult rat after focal cerebral ischemia. Neuroscience **105**, 33–41 (2001)

36. Zhao, T., Zhang, C.P., Liu, Z.H., Wu, L.Y., Huang, X., Wu, H.T., Xiong, L., Wang, X., Wang, X.M., Zhu, L.L., Fan, M.: Hypoxia-driven proliferation of embryonic neural stem/progenitor cells—role of hypoxia-inducible transcription factor-1alpha. FEBS J. **275**, 1824–1834 (2008)
37. Belmadani, A., Tran, P.B., Ren, D., Miller, R.J.: Chemokines regulate the migration of neural progenitors to sites of neuroinflammation. J. Neurosci. **26**, 3182–3191 (2006)
38. Xu, Q., Wang, S., Jiang, X., Zhao, Y., Gao, M., Zhang, Y., Wang, X., Tano, K., Kanehara, M., Zhang, W., Ishida, T.: Hypoxia-induced astrocytes promote the migration of neural progenitor cells via vascular endothelial factor, stem cell factor, stromal-derived factor-1alpha and monocyte chemoattractant protein-1 upregulation in vitro. Clin. Exp. Pharmacol. Physiol. **34**, 624–631 (2007)
39. Jin, K., Zhu, Y., Sun, Y., Mao, X.O., Xie, L., Greenberg, D.A.: Vascular endothelial growth factor (VEGF) stimulates neurogenesis in vitro and in vivo. Proc. Natl. Acad. Sci. USA **99**, 11946–11950 (2002)
40. Thored, P., Wood, J., Arvidsson, A., Cammenga, J., Kokaia, Z., Lindvall, O.: Long-term neuroblast migration along blood vessels in an area with transient angiogenesis and increased vascularization after stroke. Stroke **38**, 3032–3039 (2007)
41. Zhang, Z.G., Zhang, L., Jiang, Q., Zhang, R., Davies, K., Powers, C., Bruggen, N., Chopp, M.: VEGF enhances angiogenesis and promotes blood-brain barrier leakage in the ischemic brain. J. Clin. Invest. **106**, 829–838 (2000)
42. Zhang, Z.G., Zhang, L., Tsang, W., Soltanian-Zadeh, H., Morris, D., Zhang, R., Goussev, A., Powers, C., Yeich, T., Chopp, M.: Correlation of VEGF and angiopoietin expression with disruption of blood-brain barrier and angiogenesis after focal cerebral ischemia. J. Cereb. Blood Flow Metab. **22**, 379–392 (2002)
43. Teng, H., Zhang, Z.G., Wang, L., Zhang, R.L., Zhang, L., Morris, D., Gregg, S.R., Wu, Z., Jiang, A., Lu, M., Zlokovic, B.V., Chopp, M.: Coupling of angiogenesis and neurogenesis in cultured endothelial cells and neural progenitor cells after stroke. J. Cereb. Blood Flow Metab. **28**, 764–771 (2008)
44. Ohab, J.J., Carmichael, S.T.: Poststroke neurogenesis: emerging principles of migration and localization of immature neurons. Neuroscientist **14**, 369–380 (2008)
45. Hayashi, T., Noshita, N., Sugawara, T., Chan, P.H.: Temporal profile of angiogenesis and expression of related genes in the brain after ischemia. J. Cereb. Blood Flow Metab. **23**, 166–180 (2003)
46. Craig, C.G., Tropepe, V., Morshead, C.M., Reynolds, B.A., Weiss, S., van der Kooy, D.: In vivo growth factor expansion of endogenous subependymal neural precursor cell populations in the adult mouse brain. J. Neurosci. **16**, 2649–2658 (1996)
47. Kuhn, H.G., Winkler, J., Kempermann, G., Thal, L.J., Gage, F.H.: Epidermal growth factor and fibroblast growth factor-2 have different effects on neural progenitors in the adult rat brain. J. Neurosci. **17**, 5820–5829 (1997)
48. Shen, Q., Wang, Y., Kokovay, E., Lin, G., Chuang, S.M., Goderie, S.K., Roysam, B., Temple, S.: Adult SVZ stem cells lie in a vascular niche: a quantitative analysis of niche cell-cell interactions. Cell Stem Cell **3**, 289–300 (2008)
49. Zheng, W., Nowakowski, R.S., Vaccarino, F.M.: Fibroblast growth factor 2 is required for maintaining the neural stem cell pool in the mouse brain subventricular zone. Dev. Neurosci. **26**, 181–196 (2004)
50. Mirzadeh, Z., Merkle, F.T., Soriano-Navarro, M., Garcia-Verdugo, J.M., Alvarez-Buylla, A.: Neural stem cells confer unique pinwheel architecture to the ventricular surface in neurogenic regions of the adult brain. Cell Stem Cell **3**, 265–278 (2008)
51. Tavazoie, M., Van der Veken, L., Silva-Vargas, V., Louissaint, M., Colonna, L., Zaidi, B., Garcia-Verdugo, J.M., Doetsch, F.: A specialized vascular niche for adult neural stem cells. Cell Stem Cell **3**, 279–288 (2008)
52. Bovetti, S., Hsieh, Y.C., Bovolin, P., Perroteau, I., Kazunori, T., Puche, A.C.: Blood vessels form a scaffold for neuroblast migration in the adult olfactory bulb. J. Neurosci. **27**, 5976–5980 (2007)

53. Burns, T.C., Verfaillie, C.M., Low, W.C.: Stem cells for ischemic brain injury: a critical review. J. Comp. Neurol. **515**, 125–144 (2009)
54. Kelly, S., Bliss, T.M., Shah, A.K., Sun, G.H., Ma, M., Foo, W.C., Masel, J., Yenari, M.A., Weissman, I.L., Uchida, N., Palmer, T., Steinberg, G.K.: Transplanted human fetal neural stem cells survive, migrate, and differentiate in ischemic rat cerebral cortex. Proc. Natl. Acad. Sci. USA **101**, 11839–11844 (2004)
55. Song, H., Stevens, C.F., Gage, F.H.: Astroglia induce neurogenesis from adult neural stem cells. Nature **417**, 39–44 (2002)
56. Song, H.J., Stevens, C.F., Gage, F.H.: Neural stem cells from adult hippocampus develop essential properties of functional CNS neurons. Nat. Neurosci. **5**, 438–445 (2002)
57. Eriksson, C., Bjorklund, A., Wictorin, K.: Neuronal differentiation following transplantation of expanded mouse neurosphere cultures derived from different embryonic forebrain regions. Exp. Neurol. **184**, 615–635 (2003)
58. Herrera, D.G., Garcia-Verdugo, J.M., Alvarez-Buylla, A.: Adult-derived neural precursors transplanted into multiple regions in the adult brain. Ann. Neurol. **46**, 867–877 (1999)
59. Winkler, C., Fricker, R.A., Gates, M.A., Olsson, M., Hammang, J.P., Carpenter, M.K., Bjorklund, A.: Incorporation and glial differentiation of mouse EGF-responsive neural progenitor cells after transplantation into the embryonic rat brain. Mol. Cell Neurosci. **11**, 99–116 (1998)
60. Pluchino, S., Zanotti, L., Rossi, B., Brambilla, E., Ottoboni, L., Salani, G., Martinello, M., Cattalini, A., Bergami, A., Furlan, R., Comi, G., Constantin, G., Martino, G.: Neurosphere-derived multipotent precursors promote neuroprotection by an immunomodulatory mechanism. Nature **436**, 266–271 (2005)
61. Yandava, B.D., Billinghurst, L.L., Snyder, E.Y.: "Global" cell replacement is feasible via neural stem cell transplantation: evidence from the dysmyelinated shiverer mouse brain. Proc. Natl. Acad. Sci. USA **96**, 7029–7034 (1999)
62. Wurmser, A.E., Nakashima, K., Summers, R.G., Toni, N., D'Amour, K.A., Lie, D.C., Gage, F.H.: Cell fusion-independent differentiation of neural stem cells to the endothelial lineage. Nature **430**, 350–356 (2004)
63. Locatelli, F., Bersano, A., Ballabio, E., Lanfranconi, S., Papadimitriou, D., Strazzer, S., Bresolin, N., Comi, G.P., Corti, S.: Stem cell therapy in stroke. Cell Mol. Life Sci. **66**, 757–772 (2009)
64. Englund, U., Bjorklund, A., Wictorin, K., Lindvall, O., Kokaia, M.: Grafted neural stem cells develop into functional pyramidal neurons and integrate into host cortical circuitry. Proc. Natl. Acad Sci. USA **99**, 17089–17094 (2002)
65. Park, K.I., Teng, Y.D., Snyder, E.Y.: The injured brain interacts reciprocally with neural stem cells supported by scaffolds to reconstitute lost tissue. Nat. Biotechnol. **20**, 1111–1117 (2002)
66. Conti, L., Reitano, E., Cattaneo, E.: Neural stem cell systems: diversities and properties after transplantation in animal models of diseases. Brain Pathol. **16**, 143–154 (2006)
67. Ourednik, J., Ourednik, V., Lynch, W.P., Schachner, M., Snyder, E.Y.: Neural stem cells display an inherent mechanism for rescuing dysfunctional neurons. Nat. Biotechnol. **20**, 1103–1110 (2002)
68. Lee, J.P., Jeyakumar, M., Gonzalez, R., Takahashi, H., Lee, P.J., Baek, R.C., Clark, D., Rose, H., Fu, G., Clarke, J., McKercher, S., Meerloo, J., Muller, F.J., Park, K.I., Butters, T.D., Dwek, R.A., Schwartz, P., Tong, G., Wenger, D., Lipton, S.A., Seyfried, T.N., Platt, F.M., Snyder, E.Y.: Stem cells act through multiple mechanisms to benefit mice with neurodegenerative metabolic disease. Nat. Med. **13**, 439–447 (2007)
69. Brundin, P., Karlsson, J., Emgard, M., Schierle, G.S., Hansson, O., Petersen, A., Castilho, R.F.: Improving the survival of grafted dopaminergic neurons: a review over current approaches. Cell Transplant. **9**, 179–195 (2000)
70. Sortwell, C.E., Pitzer, M.R., Collier, T.J.: Time course of apoptotic cell death within mesencephalic cell suspension grafts: implications for improving grafted dopamine neuron survival. Exp. Neurol. **165**, 268–277 (2000)

71. Zawada, W.M., Zastrow, D.J., Clarkson, E.D., Adams, F.S., Bell, K.P., Freed, C.R.: Growth factors improve immediate survival of embryonic dopamine neurons after transplantation into rats. Brain Res. **786**, 96–103 (1998)
72. Takahashi, K., Yasuhara, T., Shingo, T., Muraoka, K., Kameda, M., Takeuchi, A., Yano, A., Kurozumi, K., Agari, T., Miyoshi, Y., Kinugasa, K., Date, I.: Embryonic neural stem cells transplanted in middle cerebral artery occlusion model of rats demonstrated potent therapeutic effects, compared to adult neural stem cells. Brain Res. **1234**, 172–182 (2008)
73. Dihne, M., Bernreuther, C., Hagel, C., Wesche, K.O., Schachner, M.: Embryonic stem cell-derived neuronally committed precursor cells with reduced teratoma formation after transplantation into the lesioned adult mouse brain. Stem Cells **24**, 1458–1466 (2006)
74. Kim, J.B., Sebastiano, V., Wu, G., Arauzo-Bravo, M.J., Sasse, P., Gentile, L., Ko, K., Ruau, D., Ehrich, M., van den Boom, D., Meyer, J., Hubner, K., Bernemann, C., Ortmeier, C., Zenke, M., Fleischmann, B.K., Zaehres, H., Scholer, H.R.: Oct4-induced pluripotency in adult neural stem cells. Cell **136**, 411–419 (2009)
75. Wianny, F., Bernat, A., Huissoud, C., Marcy, G., Markossian, S., Cortay, V., Giroud, P., Leviel, V., Kennedy, H., Savatier, P., Dehay, C.: Derivation and cloning of a novel rhesus embryonic stem cell line stably expressing tau-green fluorescent protein. Stem Cells **26**, 1444–1453 (2008)
76. Johe, K.K., Hazel, T.G., Muller, T., Dugich-Djordjevic, M.M., McKay, R.D.: Single factors direct the differentiation of stem cells from the fetal and adult central nervous system. Genes Dev. **10**, 3129–3140 (1996)
77. Reynolds, B.A., Weiss, S.: Generation of neurons and astrocytes from isolated cells of the adult mammalian central nervous system. Science **255**, 1707–1710 (1992)
78. Weiss, S., Dunne, C., Hewson, J., Wohl, C., Wheatley, M., Peterson, A.C., Reynolds, B.A.: Multipotent CNS stem cells are present in the adult mammalian spinal cord and ventricular neuroaxis. J. Neurosci. **16**, 7599–7609 (1996)
79. Richards, L.J., Kilpatrick, T.J., Bartlett, P.F.: De novo generation of neuronal cells from the adult mouse brain. Proc. Natl. Acad. Sci. USA **89**, 8591–8595 (1992)
80. Palmer, T.D., Ray, J., Gage, F.H.: FGF-2-responsive neuronal progenitors reside in proliferative and quiescent regions of the adult rodent brain. Mol. Cell Neurosci. **6**, 474–486 (1995)
81. Kelly, C.M., Zietlow, R., Dunnett, S.B., Rosser, A.E.: The effects of various concentrations of FGF-2 on the proliferation and neuronal yield of murine embryonic neural precursor cells in vitro. Cell Transplant. **12**, 215–223 (2003)
82. Zhou, F.C., Kelley, M.R., Chiang, Y.H., Young, P.: Three to four-year-old nonpassaged EGF-responsive neural progenitor cells: proliferation, apoptosis, and DNA repair. Exp. Neurol. **164**, 200–208 (2000)
83. Ray, J., Peterson, D.A., Schinstine, M., Gage, F.H.: Proliferation, differentiation, and long-term culture of primary hippocampal neurons. Proc. Natl. Acad. Sci. USA **90**, 3602–3606 (1993)
84. Robertson, M.J., Gip, P., Schaffer, D.V.: Neural stem cell engineering: directed differentiation of adult and embryonic stem cells into neurons. Front. Biosci. **13**, 21–50 (2008)
85. Gomes, W.A., Mehler, M.F., Kessler, J.A.: Transgenic overexpression of BMP4 increases astroglial and decreases oligodendroglial lineage commitment. Dev. Biol. **255**, 164–177 (2003)
86. Gritti, A., Bonfanti, L., Doetsch, F., Caille, I., Alvarez-Buylla, A., Lim, D.A., Galli, R., Verdugo, J.M., Herrera, D.G., Vescovi, A.L.: Multipotent neural stem cells reside into the rostral extension and olfactory bulb of adult rodents. J. Neurosci. **22**, 437–445 (2002)
87. Gritti, A., Frolichsthal-Schoeller, P., Galli, R., Parati, E.A., Cova, L., Pagano, S.F., Bjornson, C.R., Vescovi, A.L.: Epidermal and fibroblast growth factors behave as mitogenic regulators for a single multipotent stem cell-like population from the subventricular region of the adult mouse forebrain. J. Neurosci. **19**, 3287–3297 (1999)

88. Whittemore, S.R., Morassutti, D.J., Walters, W.M., Liu, R.H., Magnuson, D.S.: Mitogen and substrate differentially affect the lineage restriction of adult rat subventricular zone neural precursor cell populations. Exp. Cell Res. **252**, 75–95 (1999)
89. Goetz, A.K., Scheffler, B., Chen, H.X., Wang, S., Suslov, O., Xiang, H., Brustle, O., Roper, S.N., Steindler, D.A.: Temporally restricted substrate interactions direct fate and specification of neural precursors derived from embryonic stem cells. Proc. Natl. Acad. Sci. USA **103**, 11063–11068 (2006)
90. Ma, W., Tavakoli, T., Derby, E., Serebryakova, Y., Rao, M.S., Mattson, M.P.: Cell-extracellular matrix interactions regulate neural differentiation of human embryonic stem cells. BMC Dev. Biol. **8**, 90 (2008)
91. Flanagan, L.A., Rebaza, L.M., Derzic, S., Schwartz, P.H., Monuki, E.S.: Regulation of human neural precursor cells by laminin and integrins. J. Neurosci. Res. **83**, 845–856 (2006)
92. Tate, M.C., Garcia, A.J., Keselowsky, B.G., Schumm, M.A., Archer, D.R., LaPlaca, M.C.: Specific beta1 integrins mediate adhesion, migration, and differentiation of neural progenitors derived from the embryonic striatum. Mol. Cell Neurosci. **27**, 22–31 (2004)
93. Shetty, A.K., Turner, D.A.: In vitro survival and differentiation of neurons derived from epidermal growth factor-responsive postnatal hippocampal stem cells: inducing effects of brain-derived neurotrophic factor. J. Neurobiol. **35**, 395–425 (1998)
94. Benoit, B.O., Savarese, T., Joly, M., Engstrom, C.M., Pang, L., Reilly, J., Recht, L.D., Ross, A.H., Quesenberry, P.J.: Neurotrophin channeling of neural progenitor cell differentiation. J. Neurobiol. **46**, 265–280 (2001)
95. Deng, J., Petersen, B.E., Steindler, D.A., Jorgensen, M.L., Laywell, E.D.: Mesenchymal stem cells spontaneously express neural proteins in culture and are neurogenic after transplantation. Stem Cells **24**, 1054–1064 (2006)
96. Lee, J., Kuroda, S., Shichinohe, H., Ikeda, J., Seki, T., Hida, K., Tada, M., Sawada, K., Iwasaki, Y.: Migration and differentiation of nuclear fluorescence-labeled bone marrow stromal cells after transplantation into cerebral infarct and spinal cord injury in mice. Neuropathology **23**, 169–180 (2003)
97. Mezey, E., Chandross, K.J.: Bone marrow: a possible alternative source of cells in the adult nervous system. Eur. J. Pharmacol. **405**, 297–302 (2000)
98. Mezey, E., Nagy, A., Szalayova, I., Key, S., Bratincsak, A., Baffi, J., Shahar, T.: Comment on "Failure of bone marrow cells to transdifferentiate into neural cells in vivo". Science **299**, 1184; author reply 1184 (2003)
99. Garbuzova-Davis, S., Willing, A.E., Zigova, T., Saporta, S., Justen, E.B., Lane, J.C., Hudson, J.E., Chen, N., Davis, C.D., Sanberg, P.R.: Intravenous administration of human umbilical cord blood cells in a mouse model of amyotrophic lateral sclerosis: distribution, migration, and differentiation. J. Hematother. Stem Cell Res. **12**, 255–270 (2003)
100. Yu, J., Vodyanik, M.A., Smuga-Otto, K., Antosiewicz-Bourget, J., Frane, J.L., Tian, S., Nie, J., Jonsdottir, G.A., Ruotti, V., Stewart, R., Slukvin, II, Thomson, J.A.: Induced pluripotent stem cell lines derived from human somatic cells. Science **318**, 1917–1920 (2007)
101. Kim, J.B., Greber, B., Arauzo-Bravo, M.J., Meyer, J., Park, K.I., Zaehres, H., Scholer, H.R.: Direct reprogramming of human neural stem cells by OCT4. Nature **461**, 649–653 (2009)
102. Soldner, F., Hockemeyer, D., Beard, C., Gao, Q., Bell, G.W., Cook, E.G., Hargus, G., Blak, A., Cooper, O., Mitalipova, M., Isacson, O., Jaenisch, R.: Parkinson's disease patient-derived induced pluripotent stem cells free of viral reprogramming factors. Cell **136**, 964–977 (2009)
103. Singec, I., Knoth, R., Meyer, R.P., Maciaczyk, J., Volk, B., Nikkhah, G., Frotscher, M., Snyder, E.Y.: Defining the actual sensitivity and specificity of the neurosphere assay in stem cell biology. Nat. Methods **3**, 801–806 (2006)
104. Richardson, R.M., Fillmore, H.L., Holloway, K.L., Broaddus, W.C.: Progress in cerebral transplantation of expanded neuronal stem cells. J. Neurosurg. **100**, 659–671 (2004)
105. Suslov, O.N., Kukekov, V.G., Ignatova, T.N., Steindler, D.A.: Neural stem cell heterogeneity demonstrated by molecular phenotyping of clonal neurospheres. Proc. Natl. Acad. Sci. USA **99**, 14506–14511 (2002)

106. Jensen, J.B., Parmar, M.: Strengths and limitations of the neurosphere culture system. Mol. Neurobiol. **34**, 153–161 (2006)
107. Svendsen, C.N., Caldwell, M.A., Shen, J., ter Borg, M.G., Rosser, A.E., Tyers, P., Karmiol, S., Dunnett, S.B.: Long-term survival of human central nervous system progenitor cells transplanted into a rat model of Parkinson's disease. Exp. Neurol. **148**, 135–146 (1997)
108. Sun, Y.E., Martinowich, K., Ge, W.: Making and repairing the mammalian brain—signaling toward neurogenesis and gliogenesis. Semin. Cell Dev. Biol. **14**, 161–168 (2003)
109. Cukierman, E., Pankov, R., Stevens, D.R., Yamada, K.M.: Taking cell-matrix adhesions to the third dimension. Science **294**, 1708–1712 (2001)
110. Cukierman, E., Pankov, R., Yamada, K.M.: Cell interactions with three-dimensional matrices. Curr. Opin. Cell Biol. **14**, 633–639 (2002)
111. Pedersen, J.A., Swartz, M.A.: Mechanobiology in the third dimension. Ann. Biomed. Eng. **33**, 1469–1490 (2005)
112. Reilly, G.C., Engler, A.J.: Intrinsic extracellular matrix properties regulate stem cell differentiation. J. Biomech. **43**, 55–62 (2010)
113. Nat, R., Nilbratt, M., Narkilahti, S., Winblad, B., Hovatta, O., Nordberg, A.: Neurogenic neuroepithelial and radial glial cells generated from six human embryonic stem cell lines in serum-free suspension and adherent cultures. Glia **55**, 385–399 (2007)
114. Potter, W., Kalil. R.E., Kao, W.J.: Biomimetic material systems for neural progenitor cell-based therapy. Front. Biosci. **13**, 806–821 (2008)
115. Freudenberg, U., Hermann, A., Welzel, P.B., Stirl, K., Schwarz, S.C., Grimmer, M., Zieris, A., Panyanuwat, W., Zschoche, S., Meinhold, D., Storch, A., Werner, C.: A star-PEG-heparin hydrogel platform to aid cell replacement therapies for neurodegenerative diseases. Biomaterials **30**, 5049–5060 (2009)
116. Hynes, S.R., Rauch, M.F., Bertram, J.P., Lavik, E.B.: A library of tunable poly(ethylene glycol)/poly(L-lysine) hydrogels to investigate the material cues that influence neural stem cell differentiation. J. Biomed. Mater. Res. A **89**, 499–509 (2009)
117. Mahoney, M.J., Anseth, K.S.: Three-dimensional growth and function of neural tissue in degradable polyethylene glycol hydrogels. Biomaterials **27**, 2265–2274 (2006)
118. Teixeira, A.I., Ilkhanizadeh, S., Wigenius, J.A., Duckworth, J.K., Inganas, O., Hermanson, O.: The promotion of neuronal maturation on soft substrates. Biomaterials **30**, 4567–4572 (2009)
119. Li, N., Folch, A.: Integration of topographical and biochemical cues by axons during growth on microfabricated 3-D substrates. Exp. Cell Res. **311**, 307–316 (2005)
120. Ahmed, I., Liu, H.Y., Mamiya, P.C., Ponery, A.S., Babu, A.N., Weik, T., Schindler, M., Meiners, S.: Three-dimensional nanofibrillar surfaces covalently modified with tenascin-C-derived peptides enhance neuronal growth in vitro. J. Biomed. Mater. Res. A **76**, 851–860 (2006)
121. Biran, R., Noble, M.D., Tresco, P.A.: Directed nerve outgrowth is enhanced by engineered glial substrates. Exp. Neurol. **184**, 141–152 (2003)
122. Bruder, J.M., Lee, A.P., Hoffman-Kim, D.: Biomimetic materials replicating Schwann cell topography enhance neuronal adhesion and neurite alignment in vitro. J. Biomater. Sci. Polym. Ed. **18**, 967–982 (2007)
123. Oh, J., Recknor, J.B., Recknor, J.C., Mallapragada, S.K., Sakaguchi, D.S.: Soluble factors from neocortical astrocytes enhance neuronal differentiation of neural progenitor cells from adult rat hippocampus on micropatterned polymer substrates. J. Biomed. Mater. Res. A **91**, 575–585 (2009)
124. Takeuchi, S., Ziegler, D., Yoshida, Y., Mabuchi, K., Suzuki, T.: Parylene flexible neural probes integrated with microfluidic channels. Lab on a Chip **5**, 519–523 (2005)
125. Ilkhanizadeh, S., Teixeira, A.I., Hermanson, O.: Inkjet printing of macromolecules on hydrogels to steer neural stem cell differentiation. Biomaterials **28**, 3936–3943 (2007)
126. Klaver, C.L., Caplan, M.R.: Bioactive surface for neural electrodes: decreasing astrocyte proliferation via transforming growth factor-beta1. J. Biomed. Mater. Res. A **81**, 1011–1016 (2007)

127. Winter, J.O., Cogan, S.F., Rizzo, J.F., 3rd: Neurotrophin-eluting hydrogel coatings for neural stimulating electrodes. J. Biomed. Mater. Res. B Appl. Biomater. **81**, 551–563 (2007)
128. Chen, D., Du, W.B., Liu, Y., Liu, W.S., Kuznetsov, A., Mendez, F.E., Philipson, L.H., Ismagilov, R.F.: The chemistrode: a droplet-based microfluidic device for stimulation and recording with high temporal, spatial, and chemical resolution. Proc. Natl. Acad. Sci. USA **105**, 16843–16848 (2008)
129. Willerth, S.M., Rader, A., Sakiyama-Elbert, S.E.: The effect of controlled growth factor delivery on embryonic stem cell differentiation inside fibrin scaffolds. Stem Cell Res. **1**, 205–218 (2008)
130. Jhaveri, S.J., Hynd, M.R., Dowell-Mesfin, N., Turner, J.N., Shain, W., Ober, C.K.: Release of nerve growth factor from HEMA hydrogel-coated substrates and its effect on the differentiation of neural cells. Biomacromolecules **10**, 174–183 (2009)
131. Li, X., Yang, Z., Zhang, A.: The effect of neurotrophin-3/chitosan carriers on the proliferation and differentiation of neural stem cells. Biomaterials **30**, 4978–4985 (2009)
132. Nojehdehian, H., Moztarzadeh, F., Baharvand, H., Nazarian, H., Tahriri, M.: Preparation and surface characterization of poly-L-lysine-coated PLGA microsphere scaffolds containing retinoic acid for nerve tissue engineering: in vitro study. Colloids Surf. B Biointerfaces **73**, 23–29 (2009)
133. Moore, K., Macsween, M., Shoichet, M.: Immobilized concentration gradients of neurotrophic factors guide neurite outgrowth of primary neurons in macroporous scaffolds. Tissue Eng. **12**, 267–278 (2006)
134. Choi, N.W., Cabodi, M., Held, B., Gleghorn, J.P., Bonassar, L.J., Stroock, A.D.: Microfluidic scaffolds for tissue engineering. Nat. Mater. **6**, 908–915 (2007)
135. Nakaji-Hirabayashi, T., Kato, K., Arima, Y., Iwata, H.: Oriented immobilization of epidermal growth factor onto culture substrates for the selective expansion of neural stem cells. Biomaterials **28**, 3517–3529 (2007)
136. Aizawa, Y., Leipzig, N., Zahir, T., Shoichet, M.: The effect of immobilized platelet derived growth factor AA on neural stem/progenitor cell differentiation on cell-adhesive hydrogels. Biomaterials **29**, 4676–4683 (2008)
137. Ashton, R.S., Banerjee, A., Punyani, S., Schaffer, D.V., Kane, R.S.: Scaffolds based on degradable alginate hydrogels and poly(lactide-co-glycolide) microspheres for stem cell culture. Biomaterials **28**, 5518–5525 (2007)
138. Banerjee, A., Arha, M., Choudhary, S., Ashton, R.S., Bhatia, S.R., Schaffer, D.V., Kane, R.S.: The influence of hydrogel modulus on the proliferation and differentiation of encapsulated neural stem cells. Biomaterials **30**, 4695–4699 (2009)
139. Brannvall, K., Bergman, K., Wallenquist, U., Svahn, S., Bowden, T., Hilborn, J., Forsberg-Nilsson, K.: Enhanced neuronal differentiation in a three-dimensional collagen-hyaluronan matrix. J. Neurosci. Res. **85**, 2138–2146 (2007)
140. Haile, Y., Berski, S., Drager, G., Nobre, A., Stummeyer, K., Gerardy-Schahn, R., Grothe, C.: The effect of modified polysialic acid based hydrogels on the adhesion and viability of primary neurons and glial cells. Biomaterials **29**, 1880–1891 (2008)
141. Ma, W., Fitzgerald, W., Liu, Q.Y., O'Shaughnessy, T.J., Maric, D., Lin, H.J., Alkon, D.L., Barker, J.L.: CNS stem and progenitor cell differentiation into functional neuronal circuits in three-dimensional collagen gels. Exp. Neurol. **190**, 276–288 (2004)
142. Ma, W., Tavakoli, T., Chen, S., Maric, D., Liu, J.L., O'Shaughnessy, T.J., Barker, J.L.: Reconstruction of functional cortical-like tissues from neural stem and progenitor cells. Tissue Eng. Part A **14**, 1673–1686 (2008)
143. O'Connor, S.M., Stenger, D.A., Shaffer, K.M., Ma, W.: Survival and neurite outgrowth of rat cortical neurons in three-dimensional agarose and collagen gel matrices. Neurosci. Lett. **304**, 189–193 (2001)
144. O'Connor, S.M., Stenger, D.A., Shaffer, K.M., Maric, D., Barker, J.L., Ma, W.: Primary neural precursor cell expansion, differentiation and cytosolic Ca(2+) response in three-dimensional collagen gel. J. Neurosci. Methods **102**, 187–195 (2000)

145. Ren, Y.J., Zhang, H., Huang, H., Wang, X.M., Zhou, Z.Y., Cui, F.Z., An, Y.H.: In vitro behavior of neural stem cells in response to different chemical functional groups. Biomaterials **30**, 1036–1044 (2009)
146. Wang, T.W., Spector, M.: Development of hyaluronic acid-based scaffolds for brain tissue engineering. Acta Biomater. **5**, 2371–2384 (2009)
147. Watanabe, K., Nakamura, M., Okano, H., Toyama, Y.: Establishment of three-dimensional culture of neural stem/progenitor cells in collagen Type-1 gel. Restor. Neurol. Neurosci. **25**, 109–117 (2007)
148. Willerth, S.M., Arendas, K.J., Gottlieb, D.I., Sakiyama-Elbert, S.E.: Optimization of fibrin scaffolds for differentiation of murine embryonic stem cells into neural lineage cells. Biomaterials **27**, 5990–6003 (2006)
149. Fischer, S.E., Liu, X., Mao, H.Q., Harden, J.L.: Controlling cell adhesion to surfaces via associating bioactive triblock proteins. Biomaterials **28**, 3325–3337 (2007)
150. Silva, G.A., Czeisler, C., Niece, K.L., Beniash, E., Harrington, D.A., Kessler, J.A., Stupp, S.I.: Selective differentiation of neural progenitor cells by high-epitope density nanofibers. Science **303**, 1352–1355 (2004)
151. Thonhoff, J.R., Lou, D.I., Jordan, P.M., Zhao, X., Wu, P.: Compatibility of human fetal neural stem cells with hydrogel biomaterials in vitro. Brain Res. **1187**, 42–51 (2008)
152. Bhang, S.H., Lim, J.S., Choi, C.Y., Kwon, Y.K., Kim, B.S.: The behavior of neural stem cells on biodegradable synthetic polymers. J. Biomater. Sci. Polym. Ed. **18**, 223–239 (2007)
153. Jan, E., Kotov, N.A.: Successful differentiation of mouse neural stem cells on layer-by-layer assembled single-walled carbon nanotube composite. Nano Lett. **7**, 1123–1128 (2007)
154. Levenberg, S., Burdick, J.A., Kraehenbuehl, T., Langer, R.: Neurotrophin-induced differentiation of human embryonic stem cells on three-dimensional polymeric scaffolds. Tissue Eng. **11**, 506–512 (2005)
155. Levenberg, S., Huang, N.F., Lavik, E., Rogers, A.B., Itskovitz-Eldor, J., Langer, R.: Differentiation of human embryonic stem cells on three-dimensional polymer scaffolds. Proc. Natl. Acad. Sci. USA **100**, 12741–12746 (2003)
156. Mahoney, M.J., Anseth, K.S.: Contrasting effects of collagen and bFGF-2 on neural cell function in degradable synthetic PEG hydrogels. J. Biomed. Mater. Res. A **81**, 269–278 (2007)
157. Namba, R.M., Cole, A.A., Bjugstad, K.B., Mahoney, M.J.: Development of porous PEG hydrogels that enable efficient, uniform cell-seeding and permit early neural process extension. Acta Biomater. **5**, 1884–1897 (2009)
158. Saha, K., Irwin, E.F., Kozhukh, J., Schaffer, D.V., Healy, K.E.: Biomimetic interfacial interpenetrating polymer networks control neural stem cell behavior. J. Biomed. Mater. Res. A **81**, 240–249 (2007)
159. Bible, E., Chau, D.Y., Alexander, M.R., Price, J., Shakesheff, K.M., Modo, M.: The support of neural stem cells transplanted into stroke-induced brain cavities by PLGA particles. Biomaterials **30**, 2985–2994 (2009)
160. Nakajima, M., Ishimuro, T., Kato, K., Ko, I.K., Hirata, I., Arima, Y., Iwata, H.: Combinatorial protein display for the cell-based screening of biomaterials that direct neural stem cell differentiation. Biomaterials **28**, 1048–1060 (2007)
161. Gelain, F., Lomander, A., Vescovi, A.L., Zhang, S.: Systematic studies of a self-assembling peptide nanofiber scaffold with other scaffolds. J. Nanosci. Nanotechnol. **7**, 424–434 (2007)
162. Kearns, S.M., Laywell, E.D., Kukekov, V.K., Steindler, D.A.: Extracellular matrix effects on neurosphere cell motility. Exp. Neurol. **182**, 240–244 (2003)
163. Singec, I., Quiñones-Hinojosa, A.: Neurospheres. In: Gage FH, Kempermann G, Song H (eds.) Adult Neurogenesis. Cold Spring Harbor Press, Cold Spring Harbor, NY (2008)
164. Takezawa, T., Mori, Y., Yoshizato, K.: Cell culture on a thermo-responsive polymer surface. Biotechnology (N Y) **8**, 854–856 (1990)
165. Bohanon, T., Elender, G., Knoll, W., Koberle, P., Lee, J.S., Offenhausser, A., Ringsdorf, H., Sackmann, E., Simon, J., Tovar, G., Winnik, F.M.: Neural cell pattern formation on glass and oxidized silicon surfaces modified with poly(N-isopropylacrylamide). J. Biomater. Sci. Polym. Ed. **8**, 19–39 (1996)

166. Chen, R.R., Silva, E.A., Yuen, W.W., Mooney, D.J.: Spatio-temporal VEGF and PDGF delivery patterns blood vessel formation and maturation. Pharm. Res. **24**, 258–264 (2007)
167. Oppegard, S.C., Nam, K.H., Carr, J.R., Skaalure, S.C., Eddington, D.T.: Modulating temporal and spatial oxygenation over adherent cellular cultures. PLoS One **4**, e6891 (2009)
168. Riquelme, P.A., Drapeau, E., Doetsch, F.: Brain micro-ecologies: neural stem cell niches in the adult mammalian brain. Philos. Trans. R. Soc. Lond. B Biol. Sci. **363**, 123–137 (2008)
169. Scheffler, B., Edenhofer, F., Brustle, O.: Merging fields: stem cells in neurogenesis, transplantation, and disease modeling. Brain Pathol. **16**, 155–168 (2006)
170. Alvarez-Buylla. A., Lim, D.A.: For the long run: maintaining germinal niches in the adult brain. Neuron **41**, 683–686 (2004)
171. Spradling, A., Drummond-Barbosa, D., Kai, T.: Stem cells find their niche. Nature **414**, 98–104 (2001)
172. Lathia, J.D., Patton, B., Eckley, D.M., Magnus, T., Mughal, M.R., Sasaki, T., Caldwell, M.A., Rao, M.S., Mattson, M.P., ffrench-Constant, C.: Patterns of laminins and integrins in the embryonic ventricular zone of the CNS. J. Comp. Neurol. **505**, 630–643 (2007)
173. Hagg, T.: Molecular regulation of adult CNS neurogenesis: an integrated view. Trends Neurosci. **28**, 589–595 (2005)
174. Lathia, J.D., Rao, M.S., Mattson, M.P., Ffrench-Constant, C.: The microenvironment of the embryonic neural stem cell: lessons from adult niches? Dev. Dyn. **236**, 3267–3282 (2007)
175. Parent, J.M., Yu, T.W., Leibowitz, R.T., Geschwind, D.H., Sloviter, R.S., Lowenstein, D.H.: Dentate granule cell neurogenesis is increased by seizures and contributes to aberrant network reorganization in the adult rat hippocampus. J. Neurosci. **17**, 3727–3738 (1997)
176. Gould, E., Tanapat, P.: Stress and hippocampal neurogenesis. Biol. Psychiatry **46**, 1472–1479 (1999)
177. Little, L., Healy, K.E., Schaffer, D.: Engineering biomaterials for synthetic neural stem cell microenvironments. Chem. Rev. **108**, 1787–1796 (2008)
178. Hoge, R.D., Pike, G.B.: Oxidative metabolism and the detection of neuronal activation via imaging. J. Chem. Neuroanat. **22**, 43–52 (2001)
179. Silver, I., Erecinska, M.: Oxygen and ion concentrations in normoxic and hypoxic brain cells. Adv. Exp. Med. Biol. **454**, 7–16 (1998)
180. Morrison, S.J., Csete, M., Groves, A.K., Melega, W., Wold, B., Anderson, D.J.: Culture in reduced levels of oxygen promotes clonogenic sympathoadrenal differentiation by isolated neural crest stem cells. J. Neurosci. **20**, 7370–7376 (2000)
181. Storch, A., Paul, G., Csete, M., Boehm, B.O., Carvey, P.M., Kupsch, A., Schwarz, J.: Long-term proliferation and dopaminergic differentiation of human mesencephalic neural precursor cells. Exp. Neurol. **170**, 317–325 (2001)
182. Studer, L., Csete, M., Lee, S.H., Kabbani, N., Walikonis, J., Wold, B., McKay, R.: Enhanced proliferation, survival, and dopaminergic differentiation of CNS precursors in lowered oxygen. J. Neurosci. **20**, 7377–7383 (2000)
183. Zhou, L., Del Villar, K., Dong, Z., Miller, C.A.: Neurogenesis response to hypoxia-induced cell death: MAP kinase signal transduction mechanisms. Brain Res. **1021**, 8–19 (2004)
184. Ljungkvist, A.S., Bussink, J., Kaanders, J.H., van der Kogel, A.J.: Dynamics of tumor hypoxia measured with bioreductive hypoxic cell markers. Radiat. Res. **167**, 127–145 (2007)
185. Gustafsson, M.V., Zheng, X., Pereira, T., Gradin, K., Jin, S., Lundkvist, J., Ruas, J.L., Poellinger, L., Lendahl, U., Bondesson, M.: Hypoxia requires Notch signaling to maintain the undifferentiated cell state. Dev. Cell **9**, 617–628 (2005)
186. Chen, H.L., Pistollato, F., Hoeppner, D.J., Ni, H.T., McKay, R.D., Panchision, D.M.: Oxygen tension regulates survival and fate of mouse central nervous system precursors at multiple levels. Stem Cells **25**, 2291–2301 (2007)
187. Zhu, L.L., Wu, L.Y., Yew, D.T., Fan, M.: Effects of hypoxia on the proliferation and differentiation of NSCs. Mol. Neurobiol. **31**, 231–242 (2005)
188. Milosevic, J., Maisel, M., Wegner, F., Leuchtenberger, J., Wenger, R.H., Gerlach, M., Storch, A., Schwarz, J.: Lack of hypoxia-inducible factor-1 alpha impairs midbrain neural precursor cells involving vascular endothelial growth factor signaling. J. Neurosci. **27**, 412–421 (2007)

189. Bergeron, M., Yu, A.Y., Solway, K.E., Semenza, G.L., Sharp, F.R.: Induction of hypoxia-inducible factor-1 (HIF-1) and its target genes following focal ischaemia in rat brain. Eur. J. Neurosci. **11**, 4159–4170 (1999)
190. Bernaudin, M., Tang, Y., Reilly, M., Petit, E., Sharp, F.R.: Brain genomic response following hypoxia and re-oxygenation in the neonatal rat. Identification of genes that might contribute to hypoxia-induced ischemic tolerance. J. Biol. Chem. **277**, 39728–39738 (2002)
191. Sakanaka, M., Wen, T.C., Matsuda, S., Masuda, S., Morishita, E., Nagao, M., Sasaki, R.: In vivo evidence that erythropoietin protects neurons from ischemic damage. Proc. Natl. Acad. Sci. USA **95**, 4635–4640 (1998)
192. Junk, A.K., Mammis, A., Savitz, S.I., Singh, M., Roth, S., Malhotra, S., Rosenbaum, P.S., Cerami, A., Brines, M., Rosenbaum, D.M.: Erythropoietin administration protects retinal neurons from acute ischemia-reperfusion injury. Proc. Natl. Acad. Sci. USA **99**, 10659–10664 (2002)
193. Greenberg, D.A., Jin, K.: From angiogenesis to neuropathology. Nature **438**, 954–959 (2005)
194. Chavez, J.C., LaManna, J.C.: Activation of hypoxia-inducible factor-1 in the rat cerebral cortex after transient global ischemia: potential role of insulin-like growth factor-1. J. Neurosci. **22**, 8922–8931 (2002)
195. Jacques, T.S., Relvas, J.B., Nishimura, S., Pytela, R., Edwards, G.M., Streuli, C.H., ffrench-Constant, C.: Neural precursor cell chain migration and division are regulated through different beta1 integrins. Development **125**, 3167–3177 (1998)
196. Lee, S.R., Kim, H.Y., Rogowska, J., Zhao, B.Q., Bhide, P., Parent, J.M., Lo, E.H.: Involvement of matrix metalloproteinase in neuroblast cell migration from the subventricular zone after stroke. J. Neurosci. **26**, 3491–3495 (2006)
197. Wang, L., Zhang, Z.G., Zhang, R.L., Gregg, S.R., Hozeska-Solgot, A., LeTourneau, Y., Wang, Y., Chopp, M.: Matrix metalloproteinase 2 (MMP2) and MMP9 secreted by erythropoietin-activated endothelial cells promote neural progenitor cell migration. J. Neurosci. **26**, 5996–6003 (2006)
198. Belvindrah, R., Hankel, S., Walker, J., Patton, B.L., Muller, U.: Beta1 integrins control the formation of cell chains in the adult rostral migratory stream. J. Neurosci. **27**, 2704–2717 (2007)
199. Conover, J.C., Doetsch, F., Garcia-Verdugo, J.M., Gale, N.W., Yancopoulos, G.D., Alvarez-Buylla, A.: Disruption of Eph/ephrin signaling affects migration and proliferation in the adult subventricular zone. Nat. Neurosci. **3**, 1091–1097 (2000)
200. Andsberg, G., Kokaia, Z., Klein, R.L., Muzyczka, N., Lindvall, O., Mandel, R.J.: Neuropathological and behavioral consequences of adeno-associated viral vector-mediated continuous intrastriatal neurotrophin delivery in a focal ischemia model in rats. Neurobiol. Dis. **9**, 187–204 (2002)
201. Lippoldt, A., Reichel, A., Moenning, U.: Progress in the identification of stroke-related genes: emerging new possibilities to develop concepts in stroke therapy. CNS Drugs **19**, 821–832 (2005)
202. Naylor, M., Bowen, K.K., Sailor, K.A., Dempsey, R.J., Vemuganti, R.: Preconditioning-induced ischemic tolerance stimulates growth factor expression and neurogenesis in adult rat hippocampus. Neurochem. Int. **47**, 565–572 (2005)
203. Yan, Y.P., Sailor, K.A., Lang, B.T., Park, S.W., Vemuganti, R., Dempsey, R.J.: Monocyte chemoattractant protein-1 plays a critical role in neuroblast migration after focal cerebral ischemia. J. Cereb. Blood Flow Metab. **27**, 1213–1224 (2007)
204. Iihara, K., Sasahara, M., Hashimoto, N., Hazama, F.: Induction of platelet-derived growth factor beta-receptor in focal ischemia of rat brain. J. Cereb. Blood Flow Metab. **16**, 941–949 (1996)
205. Renner, O., Tsimpas, A., Kostin, S., Valable, S., Petit, E., Schaper, W., Marti, H.H.: Time- and cell type-specific induction of platelet-derived growth factor receptor-beta during cerebral ischemia. Brain Res. Mol. Brain Res. **113**, 44–51 (2003)

206. Acosta, M.A., Ymele-Leki, P., Kostov, Y.V., Leach, J.B.: Fluorescent microparticles for sensing cell microenvironment oxygen levels within 3D scaffolds. Biomaterials **30**, 3068–3074 (2009)
207. Alexandrova, M.L., Bochev, P.G.: Oxidative stress during the chronic phase after stroke. Free Radic. Biol. Med. **39**, 297–316 (2005)
208. Tsatmali, M., Walcott, E.C., Makarenkova, H., Crossin, K.L.: Reactive oxygen species modulate the differentiation of neurons in clonal cortical cultures. Mol. Cell Neurosci. **33**, 345–357 (2006)
209. Smith, J., Ladi, E., Mayer-Proschel, M., Noble, M.: Redox state is a central modulator of the balance between self-renewal and differentiation in a dividing glial precursor cell. Proc. Natl. Acad. Sci. USA **97**, 10032–10037 (2000)
210. Lampe, K.J., Namba, R.M., Silverman, T.R., Bjugstad, K.B., Mahoney, M.J.: Impact of lactic acid on cell proliferation and free radical-induced cell death in monolayer cultures of neural precursor cells. Biotechnol. Bioeng. **103**, 1214–1223 (2009)
211. Nakaji-Hirabayashi, T., Kato, K., Iwata, H.: Self-assembling chimeric protein for the construction of biodegradable hydrogels capable of interaction with integrins expressed on neural stem/progenitor cells. Biomacromolecules **9**, 1411–1416 (2008)
212. Leipzig, N.D., Shoichet, M.S.: The effect of substrate stiffness on adult neural stem cell behavior. Biomaterials **30**, 6867–6878 (2009)
213. Recknor, J.B., Sakaguchi, D.S., Mallapragada, S.K.: Directed growth and selective differentiation of neural progenitor cells on micropatterned polymer substrates. Biomaterials **27**, 4098–4108 (2006)

Biomaterial Applications in the Adult Skeletal Muscle Satellite Cell Niche: Deliberate Control of Muscle Stem Cells and Muscle Regeneration in the Aged Niche

Eric Jabart and Irina Conboy

Abstract Skeletal muscle stem cells known as satellite cells are responsible for muscle regeneration. Upon muscle injury, previously quiescent satellite cells become activated as proliferating myogenic precursors, differentiate into myoblasts, and ultimately fuse into new, multinucleated myofibers. Unfortunately, this paradigm breaks down with aging and instead of becoming activated upon injury, satellite cells remain quiescent. Recent work, however, has shed light on the mechanisms behind this impaired regeneration and these findings suggest several therapeutic avenues. Skeletal muscle tissue engineering aims to create functional muscle in vitro followed by engraftment in vivo for the replacement or repair of missing or pathological tissue in various dystrophies or myopathies. Biomaterials have rapidly become central to these regeneration efforts and numerous repair strategies already exist. However, optimization of these biomaterial platforms in order to more fully mimic the in vivo adult skeletal muscle niche is still necessary. More importantly perhaps, the effects of aged or pathological environments on skeletal muscle engraftment has yet to be fully characterized. Furthermore, debate still remains over whether or not all satellite cells (considered to be heterogeneous in genetic markers and functional properties) are in fact stem cells, and what implications this could have on in vitro regeneration strategies. Novel uses and advances in biomaterials show promise in tackling these problems. Therefore, after a discussion of muscle regeneration in both the 'young' and 'aged' niches, this chapter will examine the most up-to-date strategies for in vitro skeletal muscle regeneration and will discuss how current efforts in biomaterial technologies might

E. Jabart and I. Conboy (✉)
Department of Bioengineering, University of California Berkeley, B107B Stanley Hall, Berkeley CA, 94720, USA
e-mail: iconboy@berkeley.edu

E. Jabart
e-mail: eric_jabart@berkeley.edu

Stud Mechanobiol Tissue Eng Biomater (2010) 2: 275–308
DOI: 10.1007/8415_2010_8

Published online: 18 June 2010

be used to accurately determine the significance of satellite cells as muscle stem cells, and control robust production of myofibers in vitro by mimicking the in vivo niche.

Abbreviations

TGF-β1	Transforming growth factor beta 1
BMP	Bone morphogenetic protein
ECM	Extracellular matrix
DMD	Duchenne's muscular dystrophy
PGA	Poly(glycolic acid)
PLA	Poly(lactic acid)
PLGA	Poly(lactic-co-glycolic acid)
PDMS	Poly(dimethylsiloxane)
MBCPs	Multi-block copolymers

1 Introduction

After exercise or upon injury our muscles regenerate, restoring functionality and maintaining our structural integrity [1]. Adult skeletal muscle stem cells, known as satellite cells are responsible for this regeneration. Discovered in frog tibialis anticus muscle fibers in 1961 [1, 2], satellite cells were named based on their physiological location, between the basement membrane and the plasma membrane of the muscle fiber (Fig. 1). Their purpose was already hinted at back then, but it was not until ten years later that their role as myogenic precursors was

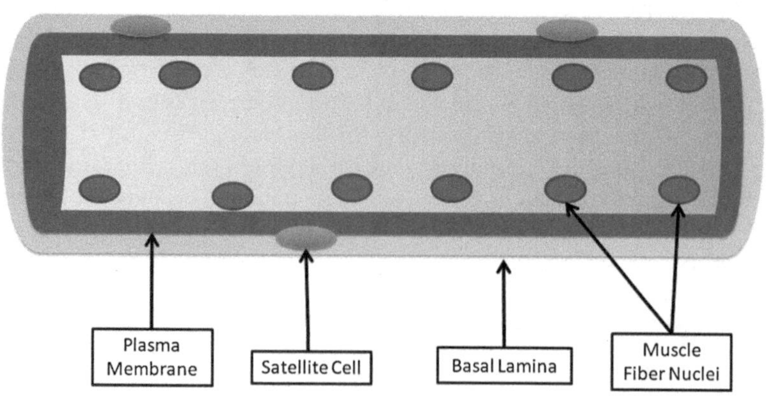

Fig. 1 The skeletal muscle fiber. Skeletal muscle satellite cells reside between the plasma membrane of myofibers and the basal lamina. Mature muscle nuclei are located towards the periphery of the fiber

definitely established [3]. Satellite cells with thymidine-labeled nuclei were observed to undergo cell division and contribute to elongation of muscle fibers. A few years later, proof that satellite cells generate fusion-competent myoblasts was determined through isolated muscle fiber analysis [4–6]. Furthermore, not only did these satellite cells differentiate into myoblasts, they also replenished the satellite cell population [7, 8] identifying them as adult skeletal muscle stem cells.

The skeletal muscle regeneration process is now reasonably understood. Upon muscle injury, quiescent satellite cells are activated by Notch/Delta signaling [9] and proliferate. Subsequent signaling through the Wnt pathway induces their differentiation and fusion into new myofibers [10] (Fig. 2). Importantly, recent studies have shown that in the 'aged' skeletal muscle niche, this signaling cascade is interrupted by increased TGF-β1 signaling and satellite cells remain quiescent upon injury [11]. Promising studies designed to combat the effects of aging on muscle regeneration have been recently conducted [11, 12].

Tissue engineering methods and biomaterial approaches would thus benefit from considering the effects of the 'aged' and pathological niches on muscle regeneration when developing applications for enhanced health and performance of skeletal muscle. By the same token, however, highly tunable, modular

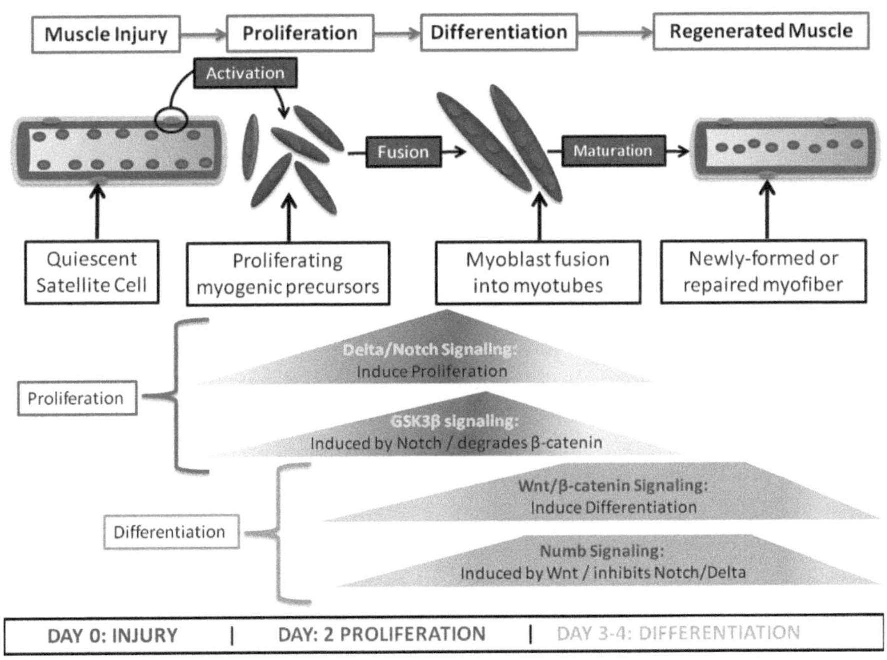

Fig. 2 Regeneration timeline and signaling in normal skeletal muscle repair. Upon muscle injury, quiescent satellite cells become activated and proliferate in response to Delta/Notch signaling. Downregulation of Delta/Notch signaling via Numb inhibition leads to increased Wnt signaling and differentiation of myogenic precursors. Fusion of myoblasts into myotubes is followed by maturation into myofibers with centrally located nuclei

biomaterial systems are uniquely positioned to tackle the problem of physiological down-regulation of muscle regenerative capacity with advancing age. Therefore, after a detailed description of 'young' and 'aged' skeletal muscle regeneration, this chapter will discuss previous and current efforts to combat the effects of aging on regeneration, and look at how the arsenal of novel biomaterials can provide effective solutions to the problem of decline in muscle regeneration in aged or pathological muscle niches.

2 Skeletal Muscle is Regenerated and Maintained by Muscle Stem Cells

2.1 Delta/Notch Signaling Leads to Activation and Proliferation of Satellite Cells

As many stem cell regenerative pathways—including muscle regeneration—are known to recapitulate embryonic organogenesis [13] and Notch signaling was known to affect cell proliferation and differentiation during embryogenesis as well as tissue repair [14–17], Notch was a logical target of study in skeletal muscle regeneration [9]. While Notch signaling can be involved in differentiation [18–20] or cell proliferation, Notch signaling in murine C2C12 myoblasts was shown to lengthen the time myoblasts remained undifferentiated, observed through lack of expression of differentiation markers MyoD and myogenin [14, 21]. After initial studies to induce satellite cell activation ex vivo [22] further work demonstrated that during satellite cell proliferation and activation Notch is active, and that the Notch antagonist Numb may play an important role in signaling for myogenic precursor differentiation [9]. Activated Notch levels were elevated during the first few days after injury, when satellite cells are highly proliferative, whereas Numb levels initially decreased. As Numb levels increased again, loss of the satellite cell marker CD34 and higher expression of the myoblast markers desmin and M-cadherin were observed. This work thus showed that Notch signaling is important in the early days of skeletal muscle regeneration by contributing to the rapidly increasing population of myogenic progenitors.

2.2 Wnt Signaling Cues Myogenic Progenitor Cells to Differentiate

The Notch pathway interacts with many other signaling pathways [e.g., transforming growth factor (TGF)/bone morphogenetic protein (BMP), Wnt] that are implicated in tissue regeneration [14, 23–26]. For skeletal muscle regeneration specifically, the balance between the Notch and Wnt signaling networks defines the transition from proliferation of myogenic precursors to the differentiation into fusion-competent myoblasts and de novo multinucleated muscle fiber formation [10].

In the same way it was known that Notch signaling was implicated in myogenesis and postnatal repair mechanisms, previous studies had shown that Wnt signaling also played a major part in muscle formation and myogenesis [10, 27–30]. Mononucleated muscle cells from regenerating muscle fibers isolated from TOPGAL mice, which express beta-galactosidase based on LEF/TCF or beta-catenin activity [31], demonstrated Wnt signaling activity through beta-galactosidase expression at days 2 and 5 of regeneration [10]. More convincingly, a study of the downstream Wnt regulator GSK3β and its phosphorylation profile at tyrosine 216 revealed a dynamic regulation that works in concert with Notch signaling. A dephosphorylated tyrosine 216 allows GSK3β to phosphorylate β-catenin, a hallmark of canonical Wnt signaling [32, 33]. Tyrosine 216 phosphorylation was high early on in muscle regeneration (when Notch signaling was high [9]), but decreased at later times, indicative of activation by Wnt. qRT-PCR studies on 4-day-cultured myofiber explants showed increased transcription of the Wnt3a ligand, and the Wnt receptors Frizzled-1 and Frizzled-2, in addition to the downstream target Axin-2.

In order to study the role of Wnt signaling in muscle regeneration, mononucleated cells isolated from single muscle fibers were treated with recombinant Wnt3a or a GSK3β inhibitor. Upon treatment of myogenic progenitors, increased β-catenin activation and nuclear localization was observed, in addition to a close to 70% increase in the number of myogenic precursors expressing desmin after 2 days in culture over untreated cells [10].

In vivo work using recombinant Wnt3a treatment on regenerating muscle led to premature differentiation of myogenic precursors into de novo myofibers. This was accompanied by an increase in the size of the regenerated myotubes in addition to an increased number of nascent myotubes early on in regeneration (days 2–3). However by days 4–5, the number of nascent myotubes in untreated samples was still increasing, whereas Wnt3a-treated samples showed no new myotube formation, and remaining injured areas of muscle, due to a depletion in myogenic progenitors [10].

Conversely, inhibition of Wnt signaling during the differentiation phase, both in vitro and in vivo, resulted in decreased number and size of de novo myotubes, confirming the importance of Wnt signaling in cell fate commitment of myogenic progenitors.

A deeper analysis of the crosstalk between the Notch and Wnt pathways showed that inhibition of Notch led to an earlier progression towards terminal myogenic differentiation, evidenced by an early increase in Wnt signaling. Inhibition of Notch during the differentiation phase had little effect. However, if Notch was exogenously activated during the differentiation phase, a decrease in nuclear β-catenin was observed, indicating inhibition of Wnt signaling. The intimacy of the Notch/Wnt crosstalk was then confirmed to occur via GSK3β, which is maintained in an active form by Notch but is inhibited by Wnt.

Skeletal muscle regeneration therefore depends on an intricately regulated crosstalk between the Notch and Wnt signaling pathways. Myogenic precursors rapidly proliferate in response to a perceived injury under the effects of Notch signaling, and then undergo differentiation and fusion into de novo myofibers due to both an activation of Wnt signaling and an inhibition of Notch signaling (itself mediated through Wnt activation).

3 The Aged Skeletal Muscle Niche Impairs Normal Regeneration: TGF-β1 Signaling Maintains Satellite Cell Quiescence and Leads to Scar Tissue Formation

There is some evidence that the number of skeletal muscle satellite cells in different muscles and species decreases with aging [34–39]. However, the regenerative potential of 'aged' satellite cells in vitro is robustly maintained [1, 11, 12, 40]. Nevertheless, in an aged niche, satellite cells fail to regenerate upon muscle injury and scar tissue is formed in the place of new myofibers [37, 41]. Transforming growth factor beta 1 (TGF-β1), a cytokine involved in many cell functions such as growth, proliferation, differentiation, and apoptosis, disrupts the dynamic Notch/Wnt crosstalk essential to myogenic proliferation and differentiation and leads to this decline in muscle regeneration in aging [11, 42–44] (Fig. 3a).

The regenerative potential of skeletal muscle satellite cells is very robust. Early on, Studitsky showed that functional, new muscle could be formed from minced, explanted muscle [45]. Cycles of degeneration and regeneration of muscle using toxins further showed that satellite cells are able to heartily contribute to the formation of new muscle even after 50 sets of induced injury-regeneration cycles [46, 47].

Although it is debated whether the satellite cell population may decline with aging [41, 48–52], the lack of muscle regeneration with aging is likely not due to exhaustion of the satellite cell population [1]. The balance of Notch/Wnt signaling is central to muscle regeneration and shifting the balance via Notch/Wnt inhibition or activation has been shown to have significant impacts on the efficacy of regeneration [9, 10, 30, 49]. Slight variations in the skeletal muscle niche that affect these pathways and their crosstalk can have a major effect on regeneration.

We saw earlier that Notch activation after injury was essential for proliferation of myogenic precursors [49] and that inhibition of Notch led to fewer nascent myotubes and overall poorer regeneration. In the aged environment it was shown that satellite cell proliferation was reduced, but that exogenous Notch activation re-established the satellite cells' proliferative capacity (similar to a young environment). In recent work, Notch activation was shown to be important for the regenerative capacity of human satellite cells and to decline in old human muscle [53]. Notably, activation of Notch by its ligand Delta is likely to be positively regulated through MAPK and interestingly, both MAPK and active Notch become diminished with age in the human muscle compartment [53]. In addition to

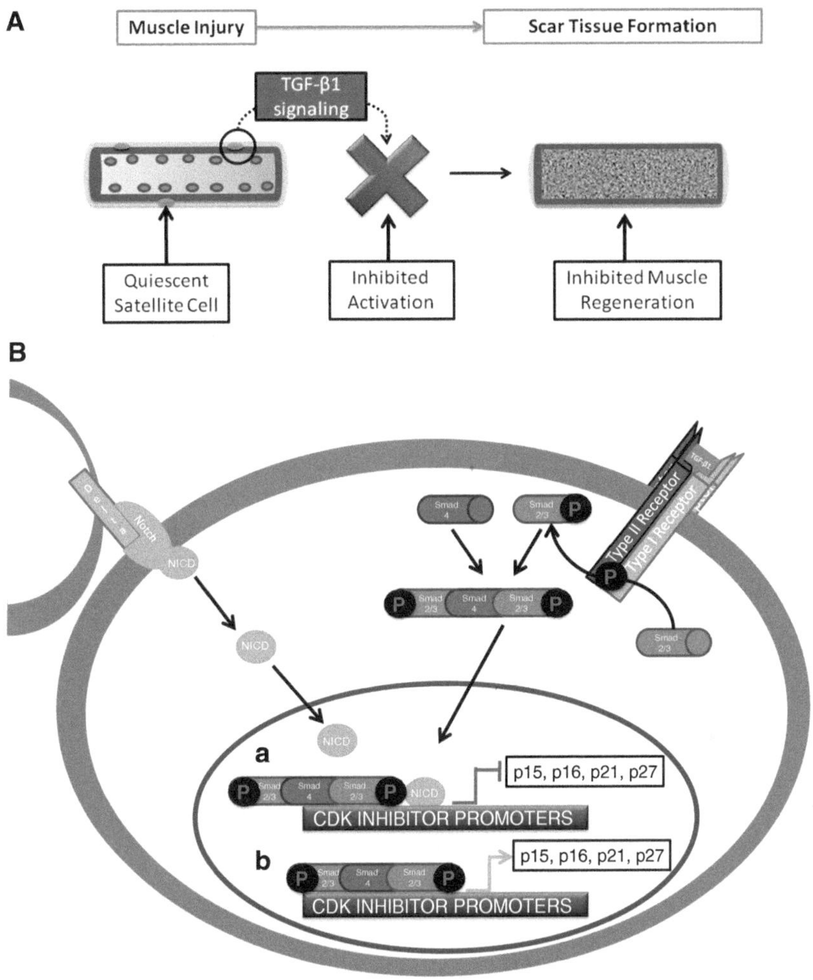

Fig. 3 **a** Increased TGF-β1 signaling inhibits proper muscle regeneration in the aged niche. Upon injury to muscle in the 'aged' niche, increased active TGF-β1 levels inhibit activation of quiescent satellite cells. Proliferation and differentiation steps of normal muscle regeneration are inhibited and few new myotubes are formed, leading to scar tissue formation in lieu of new muscle formation. **b** Notch and TGF-β1 signaling cross-talk in the aged niche. In young muscle, interaction of Delta with Notch receptor leads to release of Notch Intracellular Domain (NICD) which enters the nucleus and promotes satellite cell proliferation. TGF-β1 signaling through Type II and Type I TGF-β receptor heterotetramer leads to phosphorylation of Smad2/3 (pSmad3) and subsequent complex formation with Smad4 (2 Smad2/3s per each Smad4). This complex enters the nucleus to affect transcription of various genes. In the aged muscle niche, increased active levels of TGF-β1 (and decreased levels of active Notch) lead to increased pSmad3 signaling and expression of cyclin-dependent-kinase inhibitors p15, p16, p21, and p27 which maintain satellite cell quiescence (**b**). In the young niche TGF-β1 levels are low and Delta levels are high, so satellite cell activation is not maintained. However, forced activation of Notch in the aged niche (high active TGF-β1 levels) restores skeletal muscle regeneration after injury through antagonism of TGF-β1/pSmad3 signaling (**a**)

assessing that in this case satellite cell numbers remained constant during a lifetime, the report demonstrated that the Notch ligand Delta failed to become upregulated in aged satellite cells, thus leading to the observed decrease in proliferation. But was this due to an intrinsic aging of satellite cells? Or was it caused by other extrinsic effects?

Earlier publications had shown that transplantation of young muscle into an aged niche impaired regeneration, but that transplantation of aged muscle into a young niche allowed regeneration [54, 55]. It was therefore clear that the muscle stem cells present in the aged niche were still capable of regeneration, but that the niche itself was somehow preventing their activation. This led researchers to question what elements of the aged niche were contributing to impairment of muscle regeneration.

Using a method called heterochronic parabiosis [56], where the circulatory systems of young and aged rats are surgically linked, it was further confirmed that the aged niche caused the decline in satellite cell regeneration and that exposure of 'old' satellite cells to a young circulatory system can restore their regenerative capacity [54]. Delta signaling of Notch in old satellite cells was upregulated, leading to upregulation of Notch signaling, and enhanced proliferation in vitro. The cause of muscle regeneration impairment was thus narrowed down to either activating factors in the young circulation, inhibiting factors in the aged circulation, or both. Furthermore, this work supported the idea that 'aged' satellite cells may be able to maintain muscle homeostasis if the niche can be properly regulated [57].

In order to determine whether positive or negative regulation of muscle stem cell growth was occurring, further work demonstrated that the inhibition of young satellite cell activation or the activation of 'aged' satellite cells could be achieved in vitro with old and young serum alone, respectively [40]. More importantly, the study showed that culture of young satellite cells in old sera led to inhibition of regeneration, pointing towards the existence of a negative regulator of muscle regeneration in aged sera. A year later transcription growth factor beta 1 (TGF-β1) was identified as the inhibitory culprit [11]. Specifically, TGF-β1 levels were shown to be higher in the aged niche, leading to increased signaling via pSmad3, and antagonizing of Notch signaling. Increased pSmad3 signaling upregulated expression of the cyclin-dependent kinase inhibitors p15, p16, p21, and p27, thus preventing satellite cell activation and proliferation (Fig. 3b). Inhibition of TGF-β1 activity via antibody directed to its receptor allowed satellite cells in an 'aged' niche to regenerate muscle similarly to 'young' satellite cells. A subsequent paper dug deeper into the TGF-β1 signaling pathway and established its evolutionary conserved role (between mouse and human) in the inhibition of muscle repair [12].

In sera, TGF-β1 is present in both active and inactive forms [12, 58, 59] while in plasma only the inactive form is present. Total and active TGF-β1 on the other hand is secreted by platelets and $CD4^+$ T lymphocytes [58]. According to a recent study [12] in vitro inhibition of muscle regeneration is likely caused by TGF-β1 released from platelet activation during sera collection. However, neither systemically administered TGF-β neutralizing antibodies nor TGF-β decoys were able to restore muscle regeneration to aged mice. Both TGF-β levels and TGF-β receptor expression (3–4 fold) are increased in aged satellite cells [11], but it is unlikely that TGF-β1 acts as an endocrine inhibitor of satellite cells. Aged satellite cells expressing a dominant-negative TGF-β receptor [60, 61] were activated, proliferated, and differentiated in the presence of old serum. In vivo, systemic TGF-β receptor I kinase inhibitor [62, 63] was able to promote muscle regeneration in the aged niche, whereas TGF-β neutralizing antibody or the extracellular component of TGF-β receptor II proved ineffective [12].

In summary, skeletal muscle regeneration is an intricately regulated process that depends on the crosstalk between several essential signaling pathways (Notch/Wnt/TGF-β1). In the 'young' niche, muscle stem (satellite) cells, respond rapidly to injury and proliferate in response to Notch signaling, before differentiating and fusing into new myotubes under Wnt signaling. In the 'aged' niche, TGF-β1 signaling antagonizes Notch signaling and inhibits satellite cell activation, leading to scar tissue creation in lieu of de novo myofiber formation. Restoring youthful regeneration capacity to 'aged' satellite cells is an attractive prospect for myogenic therapies. While many advances have been made in in vitro skeletal muscle regeneration strategies, very few have yet to take into account the effects of aging on the satellite cell niche in vivo. The next two sections will thus analyze what has been accomplished so far in this regard, analyzing either purely biological or biomaterial methods to promote efficient regeneration of skeletal muscle satellite cells under adverse extrinsic conditions.

4 Toolbox to Combat TGF-β1-induced Aging of Satellite Cell Niche

Now that some of the signaling pathways behind muscle regeneration have been elucidated, including the inhibitory regulation, it is possible to deconstruct and modify these regulatory pathways in order to enhance muscle tissue engineering constructs and combat the effects of aging and pathology on the satellite cell niche. As described previously, forced activation of Notch [49] or inhibition of TGF-β1 signaling—either through blocking the interaction of TGF-β1 with its receptor, inhibiting the receptor itself, or through siRNA against Smad3 [11, 12, 64]—can lead to 'young' regeneration of satellite cells in an aged niche. However, TGF-β-neutralizing antibody or siRNA against Smad3 might require multiple daily injections, raising concerns with efficiency and safety if these techniques were ever to be considered in clinical applications.

Alternative methods to regulate TGF-β1 levels in the aged niche are currently being pursued in the hopes of improving efficiency (fewer injections) and tuning the TGF-β1 levels precisely. Not only will changes in TGF-β1 levels affect other signaling pathways, but overly long-term downregulation or complete elimination of TGF-β1 has been recently shown to negatively affect regeneration [12]. In fact, there is a defined window of active TGF-β1 levels that are necessary for proper regeneration before hitting a threshold above which active TGF-β1 levels become inhibitory.

The calibration of TGF-β signaling to young levels has broader implications in tissue engineering and regenerative medicine and raises an issue that has still been poorly addressed—although we may eventually be able to engineer tissues or organs in vitro, they will someday be engrafted into humans. However, the aged or elderly population, which will most likely have greater need for these tissues and organs, are unfortunately hosts to 'aged' tissue and organ niches. Few in vitro and in vivo tissue engineering constructs take the implications of an 'aged' niche into consideration and this could undermine performance of methods that have worked well in the non-aged environments. As has been observed in the adult muscle stem cell niche, the potential of the stem cells to regenerate does not decrease over the lifetime of the organism, but the niche conditions of the aged host prevent the satellite cells from proper function [40, 54, 57, 65, 66]. Therefore it is necessary not only to succeed in creating tissues and organs, but to also deal with the issues inherent in aged niches that may prevent proper stem cell regeneration.

When referring to the aged muscle stem cell niche, it was shown that TGF-β1 sera levels are increased over young [11, 12], indicating a possible first location to target TGF-β1 inhibition. However, systemic TGF-β1 is mostly confined to platelets and exists mainly in its latent form [12, 67–69]. Furthermore, a recent study has shown that endocrine TGF-β1 likely has no anti-myogenic activity in vivo [12]. Systemic targeting of TGF-β1 for inhibition may not therefore be a viable application. It is possible that endogenous TGF-β1 secretion by the aged tissue (such as myofibers) is responsible for the impaired regeneration [12].

In the adult skeletal muscle niche, activation of TGF-β1 occurs in the extracellular matrix (ECM) around muscle fibers [11, 70] and TGF-β1 co-localizes specifically with the laminin component of the ECM [71]. This is therefore an area in which TGF-β1 can be targeted for inhibition. Specifically, it is possible to inhibit the signaling of active TGF-β1 in the muscle niche through local addition of anti-TGF-β1 neutralizing antibody, TGF-β receptor I kinase inhibitor (systemically or locally administrated), or through a decoy [11, 12].

However, it is likely that greater efficiency in TGF-β1 regulation can be achieved and long-term therapies would be required to tackle chronic 'conditions' such as those imposed by aging of the muscle stem cell niche. Therefore, future work may consider methods to target TGF-β1 inhibition either systemically or in the local muscle niche (or both), in order to improve adult muscle stem cell regeneration in the aged niche. Several purely biological methods have already shown that downregulating TGF-β1 levels below a certain threshold (~1–5 ng/mL, [12]) has significant effects on regeneration in vitro. The use of biomaterials may facilitate the development of more effective methods to rejuvenate the muscle niche. We will analyze some of the current biomaterial methods being used to modify the adult skeletal muscle niche and will suggest a few directions which may benefit muscle regeneration further.

5 Biomaterials to the Rescue: Proposed Strategies for Adult Skeletal Muscle Regeneration

Autologous muscle transplants result in donor site morbidity and often lack proper functionality at the site of engraftment [72]. We have already suggested that biomaterials may be uniquely suited to addressing some of the issues with adult skeletal muscle regeneration due to various myopathies or to the effects of aging. In this section we will outline three different areas of application of biomaterials to the muscle stem cell niche. First we will discuss the advancement of biomaterial scaffolds for the in vitro growth of muscle and how these could be further developed to combat muscle dystrophies or aging. Next we will look at biomaterial methods to specifically remedy the inhibitory effects of TGF-β1 in the aged niche, starting with gene and drug delivery methods, followed by some novel molecular targeting strategies. Finally, we will discuss a novel biomaterial method to better define satellite cell heterogeneity and optimize selection of muscle stem cells to be used in tissue-regenerative therapies.

5.1 Engineering an In Vitro Niche for Robust Skeletal Muscle Regeneration

Effectively mimicking in vivo niches for in vitro tissue engineering is an important research tool to create possible regenerative therapies. Much in vitro work is done in two dimensions however and thus does not adequately reflect the in vivo environment of the skeletal muscle niche. Subsequently, culture systems that utilize in vitro three-dimensional niches for proliferation, differentiation, and fusion of muscle fiber progenitor cells are more likely to generate more suitable tissues for in vivo muscle repair therapies [73–75].

Indeed, it is clearly not an easy thing to mimic the native satellite cell niche ex vivo. First of all, the satellite cell is intimately associated with a muscle fiber which allows for better regeneration if implanted with intact satellite cells [8, 76]. In vitro satellite cell culture leads to a loss in proliferative activity of satellite cells as well, in addition to loss of self-renewal, as they rapidly differentiate into myoblasts; this loss in regenerative capacity is likely due to the removal of satellite cells from their in vivo niche [77]. The satellite cell niche supplies mechanical (adhesion/stiffness), structural, and chemical (small molecules) signals to the satellite cells [77, 78]. The basement membrane, below which satellite cells reside, is composed of entactin-linked networks of laminin and collagen type IV with a high number of integrin and proteoglycan binding sites [76, 77] and provides a physiological elasticity of around 21 kPa [79]. Changes in mechanical or chemical cues due to pathology or aging can significantly affect the regeneration potential of satellite cells [78]; for example young, healthy skeletal muscle tissue has been reported to have a stiffness of around 12 kPa [79, 80] that increases to over 18 kPa in some disease [80, 81] or aged states [82, 83]. Altered signaling through focal adhesion complexes can then lead to changes in cell fate [84, 85]. For example, polyacrylamide gels that are softer or stiffer than the optimal in vitro C2C12 stiffness or in vivo primary myoblast niche stiffness have been shown to impair cell proliferation [76, 80].

Biomaterial scaffolds have long been used to create three-dimensional niches for tissue regeneration ex vivo [86, 87]. For skeletal muscle regeneration, the scaffold should allow muscle progenitor adherence, growth, and proliferation, followed by fusion of myoblasts into multinucleated muscle fibers. Not only that, muscle fibers that are large, strong, aligned, and properly oxygenated and nourished will be more likely to be viable after engraftment. Generating functional, contracting muscle is one of the most important goals of regenerative medicine [72].

Choice of the appropriate cell type to use also has to be determined, but is often limited to myoblasts, since satellite cells cannot be maintained in culture [88–90]. Furthermore, myoblasts themselves may not be adequate for human transplantation as they have shown poor migration abilities after engraftment into non-human primates [72, 91]. A recent review provides a comprehensive list of cell lines (muscle-derived or other) used in skeletal muscle tissue engineering [92]. Once adequately tested, functional in vitro regeneration systems may then be applied to in vivo models of genetic myopathies [e.g., Duchenne's muscular dystrophy (DMD)] where muscle repair is inefficient due to the functional exhaustion of resident satellite cells [93–95]. The success of in vivo muscle regeneration systems to date has been low, due to cell death and immune rejection of the transplanted cell types (most often myoblasts) [96–98] and is a topic of intense research.

The first biomaterial scaffolds for in vitro skeletal muscle regeneration aimed only to create functional myotubes from seeded myoblasts; over time, numerous substrates, both artificial (biodegradable or non-biodegradable) and natural (or combinations of both) have been used to test muscle regeneration [98–110]: polymers such as poly(glycolic acid) (PGA) [111], poly(lactic acid) (PLA) [104], the PLGA co-polymer of the two [112, 113], poly(dimethylsiloxane) (PDMS) [114], poly(ethylene glycol) (PEG) [115], self-assembling peptide nanofibers [116], and many more [106, 108, 117, 118]. Biological substrates made up of various tissue extracellular matrix proteins such as collagen, hyaluronic acid [100, 101, 119–123], or other biologically active molecules [124, 125], and micropatterned surfaces including various ECM proteins or signaling molecules [126–131] have also garnered significant interest. However, many of these scaffold systems had drawbacks such as difficulty of fabrication [114] or high elastic modulus which did not adequately mimic the in vivo muscle niche [106] (Fig. 4a).

Therefore, to more efficiently and accurately tissue engineer skeletal muscle in vitro, the focus has shifted to more closely mimicking the in vivo niche in biomaterial scaffolds. Specifically, researchers have moved towards simpler scaffolding systems with controllable stiffnesses that are able to promote generation of aligned myofibers into densely packed parallel bundles and mechanically and electrically stimulate myofiber contraction [132, 133]. We will look at some of

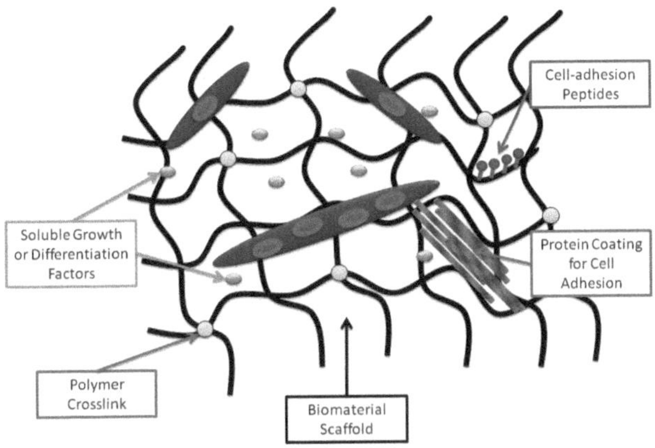

Fig. 4 Biomaterial scaffolds for skeletal muscle tissue engineering. Engineered biomaterials aim to recreate the skeletal muscle niche in vitro. Substrate stiffness can be varied to promote adhesion, spreading, proliferation, or differentiation (e.g. varying polymer scaffold crosslinking). Natural extracellular matrix protein coatings (e.g., collagen, elastin) or peptides can be displayed to promote cell adhesion and signaling. Soluble growth and differentiation factors can also be incorporated to affect cell fate

the more recent work done to optimize these scaffolds and then focus on some of the most advanced systems to date and suggest what the future may hold.

5.1.1 Alignment of In Vitro Skeletal Muscle Fibers

Randomly oriented fibers generated in vitro would not be useful for in vivo muscle transplantation therapies as aligned fibers are required for proper force generation [72]. Therefore, the initial work done in in vitro skeletal muscle regeneration which focused only on myofiber formation is now focused on scaffolds to generate parallel arrays of aligned fibers that could then be easily engrafted.

Previous work has looked at arranged muscle fibers in vitro from rat satellite cells grown on an aligned collagen gel system [110, 134] which could then be transplanted in vivo. Collagen sponges consisting of collagen-I with parallel pores were also successfully used to culture and fuse C2C12 myoblasts [100]. A more recent study has proposed the use of a cell/hydrogel system cast inside poly-dimethylsiloxane (PDMS) molds decorated with elongated posts that were able to organize the regeneration of both mouse and rat skeletal muscle tissue in such a fashion [132]. Their method, in which the post size and dimensions could be varied, allowed control of muscle fiber size, thickness, alignment, and porosity (which is important for oxygen transport). Another group using PDMS created laminin-coated microtopographically patterned wavy PDMS substrates [127]. This allows one to first align the growing myotubes on the substrate, and then organize them spatially in three-dimensions by transferring them to a degradable hydrogel. This method circumvents the issue of using rigid scaffolds alone, which promote myofiber alignment, but inhibit mechanical function of the generated fiber.

5.1.2 Effects of Synthetic Niche Stiffness on Skeletal Muscle Regeneration

Alignment of tissue-engineered myofibers is essential to generating functional skeletal muscle for subsequent engraftment. However as noted previously, rigid scaffolds may be used to promote alignment, but also inhibit myofiber contractility. The stiffness of the biomaterial scaffold is thus intricately related to the allowable contractility of the nascent myofibers. Substrate stiffness has been known to influence cell survival, adhesion, spreading, growth, proliferation, and differentiation [76, 135, 136]. For skeletal muscle, substrate stiffness affects myogenic precursor proliferation, while differentiation is controlled by both stiffness and adhesion site type and number [76]. In one study, an optimal stiffness of 21 kPa was determined for proliferation which not surprisingly corresponds to the in vivo elasticity of skeletal muscle [76, 79]. Other studies have shown that increasing substrate stiffness leads to increased proliferation of myoblasts [136] though the time-course of experiments must be taken into account [76]. Additionally important in skeletal muscle regeneration is allowing myotubes to spontaneously contract. On laminin- and poly-D-lysine-coated substrates, spontaneous contraction of myotubes was often observed, but not on collagen type IV-, Matrigel-, or entactin-collagen-laminin-coated substrates [76]. As such, two- and three-dimensional peptide-modified alginate scaffolds [102, 136] demonstrated that mouse myoblasts in non-degradable, stiff gels proliferated more, but fused less than cells seeded onto degradable gels [137]. This work is important in determining the characteristics of future scaffolds to be used for in vivo muscle regeneration.

5.1.3 Electrical Stimulation of Tissue-engineered Skeletal Muscle

While mechanical stimulation of myotubes can allow for contraction, external electrical stimulation has also been looked into as a means of inducing myotube contraction. Skeletal myoblasts cultured on aligned electrospun polyurethane fibers and subjected to external electric stimuli formed more aligned, contractile myotubes, than those cultured on unaligned fibers without stimulation [138] (Fig. 5). A quantitative study of the effects of electrical pulse frequency on myotube formation was recently completed and determined the proper frequency for synchronous contractility of myotubes [139]. Another study showed that it is possible to also locally stimulated myoblasts using a microporous alumina membrane on a PDMS substrate [140]. Finally, a recent study has shown that electrical stimulation can even encourage myofiber alignment [141]. Previous studies had shown that myoblasts could be coaxed to fuse into aligned myofibers based on the growth substrate itself, such as elastic membranes or micropatterned scaffolds [142, 143], or if subjected to external mechanical stimuli such as stretching [144]. The latter method recapitulates the in vivo passive stretching of muscle [145] and is a reminder that closely mimicking in vivo regenerative processes in vitro by manipulating synthetic biomaterial systems can lead to effective myofiber synthesis methods for therapeutic purposes [146].

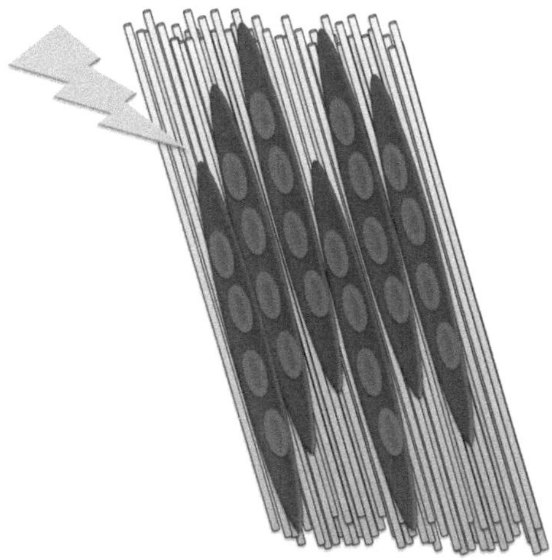

Fig. 5 Electrospun nanofibers for generation of aligned myotubes. Parallel bundles of electrospun polymeric nanofibers can generate groups of aligned myotubes. Conductive polymers can be used to generate contracting fibers through electrical stimulation

Some of the most advanced biomaterial niches seek to optimize substrate stiffness, myofiber alignment, and electrical stimulation all at once. Electrospinning of poly(caprolactone)/collagen nanofibers was recently used to create a scaffold for generating human skeletal muscle in vitro with controlled orientation [147]. Unlike randomly oriented nanofibers, parallel electrospun nanofibers were able to induce aligned myotube formation and generated longer, more native-like myofibers. A subsequent, more pointed study [148] electrospun fibers of poly(L-lactide-co-3-caprolactone) blended with conductive polyaniline [149] in order to generate myotubes through electrical stimulation of C2C12 myoblasts.

5.1.4 Vascularization of Tissue-engineered Skeletal Muscle

Tissue-engineered constructs require proper oxygenation and nutrient delivery to survive and integrate with the host tissue upon engraftment. In order for that to be possible, no tissue can be located at more than approximately 150 μm from supply sources [142]. This issue has been considered as well for tissue-engineered skeletal muscle. The foreign body reaction that occurs upon transplantation of tissue leads to some viable vascularization [92, 150–152]. Other methods involve prevascularizing the tissue engineering construct [105, 153] before implantation; the method by Levenberg et al. co-cultures endothelial cells and embryonic fibroblasts with the myoblasts leading to improved integration with the host vasculature upon transplantation. However, few vascularization methods have been developed to date and none was applied in humans.

To conclude this section on synthetic biomaterial niches, we have seen that they are highly modifiable, adaptable systems and are currently being optimized to more effectively mimic the in vivo skeletal muscle niche. The initial steps of generating myofibers, aligning them, and stimulating their contraction electromechanically are fairly well understood and developed. However, the issue of providing proper

Table 1 Biomaterial strategies for skeletal muscle regeneration

Strategy	Method	Materials	References
Myofiber alignment	Aligned collagen gel	Collagen	[103, 126]
	Collagen sponge	Collagen	[93]
	Hydrogel/mold with elongated posts	PDMS	[124]
	Micropatterned surface	Laminin on wavy PDMS substrate	[119]
	Micropatterned hydrogel	Micro-contact printing of ECM proteins	[148]
	Electrospinning of parallel fibers	Poly(caprolactone)/collagen or poly(caprolactone)/ polyaniline	[139, 141]
	Electrical stimulation	Micropatterned poly-(L-lactic acid) membranes	[133]
Substrate stiffness	Selection of appropriate stiffness substrate that allows myofiber contraction and mimics in vivo stiffness	Laminin	[76]
		Poly-D-lysine	[76]
		2- and 3D peptide-modified alginate scaffolds	[95, 128, 129]
Electrical stimulation	External electric stimuli	Electrospun polyurethane fibers	[130, 131]
		Microporous alumina membrane on PDMS substrate	[132]
		ECM-coated conductive polypyrrole substrate	[92]
Vascularization	Prevascularization of tissue engineering construct	Co-culture system within highly porous, biodegradable polymer scaffold	[98, 145]

vascularization once engrafted has yet to be fully addressed, especially for human transplantation. A summary of the in vitro biomaterial strategies for skeletal muscle regeneration can be found in Table 1. In addition, issues of biocompatibility and immune response to foreign constructs and the likelihood of inflammation upon engrafting still remain. In attempts to circumvent some of these issues, attention for skeletal muscle tissue engineering has been shifting to in vitro niches that are more biologically natural in composition, and are the subject of our next section.

5.1.5 Natural Skeletal Muscle Niches: Mimicking the In Vivo Environment

While synthetic materials used for three-dimensional skeletal muscle scaffolds may be very versatile and tunable, they may not recapitulate the in vivo muscle niche as well as natural biomaterials already present in the niche. Specifically the extracellular matrix, composed of collagen, laminin, fibronectin, among other components [154], provides both three-dimensional structure and cell-adhesion sites for organizing muscle precursors into functional tissue. Natural biomaterials, including such

ECM proteins, are also being heavily researched for in vitro muscle regeneration. A recent study used collagen in combination with Matrigel, an ECM extract from an Engelbreth Holm-Swarm murine sarcoma, to form rat skeletal muscle fibers from myoblasts in vitro before injecting the gel-like cell/matrix structure in vivo [155]. Other work with 3D collagen matrices used a co-culture system of human myoblasts and fibroblasts to determine that the non-myogenic cells were essential for differentiation, matrix remodeling, and force generation [73]. This study also confirms that the in vivo niche provides not only adhesion and organization of cells into 3D tissues, but also provides soluble factors to the cells that influence growth, proliferation, and differentiation. Another group studied the proliferation and fusion of myogenic precursors from wild type and MDX mice (a model for DMD) on hydrogels micro-contact-printed with various ECM proteins for selective adhesion of satellite cells [156]. Using their micropatterning technique to mimic ECM adhesion sites, they generated nicely aligned muscle fibers from seeded satellite cells. A more recent study went further by using both a conductive polymer substrate in combination with the coating of ECM proteins (in this case hyaluronic acid and chondroitin sulfate) for adhesion, proliferation, and differentiation of both murine C2C12s and mouse primary myoblasts [99].

In a very interesting study ECM was extracted from rats and used as an in vitro substrate for skeletal muscle growth and differentiation [157]. This novel substrate allowed greater proliferation and differentiation of myoblasts than uncoated surfaces or collagen-coated surfaces. Importantly, this study also showed that collagen is the major component of the in vivo skeletal muscle niche, and that the extracted ECM must also contain additional components, as treatment with collagenase did not obviate adhesion and proliferation of myogenic cells. This study thus provided an important step in deconstructing the in vivo skeletal muscle niche in order to create enhanced in vitro niches.

The Conboy lab has also done work in this area of natural three-dimensional matrices for skeletal muscle regeneration, specifically focusing on reconstructing an in vitro skeletal muscle niche that highly mimics the in vivo niche [126]. In a first series of experiments, myogenic cell fate was intricately controlled using a combination of known myogenic growth and differentiation factors embedded in Matrigel [9, 70, 158–160]. A related delivery mechanism has just been developed that is used to pattern nanoliter amounts of reagents (e.g., growth/differentiation factors) that will stay in a defined space, even in the presence of a second aqueous phase (e.g., cell culture medium) [161]. In the system developed by de Juan-Pardo et al., delivery of the factors was manipulated spatially so as to locally promote growth or differentiation of myoblasts even in the presence of differentiation or growth medium, respectively. By having the power to control cell fate regardless of the external medium, this artificial niche could be a great tool in promoting muscle regeneration in various myopathic conditions caused by changes in the in vivo niche. This was enabled by the unique method of incorporating growth and differentiation factors into the ECM gel, and allowing both proliferation and differentiation to occur simultaneously in the same dish. As expected, cells in growth medium alone proliferated, while cells in differentiation medium alone fused into myotubes. However cells in growth medium exposed to embedded differentiation factors preferentially differentiated, while cells in differentiation medium exposed to growth factors preferentially proliferated. In addition, cells cultured in neutral medium (which promotes neither proliferation nor differentiation) could be spatially directed to proliferate or differentiate if locally exposed to growth or differentiation factors respectively.

In the future, a completely reverse-engineering in vivo skeletal muscle niche, reconstructed from defined components and factors may provide an optimal solution as an in vitro skeletal muscle regeneration platform.

Section 5.1 looked at the current state of biomaterial niches for generating functional muscle in vitro for in vivo transplantation. The use of biomaterials has rapidly advanced the work by moving to recapitulate the in vivo skeletal muscle niche ex vivo (e.g., conductive polymers for electrical stimulation or the use of extracellular matrix components for adhesion). In addition, biomaterials have provided the tools to generate large, aligned, and strong muscle fibers which are more suitable for transplantation. And although synthetic biomaterials are highly modular and modifiable, the move now seems to be towards simpler niches, which more greatly mimic the in vivo niche through use of extracellular matrix components and signaling factors. Natural niches (e.g., collagen, Matrigel) provide more numerous and higher affinity binding sites for skeletal muscle and better mimic the in vivo niche.

What comes next? While there are therefore countless possible scaffolds and methods for preparation of muscle progenitors or myotubes ex vivo, which could then be used for muscle regeneration or replacement for various myopathies or dystrophies, no one has yet to address the task of muscle regeneration where the in vivo niche inhibits or prevents regeneration. Specifically, skeletal muscle regeneration in an aged niche is inhibited by increased TGF-β1 signaling and may therefore impede proper functioning of any of these systems upon implantation. Future work in this area may look at the incorporation of TGF-β1 inhibitors into such scaffolds or of the Notch ligand Delta to promote proliferation of implanted cells. However, as we have seen previously, satellite cells in an aged niche are still very capable of regeneration, and "rejuvenating" the niche itself may be the only necessity to restoring youthful muscle regeneration. The next section will analyze the use of biomaterial methods to address this issue of regeneration of skeletal muscle in aging.

5.2 Biomaterial Strategies to Combat Aging of the Muscle Stem Cell Niche

5.2.1 Gene and Drug Delivery Methods to Promote Skeletal Muscle Regeneration

Biomaterials have long been used in drug and gene delivery methods in order to protect cargo, target specific cells or tissues, and provide controlled, sustained release for treatment [162]. Gene delivery methods to skeletal muscle could provide possible treatment for degenerative myopathies such as Duchenne's muscular dystrophy, in which the protein dystrophin, which is essential in maintaining interactions between the extracellular matrix and the muscle fiber cytoskeleton, is not expressed [163]. Typical gene delivery methods include naked plasmid DNA, plasmid DNA delivered in polymeric particles [164], or through viruses [165]. However, naked DNA transfection is very inefficient and viral transfection, although very efficient, raises several concerns about oncogenicity, immune response, and safety [165]. Polymeric vehicles which release DNA based on diffusion and polymer degradation [164], allow long-term controlled and sustained release of DNA [166]. Therefore, in past years research focused on optimizing polymeric delivery vehicles in order to enhance gene delivery without the viral side effects. Various polymeric vehicles have been developed [167–170] and 'smart' polymers that are pH or temperature-sensitive have become popular [171–176]. Temperature-sensitive polymers, which are liquid at room temperature, but gel in the body at 37°C provide ease of use and injectability, in addition to the controlled, sustained release of a drug or gene payload [177]. Recent work has been done to enhance these methods of delivery to skeletal muscle cells [178] and is thus promising for possible therapies. Specifically, in vitro and in vivo delivery of plasmid DNA was mediated via multi-block copolymers (MBCPs) of pluronic® and di-(ethylene glycol) divinyl ether [179]. Importantly, the MBCPs were much more effective at in vivo (although not in vitro) transfection than both naked DNA or polymeric/DNA complexes that incorporated poly(ethylenimine), a standard cationic polymer used in drug delivery applications that had been previously shown ineffective as a gene delivery vehicle in muscle [180, 181].

As explained above, gene delivery systems could be useful in delivering plasmid DNA to skeletal muscle cells. This could obviously apply in cases of genetic disorders, many of which lead to muscular dystrophies such as DMD, facioscapulohumeral dystrophy, Limb-Girdle muscular dystrophy, etc. where replacing deleted or mutated genes with healthy ones could be used to restore proper muscle physiology. In terms of promoting muscle regeneration in the aged niche, gene delivery of the Notch ligand Delta or the TGF-β1 receptor kinase inhibitor could be considered as possible therapeutics for restoring 'youthful' skeletal muscle regeneration, either through activation of the Notch pathway or through inhibition of TGF-β1 signaling, respectively. The next section will discuss novel biomaterial applications to the regulation of TGF-β1 signaling in the aged niche.

5.2.2 Novel Targeting Strategies for TGF-β1 Inhibition

A biomaterial platform for regulating TGF-β1 levels to 'young' levels in the aged niche

As described previously [11], attenuation of TGF-β1 signaling in the aged muscle niche promotes restoration of skeletal muscle regeneration. However, inhibition of TGF-β1 signaling required twice-a-day injection of TGF-β1 receptor I kinase inhibitor or Notch, or lentiviral transfection of shRNA against pSmad3, a downstream effector of TGF-β1 signaling. Complete silencing of signaling using viral transfection would be harmful as TGF-β1 is involved in many other signaling pathways [42], and non-viral transfection methods, which are less efficient than viral methods, would need to be optimized. In addition, were these strategies to be considered for clinical use, fewer injections, and a more efficient delivery system would be desired.

As such, novel biomaterials constructs may pave the way for some solutions. Future work will generate a wide array of biodegradable, biocompatible, nanoparticulate platform technologies that allow the display of a wide array of peptides or proteins at the particle surface through site-selective conjugation. These particles can be engineered to both encapsulate drugs or proteins of interest, and target specifically based on what molecule is used to decorate them. Unlike most drug delivery particles however, the site-specific conjugation technique provides controlled and consistent orientation of the displayed molecule of choice. This allows the display of the desired portion of surface molecule (i.e. actively targeting portion), and a significant increase in efficiency, as none of the displayed molecules will be randomly oriented. Such particles could be used in a variety of ways to combat the effects of aging in the muscle niche. Specifically, activation of Notch signaling through long-term sustained delivery of Delta [49] encapsulated in nanoparticles may promote youthful muscle regeneration and effectively reduce the amount of injections. Tailoring protein release rate, particle degradation rate, and amount of protein loaded will all be important in determining the efficacy of such a treatment. In addition, determining which possible targeting molecules to use or generating targeting peptides will have to be done. However, a systemic delivery of Delta may not be the most efficient. Nanoparticles will not exit blood vessels into tissues and may be rapidly cleared from circulation [182] (although the delivery time would greatly exceed that of Delta alone). Delta is also involved in many other signaling pathways so a non-localized delivery of Delta could generate unwanted effects. Nevertheless, a localized, controlled, and sustained release of Delta would be beneficial in enhancing muscle repair, but intramuscular injections would be difficult and injections would need to be timed to moments of injury (or before going to exercise). A more precise targeting strategy would therefore need to be established in order to consider this a viable therapy.

Similarly to these targeted Delta delivery strategies, delivery of TGF-β1 receptor kinase inhibitor, which was also effectively used to regulate TGF-β1 signaling in the aged muscle niche could be used [11, 12] In this case, however, as aged mice treated with systemically injected TGF-β1 receptor kinase inhibitor showed a marked improvement in muscle regeneration, a circulating particle loaded with TGF-β1 receptor kinase inhibitor may be suitable [12].

As a parallel strategy, the same nanoparticulate platforms could be used to directly inhibit the action of TGF-β1. Specifically, the nanoparticles could be decorated with TGF-β1-binding molecules such as antibodies, antibody fragments, etc. Unlike in the Delta or TGF-β1 receptor kinase inhibitor cases, the decorated nanoparticles will not be delivering a payload, but will instead serve as TGF-β1 scavengers that will bind and inhibit the action of TGF-β1 in the aged niche. These multivalent TGF-β1 'sponges' could collect TGF-β1 molecules and could greatly improve efficiency of TGF-β1 inhibition (over free antibody injections). Regulating 'aged' TGF-β1 levels to the precise 'young' levels will require manipulation of the number of TGF-β1-binding sites on the particles and the number of particles to be used; too few particles will have little or no effect on the TGF-β1 inhibition of muscle regeneration; too many particles will also have deleterious effects, as a certain level of TGF-β1 signaling must be maintained [12, 42]. As in the previous case with delivery of Delta, the immune clearance of nanoparticles will occur. However, in this case, this removal of TGF-β1-decorated particles by immune cells will be beneficial as it will decrease active TGF-β1 signaling levels, thus lowering the block on muscle regeneration. The remaining issue is that of TGF-β1 localization. It is known that TGF-β1 can be internal to circulating platelets or secreted by various cell types [12, 42] (though some TGF-β1 has been observed systemically in diseased states [183]); upon platelet activation, TGF-β1 is released, crosses into tissue, and becomes active in the local extracellular matrix [12, 42]. Therefore the issue of targeting TGF-β1 locally (in the muscle niche) arises, and consequently the same concerns as in the Delta delivery situation. In any case, this multivalent nanoparticulate platform can provide greater local regulation of TGF-β1 levels than a simple free antibody-driven system and may at least be very useful as a model for in vitro studies, where niche targeting is not an issue.

5.3 Satellite Cells and Muscle Stem Cells: Biomaterials to Help Determine Who is Who

Skeletal muscle satellite cells make up a small portion of the cells on muscle fibers [184]. Correctly identifying adult muscle stem cells in order to generate new muscle is therefore essential for therapeutic purposes. However, whether or not satellite cells are the only myogenic precursor cells, and whether all satellite cells are in fact muscle stem cells is still a topic of debate [7, 38, 78]. Several other cell lines, including hematopoietic cells [185–188], mesoangioblasts [189], and pericytes [190] have shown myogenic potential [78]. Satellite cells may even differ between muscle types [1, 191]. In fact disagreement still remains over which markers define satellite cells and which characterize actual muscle stem cells [7]. Finally there is the disputed place of reserve cells which arise during myoblast differentiation in vitro and may contribute to replenishing the satellite cell pool [38, 48, 192–194]. These problems are exacerbated by the fact that there are so few satellite cells per fiber, which makes identifying surface markers with conventional methods such as flow cytometry or immunostaining less reliable. Novel uses of biomaterials may provide the answers to some of these questions and after discussing the current satellite cell classification, we will suggest one such method.

Originally, satellite cells were identified solely based on their biological location beneath the basal lamina of skeletal muscle fibers [2]. The most often cited current satellite cell marker (and muscle stem cell marker) is the paired box transcription factor Pax7 [1, 7, 38, 40, 48, 57, 90, 192, 194, 195]. However, a recent publication suggests that this Pax7 may not be a unique identifier and that it is only a subpopulation of the Pax7-positive satellite cell population (i.e., satellite cells that are $Pax7^{+}/Myf5^{+}$) that actually contributes to muscle regeneration. The $Pax7^{+}/Myf5^{-}$ population was found to serve as a muscle stem cell reservoir to replenish satellite cells after a cycle of regeneration [7].

Current satellite cell isolation and sorting strategies are mostly limited to FACSs sorting based on specific markers [7]. One such example defines $CD34^{+}$, $\alpha7$-integrin^{+}, $CD31^{-}$, $CD45^{-}$, $CD11b^{-}$, $Sca1^{-}$ as satellite cells [7] (>95% satellite cells) while another includes $CD45^{-}$, Sca-1^{-}, Mac-1^{-}, $CXCR4^{+}$, $\beta1$-integrin^{+} as part of the satellite cell pool [196]. Overall, surface markers that have been considered as identifying of quiescent satellite cells include: $CXCR4^{+}$ (fusin), $CD34^{+}$ (not specific, but useful for sorting from heterogeneous populations of cells) [1, 38, 57, 191, 197], $\alpha7$-integrin^{+}, $CD31^{-}$ (PECAM-1), $CD45^{-}$ (PTPRC), $CD11b^{-}$ (integrin αM, Mac-1,CR3A), $Sca1^{-}$ (ATX1, D6S504E), Syndecan-3^{+} [38, 198], Syndecan-4^{+} [7, 38, 48, 57, 191, 195, 197–200], cMet^{+} [hepatocyte growth factor receptor (HGFR)] [38, 201, 202], caveolin1^{+} [202], Fzd7^{+} [7], M-cadherin (CDH15) [38, 57, 201, 203], β1 integrin (CD29) [57, 204], VCAM-1^{+} [38, 205], NCAM^{+} (CD56, leu-19) [38, 206, 207], ABGC2 [78, 208], and the antigen for the SM/C-2.6 monoclonal antibody [78, 209], while internal markers include: MyoD [48, 199], Myf5 [38, 57, 67, 191, 195, 197, 201] (though this is now debated by Le Grand's work), MNF (myocyte nuclear factor) [38, 210], myostatin (growth differentiation factor 8) [38, 198, 211, 212], IRF-2 (interferon regulatory factor-2) [38, 205], Msx1 (hox7, hyd1) [38, 198].

However there is still a lack of agreement on which markers completely define satellite cells, but maybe more importantly, which define the muscle stem cell population. In order to work on isolating these markers from such minute populations of cells, novel biomaterial-based microfluidic devices could be used to effectively sort small numbers of cells based on their surface markers (based on past work, [213]). These devices could be used both identify the type and number of surface markers on a cell. Furthermore, unlike flow cytometry or immunostaining, these devices employ a label-free technique which allows subsequent collection and culture of the cells that pass through. Myogenicity of these small populations of purified cells can then be determined and identification of muscle stem cell populations. Finally incorporation of other microfluidic technologies such as single-cell protein analysis techniques could push this work further by allowing determination of gene expression at the single cell level [214]. This will allow detection of differences between these small populations of cells and identification of subpopulations of cells within the satellite cell niche. Together, this work has promise in shedding light on the characteristics of skeletal muscle satellite cells, and the determination of the subpopulation of muscle stem cells.

5.4 Use of Biomaterials in Tissue Engineering Applications

As we have seen, biomaterials have many applications in tissue regenerative therapies for adult skeletal muscle. However, the possible negative impact of biomaterials must still be kept in mind. Biocompatibility of materials, the effects of biomaterial degradation and degradation products on the niche, the host response to non-degrading implants or materials, and the inflammatory and immune response will all need to be considered once biomaterial platforms are optimized and ready for clinical trials. Nevertheless, the potential to treat myopathic conditions or the effects of aging on the adult skeletal muscle niche with biomaterials is real and perhaps not that distant.

6 Conclusion

Adult skeletal muscle regeneration is orchestrated by a complex network of signaling molecules that are still being uncovered. The onset of aging impedes the regenerative potential of muscle stem cells and much work is being done to reverse these effects. However, there are still some doubts as to whether all satellite cells are muscle stem cells and how to most efficiently enhance muscle regeneration in the old and in individuals afflicted by genetic muscle wasting. The use of biomaterials as 3D muscle satellite cell niches, delivery vehicles for genes, drugs, or pro-regenerative factors, and for further definition of the heterogeneous skeletal muscle stem cell pool will likely lead to efficient, robust, and clinically feasible enhancements to muscle regeneration strategies.

Acknowledgments The authors would like to acknowledge funding through the ARRA NIH grant AG 027252S1 awarded to I.M. Conboy. The authors would also like to offer recognition to other works that fit within the scope of this chapter but were not included.

References

1. Zammit, P.S., Partridge, T.A., Yablonka-Reuveni, Z.: The skeletal muscle satellite cell: the stem cell that came in from the cold. J. Histochem. Cytochem. **54**, 1177–1191 (2006)
2. Mauro, A.: Satellite cell of skeletal muscle fibers. J. Biophys. Biochem. Cytol. **9**, 493–495 (1961)
3. Moss, F.P., Leblond, C.P.: Satellite cells as the source of nuclei in muscles of growing rats. Anat. Rec. **170**, 421–435 (1971)
4. Bischoff, R.: Regeneration of single skeletal muscle fibers in vitro. Anat. Rec. **182**, 215–235 (1975)
5. Bischoff, R.: Proliferation of muscle satellite cells on intact myofibers in culture. Dev. Biol. **115**, 129–139 (1986)
6. Konigsberg, U.R., Lipton, B.H., Konigsberg, I.R.: The regenerative response of single mature muscle fibers isolated in vitro. Dev. Biol. **45**, 260–275 (1975)

7. Le Grand, F., Jones, A.E., Seale, V., Scime, A., Rudnicki, M.A.: Wnt7a activates the planar cell polarity pathway to drive the symmetric expansion of satellite stem cells. Cell Stem Cell **4**, 535–547 (2009)
8. Collins, C.A., Olsen, I., Zammit, P.S., Heslop, L., Petrie, A., Partridge, T.A., Morgan, J.E.: Stem cell function, self-renewal, and behavioral heterogeneity of cells from the adult muscle satellite cell niche. Cell **122**, 289–301 (2005)
9. Conboy, I.M., Rando, T.A.: The regulation of Notch signaling controls satellite cell activation and cell fate determination in postnatal myogenesis. Dev. Cell **3**, 397–409 (2002)
10. Brack, A.S., Conboy, I.M., Conboy, M.J., Shen, J., Rando, T.A.: A temporal switch from notch to Wnt signaling in muscle stem cells is necessary for normal adult myogenesis. Cell Stem Cell **2**, 50–59 (2008)
11. Carlson, M.E., Hsu, M., Conboy, I.M.: Imbalance between pSmad3 and Notch induces CDK inhibitors in old muscle stem cells. Nature **454**, 528–532 (2008)
12. Carlson, M.E., Conboy, M.J., Hsu, M., Barchas, L., Jeong, J., Agrawal, A., Mikels, A.J., Agrawal, S., Schaffer, D.V., Conboy, I.M.: Relative roles of TGF-beta1 and Wnt in the systemic regulation and aging of satellite cell responses. Aging Cell **6**, 676–689 (2009)
13. Zhao, P., Hoffman, E.P.: Embryonic myogenesis pathways in muscle regeneration. Dev. Dyn. **229**, 380–392 (2004)
14. Carlson, M.E., Conboy, I.M.: Regulating the Notch pathway in embryonic, adult and old stem cells. Curr. Opin. Pharmacol. **7**, 303–309 (2007)
15. Artavanis-Tsakonas, S., Rand, M.D., Lake, R.J.: Notch signaling: cell fate control and signal integration in development. Science (New York, NY) **284**, 770–776 (1999)
16. Bray, S.J.: Notch signalling: a simple pathway becomes complex. Nat. Rev. **7**, 678–689 (2006)
17. Kadesch, T.: Notch signaling: the demise of elegant simplicity. Curr. Opin. Genet. Dev. **14**, 506–512 (2004)
18. Cheng, P., Nefedova, Y., Miele, L., Osborne, B.A., Gabrilovich, D.: Notch signaling is necessary but not sufficient for differentiation of dendritic cells. Blood **102**, 3980–3988 (2003)
19. Schroeder, T., Kohlhof, H., Rieber, N., Just, U.: Notch signaling induces multilineage myeloid differentiation and up-regulates PU.1 expression. J. Immunol. **170**, 5538–5548 (2003)
20. Schroeder, T., Meier-Stiegen, F., Schwanbeck, R., Eilken, H., Nishikawa, S., Hasler, R., Schreiber, S., Bornkamm, G.W., Nishikawa, S., Just, U.: Activated Notch1 alters differentiation of embryonic stem cells into mesodermal cell lineages at multiple stages of development. Mech. Dev. **123**, 570–579 (2006)
21. Nofziger, D., Miyamoto, A., Lyons, K.M., Weinmaster, G.: Notch signaling imposes two distinct blocks in the differentiation of C2C12 myoblasts. Development (Cambridge, England) **126**, 1689–1702 (1999)
22. Rosenblatt, J.D., Lunt, A.I., Parry, D.J., Partridge, T.A.: Culturing satellite cells from living single muscle fiber explants. In Vitro Cell. Dev. Biol. **31**, 773–779 (1995)
23. Kamakura, S., Oishi, K., Yoshimatsu, T., Nakafuku, M., Masuyama, N., Gotoh, Y.: Hes binding to STAT3 mediates crosstalk between Notch and JAK-STAT signalling. Nat. Cell Biol. **6**, 547–554 (2004)
24. Molofsky, A.V., Pardal, R., Morrison, S.J.: Diverse mechanisms regulate stem cell self-renewal. Curr. Opin. Cell Biol. **16**, 700–707 (2004)
25. Sundaram, M.V.: The love-hate relationship between Ras and Notch. Genes Dev. **19**, 1825–1839 (2005)
26. Carlson, M.E., Silva, H.S., Conboy, I.M.: Aging of signal transduction pathways, and pathology. Exp. Cell Res. **314**, 1951–1961 (2008)
27. Cossu, G., Borello, U.: Wnt signaling and the activation of myogenesis in mammals. EMBO J. **18**, 6867–6872 (1999)
28. Anakwe, K., Robson, L., Hadley, J., Buxton, P., Church, V., Allen, S., Hartmann, C., Harfe, B., Nohno, T., Brown, A.M., Evans, D.J., Francis-West, P.: Wnt signalling regulates myogenic differentiation in the developing avian wing. Development (Cambridge, England) **130**, 3503–3514 (2003)

29. Petropoulos, H., Skerjanc, I.S.: Beta-catenin is essential and sufficient for skeletal myogenesis in P19 cells. J. Biol. Chem. **277**, 15393–15399 (2002)
30. Brack, A.S., Conboy, M.J., Roy, S., Lee, M., Kuo, C.J., Keller, C., Rando, T.A.: Increased Wnt signaling during aging alters muscle stem cell fate and increases fibrosis. Science (New York, NY) **317**, 807–810 (2007)
31. DasGupta, R., Fuchs, E.: Multiple roles for activated LEF/TCF transcription complexes during hair follicle development and differentiation. Development (Cambridge, England) **126**, 4557–4568 (1999)
32. Hagen, T., Di Daniel, E., Culbert, A.A., Reith, A.D.: Expression and characterization of GSK-3 mutants and their effect on beta-catenin phosphorylation in intact cells. J. Biol. Chem. **277**, 23330–23335 (2002)
33. Yuan, H., Mao, J., Li, L., Wu, D.: Suppression of glycogen synthase kinase activity is not sufficient for leukemia enhancer factor-1 activation. J. Biol. Chem. **274**, 30419–30423 (1999)
34. Schmalbruch, H., Hellhammer, U.: The number of nuclei in adult rat muscles with special reference to satellite cells. Anat. Rec. **189**, 169–175 (1977)
35. Gibson, M.C., Schultz, E.: Age-related differences in absolute numbers of skeletal muscle satellite cells. Muscle Nerve **6**, 574–580 (1983)
36. Charlton, C.A., Mohler, W.A., Radice, G.L., Hynes, R.O., Blau, H.M.: Fusion competence of myoblasts rendered genetically null for N-cadherin in culture. J. Cell Biol. **138**, 331–336 (1997)
37. Bockhold, K.J., Rosenblatt, J.D., Partridge, T.A.: Aging normal and dystrophic mouse muscle: analysis of myogenicity in cultures of living single fibers. Muscle Nerve **21**, 173–183 (1998)
38. Charge, S.B., Rudnicki, M.A.: Cellular and molecular regulation of muscle regeneration. Physiol. Rev. **84**, 209–238 (2004)
39. Snow, M.H.: A quantitative ultrastructural analysis of satellite cells in denervated fast and slow muscles of the mouse. Anat. Rec. **207**, 593–604 (1983)
40. Carlson, M.E., Conboy, I.M.: Loss of stem cell regenerative capacity within aged niches. Aging Cell **6**, 371–382 (2007)
41. Schultz, E., Lipton, B.H.: Skeletal muscle satellite cells: changes in proliferation potential as a function of age. Mech. Ageing Dev. **20**, 377–383 (1982)
42. Derynck, R., Zhang, Y.E.: Smad-dependent and Smad-independent pathways in TGF-beta family signalling. Nature **425**, 577–584 (2003)
43. Massague, J.: TGF-beta signal transduction. Annu. Rev. Biochem. **67**, 753–791 (1998)
44. Massague, J., Chen, Y.G.: Controlling TGF-beta signaling. Genes Dev. **14**, 627–644 (2000)
45. Studitsky, A.N.: Free auto- and homografts of muscle tissue in experiments on animals. Ann. N Y Acad. Sci. **120**, 789–801 (1964)
46. Sadeh, M., Czyewski, K., Stern, L.Z.: Chronic myopathy induced by repeated bupivacaine injections. J. Neurol. Sci. **67**, 229–238 (1985)
47. Luz, M.A., Marques, M.J., Santo Neto, H.: Impaired regeneration of dystrophin-deficient muscle fibers is caused by exhaustion of myogenic cells. Braz. J. Med. Biol. Res. (=Revista brasileira de pesquisas medicas e biologicas/Sociedade Brasileira de Biofisica) **35**, 691–695 (2002)
48. Shefer, G., Van de Mark, D.P., Richardson, J.B., Yablonka-Reuveni, Z.: Satellite-cell pool size does matter: defining the myogenic potency of aging skeletal muscle. Dev. Biol. **294**, 50–66 (2006)
49. Conboy, I.M., Conboy, M.J., Smythe, G.M., Rando, T.A.: Notch-mediated restoration of regenerative potential to aged muscle. Science (New York, NY) **302**, 1575–1577 (2003)
50. Gibson, A., Yu, O.: Biphasic non-adrenergic, non-cholinergic relaxations of the mouse anococcygeus muscle. Br. J. Pharmacol. **79**, 611–615 (1983)
51. Nnodim, J.O.: Satellite cell numbers in senile rat levator ani muscle. Mech. Ageing Dev. **112**, 99–111 (2000)
52. Roth, S.M., Martel, G.F., Ivey, F.M., Lemmer, J.T., Metter, E.J., Hurley, B.F., Rogers, M.A.: Skeletal muscle satellite cell populations in healthy young and older men and women. Anat. Rec. **260**, 351–358 (2000)

53. Carlson, M.S.C., Conboy, M.J., Aagaard, P., Mackey, A., Kjaer, M., Conboy, I.M.: Molecular aging and rejuvenation of human muscle stem cells. EMBO Mol. Med. **1**, 381–391 (2009)
54. Conboy, I.M., Conboy, M.J., Wagers, A.J., Girma, E.R., Weissman, I.L., Rando, T.A.: Rejuvenation of aged progenitor cells by exposure to a young systemic environment. Nature **433**, 760–764 (2005)
55. Carlson, B.M., Dedkov, E.I., Borisov, A.B., Faulkner, J.A.: Skeletal muscle regeneration in very old rats. J. Gerontol. **56**, B224–233 (2001)
56. McCay, C.M., Pope, F., Lunsford, W., Sperling, G., Sambhavaphol, P.: Parabiosis between old and young rats. Gerontologia **1**, 7–17 (1957)
57. Conboy, I.M., Rando, T.A.: Aging, stem cells and tissue regeneration: lessons from muscle. Cell Cycle (Georgetown, Tex) **4**, 407–410 (2005)
58. Assoian, R.K., Komoriya, A., Meyers, C.A., Miller, D.M., Sporn, M.B.: Transforming growth factor-beta in human platelets. Identification of a major storage site, purification, and characterization. J. Biol. Chem. **258**, 7155–7160 (1983)
59. Massague, J.: The TGF-beta family of growth and differentiation factors. Cell **49**, 437–438 (1987)
60. Tang, P., Jasser, S.A., Sung, J.C., Shi, Y., Steck, P.A., Yung, W.K.: Transforming growth factor-alpha antisense vectors can inhibit glioma cell growth. J. Neuro-oncol. **43**, 127–135 (1999)
61. Yu, J.H., Schaffer, D.V.: High-throughput, library-based selection of a murine leukemia virus variant to infect nondividing cells. J. Virol. **80**, 8981–8988 (2006)
62. Gellibert, F., Woolven, J., Fouchet, M.H., Mathews, N., Goodland, H., Lovegrove, V., Laroze, A., Nguyen, V.L., Sautet, S., Wang, R., Janson, C., Smith, W., Krysa, G., Boullay, V., De Gouville, A.C., Huet, S., Hartley, D.: Identification of 1,5-naphthyridine derivatives as a novel series of potent and selective TGF-beta type I receptor inhibitors. J. Med. Chem. **47**, 4494–4506 (2004)
63. Singh, J., Chuaqui, C.E., Boriack-Sjodin, P.A., Lee, W.C., Pontz, T., Corbley, M.J., Cheung, H.K., Arduini, R.M., Mead, J.N., Newman, M.N., Papadatos, J.L., Bowes, S., Josiah, S., Ling, L.E.: Successful shape-based virtual screening: the discovery of a potent inhibitor of the type I TGFbeta receptor kinase (TbetaRI). Bioorg. Med. Chem. Lett. **13**, 4355–4359 (2003)
64. Yeom, S.Y., Jeoung, D., Ha, K.S., Kim, P.H.: Small interfering RNA (siRNA) targetted to Smad3 inhibits transforming growth factor-beta signaling. Biotechnol. Lett. **26**, 699–703 (2004)
65. Zacks, S.I., Sheff, M.F.: Age-related impeded regeneration of mouse minced anterior tibial muscle. Muscle Nerve **5**, 152–161 (1982)
66. Carlson, B.M., Faulkner, J.A.: Muscle transplantation between young and old rats: age of host determines recovery. Am. J. Physiol. **256**, C1262–1266 (1989)
67. Alliston, T., Ko, T.C., Cao, Y., Liang, Y.Y., Feng, X.H., Chang, C., Derynck, R.: Repression of bone morphogenetic protein and activin-inducible transcription by Evi-1. J. Biol. Chem. **280**, 24227–24237 (2005)
68. Annes, J.P., Munger, J.S., Rifkin, D.B.: Making sense of latent TGFbeta activation. J. Cell Sci. **116**, 217–224 (2003)
69. Hyytiainen, M., Penttinen, C., Keski-Oja, J.: Latent TGF-beta binding proteins: extracellular matrix association and roles in TGF-beta activation. Crit. Rev. Clin. Lab. Sci. **41**, 233–264 (2004)
70. Husmann, I., Soulet, L., Gautron, J., Martelly, I., Barritault, D.: Growth factors in skeletal muscle regeneration. Cytokine Growth Factor Rev. **7**, 249–258 (1996)
71. Barcellos-Hoff, M.H.: Latency and activation in the control of TGF-beta. J. Mammary Gland Biol. Neoplasia **1**, 353–363 (1996)
72. Scime, A., Caron, A.Z., Grenier, G.: Advances in myogenic cell transplantation and skeletal muscle tissue engineering. Front. Biosci. **14**, 3012–3023 (2009).
73. Brady, M.A., Lewis, M.P., Mudera, V.: Synergy between myogenic and non-myogenic cells in a 3D tissue-engineered craniofacial skeletal muscle construct. J. Tissue Eng. Regen. Med. **2**, 408–417 (2008)

74. Cheema, U., Yang, S.Y., Mudera, V., Goldspink, G.G., Brown, R.A.: 3-D in vitro model of early skeletal muscle development. Cell Motil. Cytoskeleton **54**, 226–236 (2003)
75. Kosnik, P.E., Dennis, R.G., Vandenburgh, H.H.: Tissue engineering skeletal muscle. In: Guilak, F.B.D., Goldstein, S.A., et al. (eds.) In Functional Tissue Engineering, pp. 377–392. Springer, New York (2003)
76. Boonen, K.J., Rosaria-Chak, K.Y., Baaijens, F.P., van der Schaft, D.W., Post, M.J.: Essential environmental cues from the satellite cell niche: optimizing proliferation and differentiation. Am. J. Physiol. **296**, C1338–1345 (2009)
77. Boonen, K.J., Post, M.J.: The muscle stem cell niche: regulation of satellite cells during regeneration. Tissue Eng. Part B Rev. **14**, 419–431 (2008)
78. Cosgrove, B.D., Sacco, A., Gilbert, P.M., Blau, H.M.: A home away from home: challenges and opportunities in engineering in vitro muscle satellite cell niches. Differentiation **78**, 185–194 (2009)
79. Collinsworth, A.M., Zhang, S., Kraus, W.E., Truskey, G.A.: Apparent elastic modulus and hysteresis of skeletal muscle cells throughout differentiation. Am. J. Physiol. **283**, C1219–1227 (2002)
80. Engler, A.J., Griffin, M.A., Sen, S., Bonnemann, C.G., Sweeney, H.L., Discher, D.E.: Myotubes differentiate optimally on substrates with tissue-like stiffness: pathological implications for soft or stiff microenvironments. J. Cell Biol. **166**, 877–887 (2004)
81. Stedman, H.H., Sweeney, H.L., Shrager, J.B., Maguire, H.C., Panettieri, R.A., Petrof, B., Narusawa, M., Leferovich, J.M., Sladky, J.T., Kelly, A.M.: The mdx mouse diaphragm reproduces the degenerative changes of Duchenne muscular dystrophy. Nature **352**, 536–539 (1991)
82. Gao, Y., Kostrominova, T.Y., Faulkner, J.A., Wineman, A.S.: Age-related changes in the mechanical properties of the epimysium in skeletal muscles of rats. J. Biomech. **41**, 465–469 (2008)
83. Rosant, C., Nagel, M.D., Perot, C.: Aging affects passive stiffness and spindle function of the rat soleus muscle. Exp. Gerontol. **42**, 301–308 (2007)
84. Geiger, B., Spatz, J.P., Bershadsky, A.D.: Environmental sensing through focal adhesions. Nat. Rev. **10**, 21–33 (2009)
85. Lopez, J.I., Mouw, J.K., Weaver, V.M.: Biomechanical regulation of cell orientation and fate. Oncogene **27**, 6981–6993 (2008)
86. Langer, R., Vacanti, J.P.: Tissue engineering. Science (New York, NY) **260**, 920–926 (1993)
87. Mooney, D.J., Organ, G., Vacanti, J.P., Langer, R.: Design and fabrication of biodegradable polymer devices to engineer tubular tissues. Cell Transplant. **3**, 203–210 (1994)
88. Partridge, T.A., Grounds, M., Sloper, J.C.: Evidence of fusion between host and donor myoblasts in skeletal muscle grafts. Nature **273**, 306–308 (1978)
89. Peault, B., Rudnicki, M., Torrente, Y., Cossu, G., Tremblay, J.P., Partridge, T., Gussoni, E., Kunkel, L.M., Huard, J.: Stem and progenitor cells in skeletal muscle development, maintenance, and therapy. Mol. Ther. **15**, 867–877 (2007)
90. O'Connor, M.S., Carlson, M.E., Conboy, I.M.: Differentiation rather than aging of muscle stem cells abolishes their telomerase activity. Biotechnol. Prog. **25**, 1130–1137 (2009)
91. Skuk, D., Goulet, M., Roy, B., Tremblay, J.P.: Efficacy of myoblast transplantation in nonhuman primates following simple intramuscular cell injections: toward defining strategies applicable to humans. Exp. Neurol. **175**, 112–126 (2002)
92. Koning, M., Harmsen, M.C., van Luyn, M.J., Werker, P.M.: Current opportunities and challenges in skeletal muscle tissue engineering. J. Tissue Eng. Regen. Med. **3**, 407–415 (2009)
93. Chakkalakal, J.V., Thompson, J., Parks, R.J., Jasmin, B.J.: Molecular, cellular, and pharmacological therapies for Duchenne/Becker muscular dystrophies. FASEB J. **19**, 880–891 (2005)
94. Negroni, E., Butler-Browne, G.S., Mouly, V.: Myogenic stem cells: regeneration and cell therapy in human skeletal muscle. Pathol. Biol. **54**, 100–108 (2006)

95. Partridge, T.: Myoblast transplantation. Neuromuscul. Disord. **12**(Suppl 1), S3–6 (2002)
96. Mouly, V., Aamiri, A., Perie, S., Mamchaoui, K., Barani, A., Bigot, A., Bouazza, B., Francois, V., Furling, D., Jacquemin, V., Negroni, E., Riederer, I., Vignaud, A., St Guily, J.L., Butler-Browne, G.S.: Myoblast transfer therapy: is there any light at the end of the tunnel? Acta Myol. **24**, 128–133 (2005)
97. Das, M., Rumsey, J.W., Bhargava, N., Stancescu, M., Hickman, J.J.: Skeletal muscle tissue engineering: a maturation model promoting long-term survival of myotubes, structural development of the excitation-contraction coupling apparatus and neonatal myosin heavy chain expression. Biomaterials **30**, 5392–5402 (2009)
98. Saxena, A.K., Marler, J., Benvenuto, M., Willital, G.H., Vacanti, J.P.: Skeletal muscle tissue engineering using isolated myoblasts on synthetic biodegradable polymers: preliminary studies. Tissue Eng. **5**, 525–532 (1999)
99. Gilmore, K.J., Kita, M., Han, Y., Gelmi, A., Higgins, M.J., Moulton, S.E., Clark, G.M., Kapsa, R., Wallace, G.G.: Skeletal muscle cell proliferation and differentiation on polypyrrole substrates doped with extracellular matrix components. Biomaterials **30**, 5292–5304 (2009)
100. Kroehne, V., Heschel, I., Schugner, F., Lasrich, D., Bartsch, J.W., Jockusch, H.: Use of a novel collagen matrix with oriented pore structure for muscle cell differentiation in cell culture and in grafts. J. Cell. Mol. Med. **12**, 1640–1648 (2008)
101. Bardouille, C., Lehmann, J., Heimann, P., Jockusch, H.: Growth and differentiation of permanent and secondary mouse myogenic cell lines on microcarriers. Appl. Microbiol. Biotechnol. **55**, 556–562 (2001)
102. Rowley, J.A., Madlambayan, G., Mooney, D.J.: Alginate hydrogels as synthetic extracellular matrix materials. Biomaterials **20**, 45–53 (1999)
103. Huang, Y.C., Dennis, R.G., Larkin, L., Baar, K.: Rapid formation of functional muscle in vitro using fibrin gels. J. Appl. Physiol. **98**, 706–713 (2005)
104. Cronin, E.M., Thurmond, F.A., Bassel-Duby, R., Williams, R.S., Wright, W.E., Nelson, K.D., Garner, H.R.: Protein-coated poly(L-lactic acid) fibers provide a substrate for differentiation of human skeletal muscle cells. J. Biomed. Mater. Res. **69**, 373–381 (2004)
105. Levenberg, S., Rouwkema, J., Macdonald, M., Garfein, E.S., Kohane, D.S., Darland, D.C., Marini, R., van Blitterswijk, C.A., Mulligan, R.C., D'Amore, P.A., Langer, R.: Engineering vascularized skeletal muscle tissue. Nat. Biotechnol. **23**, 879–884 (2005)
106. Neumann, T., Hauschka, S.D., Sanders, J.E.: Tissue engineering of skeletal muscle using polymer fiber arrays. Tissue Eng. **9**, 995–1003 (2003)
107. Shen, J.Y., Chan-Park, M.B., Feng, Z.Q., Chan, V., Feng, Z.W.: UV-embossed microchannel in biocompatible polymeric film: application to control of cell shape and orientation of muscle cells. J. Biomed. Mater. Res. B Appl. Biomater. **77**, 423–430 (2006)
108. Riboldi, S.A., Sampaolesi, M., Neuenschwander, P., Cossu, G., Mantero, S.: Electrospun degradable polyesterurethane membranes: potential scaffolds for skeletal muscle tissue engineering. Biomaterials **26**, 4606–4615 (2005)
109. Vandenburgh, H.H., Karlisch, P., Farr, L.: Maintenance of highly contractile tissue-cultured avian skeletal myotubes in collagen gel. In Vitro Cell. Dev. Biol. **24**, 166–174 (1988)
110. Okano, T., Satoh, S., Oka, T., Matsuda, T.: Tissue engineering of skeletal muscle. Highly dense, highly oriented hybrid muscular tissues biomimicking native tissues. ASAIO J. **43**, M749–753 (1997)
111. Saxena, A.K., Willital, G.H., Vacanti, J.P.: Vascularized three-dimensional skeletal muscle tissue-engineering. Biomed. Mater. Eng. **11**, 275–281 (2001)
112. Levy-Mishali, M., Zoldan, J., Levenberg, S.: Effect of scaffold stiffness on myoblast differentiation. Tissue Eng. Part A **15**, 935–944 (2009)
113. Boldrin, L., Malerba, A., Vitiello, L., Cimetta, E., Piccoli, M., Messina, C., Gamba, P.G., Elvassore, N., De Coppi, P.: Efficient delivery of human single fiber-derived muscle precursor cells via biocompatible scaffold. Cell Transplant. **17**, 577–584 (2008)
114. Lam, M.T., Sim, S., Zhu, X., Takayama, S.: The effect of continuous wavy micropatterns on silicone substrates on the alignment of skeletal muscle myoblasts and myotubes. Biomaterials **27**, 4340–4347 (2006)

115. Lutolf, M.P., Hubbell, J.A.: Synthetic biomaterials as instructive extracellular microenvironments for morphogenesis in tissue engineering. Nat. Biotechnol. **23**, 47–55 (2005)
116. Dubois, G., Segers, V.F., Bellamy, V., Sabbah, L., Peyrard, S., Bruneval, P., Hagege, A.A., Lee, R.T., Menasche, P.: Self-assembling peptide nanofibers and skeletal myoblast transplantation in infarcted myocardium. J. Biomed. Mater. Res. **87**, 222–228 (2008)
117. Mulder, M.M., Hitchcock, R.W., Tresco, P.A.: Skeletal myogenesis on elastomeric substrates: implications for tissue engineering. J. Biomater. Sci. **9**, 731–748 (1998)
118. Ren, K., Crouzier, T., Roy, C., Picart, C.: Polyelectrolyte multilayer films of controlled stiffness modulate myoblast cells differentiation. Adv. Funct. Mater. **18**, 1378–1389 (2008)
119. Wang, W., Fan, M., Zhang, L., Liu, S.H., Sun, L., Wang, C.Y.: Compatibility of hyaluronic acid hydrogel and skeletal muscle myoblasts. Biomed. Mater. (Bristol, England) **4**(2), 25011 (2009)
120. Loken, S., Jakobsen, R.B., Aroen, A., Heir, S., Shahdadfar, A., Brinchmann, J.E., Engebretsen, L., Reinholt, F.P.: Bone marrow mesenchymal stem cells in a hyaluronan scaffold for treatment of an osteochondral defect in a rabbit model. Knee Surg. Sports Traumatol. Arthrosc. **16**, 896–903 (2008)
121. David-Raoudi, M., Tranchepain, F., Deschrevel, B., Vincent, J.C., Bogdanowicz, P., Boumediene, K., Pujol, J.P.: Differential effects of hyaluronan and its fragments on fibroblasts: relation to wound healing. Wound Repair Regen. **16**, 274–287 (2008)
122. Cui, F.Z., Tian, W.M., Hou, S.P., Xu, Q.Y., Lee, I.S.: Hyaluronic acid hydrogel immobilized with RGD peptides for brain tissue engineering. J. Mater. Sci. **17**, 1393–1401 (2006)
123. Han, F., Ishiguro, N., Ito, T., Sakai, T., Iwata, H: Effects of sodium hyaluronate on experimental osteoarthritis in rabbit knee joints. Nagoya J. Med. Sci. **62**, 115–126 (1999)
124. Gupta, P., Vermani, K., Garg, S.: Hydrogels: from controlled release to pH-responsive drug delivery. Drug Discov. Today **7**, 569–579 (2002)
125. Beier, J.P., Kneser, U., Stern-Strater, J., Stark, G.B., Bach, A.D.: Y chromosome detection of three-dimensional tissue-engineered skeletal muscle constructs in a syngeneic rat animal model. Cell Transplant. **13**, 45–53 (2004)
126. de Juan-Pardo, E.M., Hoang, M.B., Conboy, I.M.: Geometric control of myogenic cell fate. Int. J. Nanomed. **1**, 203–212 (2006)
127. Lam, M.T., Huang, Y.C., Birla, R.K., Takayama, S.: Microfeature guided skeletal muscle tissue engineering for highly organized 3-dimensional free-standing constructs. Biomaterials **30**, 1150–1155 (2009)
128. Riboldi, S.A., Sadr, N., Pigini, L., Neuenschwander, P., Simonet, M., Mognol, P., Sampaolesi, M., Cossu, G., Mantero, S.: Skeletal myogenesis on highly orientated microfibrous polyesterurethane scaffolds. J. Biomed. Mater. Res. **84**, 1094–1101 (2008)
129. Rockwood, D.N., Woodhouse, K.A., Fromstein, J.D., Chase, D.B., Rabolt, J.F.: Characterization of biodegradable polyurethane microfibers for tissue engineering. J. Biomater. Sci. **18**, 743–758 (2007)
130. Evans, D.J., Britland, S., Wigmore, P.M.: Differential response of fetal and neonatal myoblasts to topographical guidance cues in vitro. Dev. Genes Evol. **209**, 438–442 (1999)
131. Clark, P., Dunn, G.A., Knibbs, A., Peckham, M.: Alignment of myoblasts on ultrafine gratings inhibits fusion in vitro. Int. J. Biochem. Cell Biol. **34**, 816–825 (2002)
132. Bian, W., Bursac, N.: Engineered skeletal muscle tissue networks with controllable architecture. Biomaterials **30**, 1401–1412 (2009)
133. Davis, B.H., Schroeder, T., Yarmolenko, P.S., Guilak, F., Dewhirst, M.W., Taylor, D.A.: An in vitro system to evaluate the effects of ischemia on survival of cells used for cell therapy. Ann. Biomed. Eng. **35**, 1414–1424 (2007)
134. Simpson, D.G., Terracio, L., Terracio, M., Price, R.L., Turner, D.C., Borg, T.K.: Modulation of cardiac myocyte phenotype in vitro by the composition and orientation of the extracellular matrix. J. Cell. Physiol. **161**, 89–105 (1994)
135. Engler, A.J., Sen, S., Sweeney, H.L., Discher, D.E.: Matrix elasticity directs stem cell lineage specification. Cell **126**, 677–689 (2006)

136. Boontheekul, T., Hill, E.E., Kong, H.J., Mooney, D.J.: Regulating myoblast phenotype through controlled gel stiffness and degradation. Tissue Eng. **13**, 1431–1442 (2007)
137. Dhoot, N.O., Tobias, C.A., Fischer, I., Wheatley, M.A.: Peptide-modified alginate surfaces as a growth permissive substrate for neurite outgrowth. J. Biomed. Mater. Res. **71**, 191–200 (2004)
138. Liao, I.C., Liu, J.B., Bursac, N., Leong, K.W.: Effect of electromechanical stimulation on the maturation of myotubes on aligned electrospun fibers. Cell. Mol. Bioeng. **1**, 133–145 (2008)
139. Yamasaki, K., Hayashi, H., Nishiyama, K., Kobayashi, H., Uto, S., Kondo, H., Hashimoto, S., Fujisato, T.: Control of myotube contraction using electrical pulse stimulation for bio-actuator. J. Artif. Organs **12,** 131–137 (2009)
140. Ishibashi, T., Hoshino, Y., Kaji, H., Kanzaki, M., Sato, M., Nishizawa, M.: Localized electrical stimulation to C2C12 myotubes cultured on a porous membrane-based substrate. Biomed. Microdevices **11**, 413–419 (2009)
141. Flaibani, M., Boldrin, L., Cimetta, E., Piccoli, M., De Coppi, P., Elvassore, N.: Muscle differentiation and myotubes alignment is influenced by micropatterned surfaces and exogenous electrical stimulation. Tissue Eng. Part A **15**, 2447–2457 (2009)
142. Dennis, R.G., Kosnik, P.E., 2nd: Excitability and isometric contractile properties of mammalian skeletal muscle constructs engineered in vitro. In Vitro Cell. Dev. Biol. **36**, 327–335 (2000)
143. Vandenburgh, H.H., Swasdison, S., Karlisch, P.: Computer-aided mechanogenesis of skeletal muscle organs from single cells in vitro. FASEB J. **5**, 2860–2867 (1991)
144. Tatsumi, R., Hattori, A., Ikeuchi, Y., Anderson, J.E., Allen, R.E.: Release of hepatocyte growth factor from mechanically stretched skeletal muscle satellite cells and role of pH and nitric oxide. Mol. Biol. Cell **13**, 2909–2918 (2002)
145. Tatsumi, R., Liu, X., Pulido, A., Morales, M., Sakata, T., Dial, S., Hattori, A., Ikeuchi, Y., Allen, R.E.: Satellite cell activation in stretched skeletal muscle and the role of nitric oxide and hepatocyte growth factor. Am. J. Physiol. Cell Physiol. **290**, C1487–1494 (2006)
146. Desai, T.A.: Micro- and nanoscale structures for tissue engineering constructs. Med. Eng. Phys. **22**, 595–606 (2000)
147. Choi, J.S., Lee, S.J., Christ, G.J., Atala, A., Yoo, J.J.: The influence of electrospun aligned poly(epsilon-caprolactone)/collagen nanofiber meshes on the formation of self-aligned skeletal muscle myotubes. Biomaterials **29**, 2899–2906 (2008)
148. Jun, I., Jeong, S., Shin, H.: The stimulation of myoblast differentiation by electrically conductive sub-micron fibers. Biomaterials **30**, 2038–2047 (2009)
149. Jeong, S.I., Jun, I.D., Choi, M.J., Nho, Y.C., Lee, Y.M., Shin, H.: Development of electroactive and elastic nanofibers that contain polyaniline and poly(L-lactide-co-epsilon-caprolactone) for the control of cell adhesion. Macromol. Biosci. **8**, 627–637 (2008)
150. Tanaka, Y., Tsutsumi, A., Crowe, D.M., Tajima, S., Morrison, W.A.: Generation of an autologous tissue (matrix) flap by combining an arteriovenous shunt loop with artificial skin in rats: preliminary report. Br. J. Plastic Surg. **53**, 51–57 (2000)
151. Mian, R., Morrison, W.A., Hurley, J.V., Penington, A.J., Romeo, R., Tanaka, Y., Knight, K.R.: Formation of new tissue from an arteriovenous loop in the absence of added extracellular matrix. Tissue Eng. **6**, 595–603 (2000)
152. Beier, J.P., Stern-Straeter, J., Foerster, V.T., Kneser, U., Stark, G.B., Bach, A.D.: Tissue engineering of injectable muscle: three-dimensional myoblast-fibrin injection in the syngeneic rat animal model. Plast. Reconstr. Surg. **118**, 1113–1121; discussion 1122–1114 (2006)
153. Bach, A.D., Stem-Straeter, J., Beier, J.P., Bannasch, H., Stark, G.B.: Engineering of muscle tissue. Clin. Plast. Surg. **30**, 589–599 (2003)
154. Lodish, H.F., Baltimore, D., Berk, A., Zipursky, L., Matsudaira, P., Darnell, J.E.: Molecular Cell Biology, 4th edn, pp. 968–1002. Scientific American Books, New York (2000)
155. Giraud, M.N., Ayuni, E., Cook, S., Siepe, M., Carrel, T.P., Tevaearai, H.T.: Hydrogel-based engineered skeletal muscle grafts normalize heart function early after myocardial infarction. Artif. Organs **32**, 692–700 (2008)

156. Cimetta, E., Pizzato, S., Bollini, S., Serena, E., De Coppi, P., Elvassore, N.: Production of arrays of cardiac and skeletal muscle myofibers by micropatterning techniques on a soft substrate. Biomed. Microdevices **11**, 389–400 (2009)
157. Stern, M.M., Myers, R.L., Hammam, N., Stern, K.A., Eberli, D., Kritchevsky, S.B., Soker, S., Van Dyke, M.: The influence of extracellular matrix derived from skeletal muscle tissue on the proliferation and differentiation of myogenic progenitor cells ex vivo. Biomaterials **30**, 2393–2399 (2009)
158. Pola, R., Ling, L.E., Aprahamian, T.R., Barban, E., Bosch-Marce, M., Curry, C., Corbley, M., Kearney, M., Isner, J.M., Losordo, D.W.: Postnatal recapitulation of embryonic hedgehog pathway in response to skeletal muscle ischemia. Circulation **108**, 479–485 (2003)
159. Tidball, J.G.: Inflammatory processes in muscle injury and repair. Am. J. Physiol. Regul. Integr. Comp. Physiol. **288**, R345–353 (2005)
160. Wagers, A.J., Conboy, I.M.: Cellular and molecular signatures of muscle regeneration: current concepts and controversies in adult myogenesis. Cell **122**, 659–667 (2005)
161. Tavana, H., Jovic, A., Mosadegh, B., Lee, Q.Y., Liu, X., Luker, K.E., Luker, G.D., Weiss, S.J., Takayama, S.: Nanolitre liquid patterning in aqueous environments for spatially defined reagent delivery to mammalian cells. Nat. Mater. **8**, 736–741 (2009)
162. Ratner, B.D., Hoffman, A.S., Schoen, F.J., Lemons, J.E.: An introduction to materials in medicine. Biomater. Sci. (2nd edn) (2004)
163. Bushby, A.: Muscular Dystrophy, Methods and Protocols. Humana Press, New York (2001)
164. Pannier, A.K., Shea, L.D.: Controlled release systems for DNA delivery. Mol. Ther. **10**, 19–26 (2004)
165. Lodish, H.F., Baltimore, D., Berk, A., Zipursky, L., Matsudaira, P., Darnell, J.E.: Molecular Cell Biology, 4th edn, pp. 203–204. Scientific American Books, New York (2000)
166. Masotti, A., Ortaggi, G.: Chitosan micro- and nanospheres: fabrication and applications for drug and DNA delivery. Mini Rev. Med. Chem. **9**, 463–469 (2009)
167. Ziauddin, J., Sabatini, D.M.: Microarrays of cells expressing defined cDNAs. Nature **411**, 107–110 (2001)
168. Jewell, C.M., Zhang, J., Fredin, N.J., Lynn, D.M.: Multilayered polyelectrolyte films promote the direct and localized delivery of DNA to cells. J. Control Release **106**, 214–223 (2005)
169. Segura, T., Chung, P.H., Shea, L.D.: DNA delivery from hyaluronic acid-collagen hydrogels via a substrate-mediated approach. Biomaterials **26**, 1575–1584 (2005)
170. Segura, T., Shea, L.D. Surface-tethered DNA complexes for enhanced gene delivery. Bioconjug. Chem. **13**, 621–629 (2002)
171. Jeong, B., Bae, Y.H., Lee, D.S., Kim, S.W.: Biodegradable block copolymers as injectable drug-delivery systems. Nature **388**, 860–862 (1997)
172. Booth, C., Attwood, D., Price, C.: Self-association of block copoly(oxyalkylene)s in aqueous solution. Effects of composition, block length and block architecture. Phys. Chem. Chem. Phys. **8**, 3612–3622 (2006)
173. Bhardwaj, R., Blanchard, J.: Controlled-release delivery system for the alpha-MSH analog melanotan-I using poloxamer 407. J. Pharm. Sci. **85**, 915–919 (1996)
174. Wenzel, J.G., Balaji, K.S., Koushik, K., Navarre, C., Duran, S.H., Rahe, C.H., Kompella, U.B.: Pluronic F127 gel formulations of deslorelin and GnRH reduce drug degradation and sustain drug release and effect in cattle. J. Control Release **85**, 51–59 (2002)
175. Hinrichs, W.L., Schuurmans-Nieuwenbroek, N.M., van de Wetering, P., Hennink, W.E.: Thermosensitive polymers as carriers for DNA delivery. J. Control Release **60**, 249–259 (1999)
176. Li, Z., Ning, W., Wang, J., Choi, A., Lee, P.Y., Tyagi, P., Huang, L.: Controlled gene delivery system based on thermosensitive biodegradable hydrogel. Pharma. Res. **20**, 884–888 (2003)
177. Zentner, G.M., Rathi, R., Shih, C., McRea, J.C., Seo, M.H., Oh, H., Rhee, B.G., Mestecky, J., Moldoveanu, Z., Morgan, M., Weitman, S.: Biodegradable block copolymers for delivery of proteins and water-insoluble drugs. J. Control Release **72**, 203–215 (2001)

178. Namgung, R., Nam, S., Kim, S.K., Son, S., Singha, K., Kwon, J.S., Ahn, Y., Jeong, M.H., Park, I.K., Garripelli, V.K., Jo, S., Kim, W.J.: An acid-labile temperature-responsive sol-gel reversible polymer for enhanced gene delivery to the myocardium and skeletal muscle cells. Biomaterials **30**, 5225–5233 (2009)
179. Garripelli, V.K., Kim, J.K., Namgung, R., Kim, W.J., Repka, M.A., Jo, S.: A novel thermosensitive polymer with pH-dependent degradation for drug delivery. Acta Biomater. **6**, 477–485 (2009)
180. Chang, C.W., Choi, D., Kim, W.J., Yockman, J.W., Christensen, L.V., Kim, Y.H., Kim, S.W.: Non-ionic amphiphilic biodegradable PEG-PLGA-PEG copolymer enhances gene delivery efficiency in rat skeletal muscle. J. Control Release **118**, 245–253 (2007)
181. Anderson, D.G., Peng, W., Akinc, A., Hossain, N., Kohn, A., Padera, R., Langer, R., Sawicki, J.A.: A polymer library approach to suicide gene therapy for cancer. Proc. Natl. Acad. Sci. U.S.A. **101**, 16028–16033 (2004)
182. Dobrovolskaia, M.A., Aggarwal, P., Hall, J.B., McNeil, S.E.: Preclinical studies to understand nanoparticle interaction with the immune system and its potential effects on nanoparticle biodistribution. Mol. Pharmaceut. **5**, 487–495 (2008)
183. Matt, P., Schoenhoff, F., Habashi, J., Holm, T., Van Erp, C., Loch, D., Carlson, O.D., Griswold, B.F., Fu, Q., De Backer, J., Loeys, B., Huso, D.L., McDonnell, N.B., Van Eyk, J.E., Dietz, H.C.: Circulating transforming growth factor-beta in Marfan syndrome. Circulation **120**, 526–532 (2009)
184. Morgan, J.E., Partridge, T.A.: Muscle satellite cells. Int. J. Biochem. Cell Biol. **35**, 1151–1156 (2003)
185. Camargo, F.D., Green, R., Capetanaki, Y., Jackson, K.A., Goodell, M.A.: Single hematopoietic stem cells generate skeletal muscle through myeloid intermediates. Nat. Med. **9**, 1520–1527 (2003)
186. Corbel, S.Y., Lee, A., Yi, L., Duenas, J., Brazelton, T.R., Blau, H.M., Rossi, F.M.: Contribution of hematopoietic stem cells to skeletal muscle. Nat. Med. **9**, 1528–1532 (2003)
187. Doyonnas, R., LaBarge, M.A., Sacco, A., Charlton, C., Blau, H.M.: Hematopoietic contribution to skeletal muscle regeneration by myelomonocytic precursors. Proc. Natl. Acad. Sci. U.S.A. **101**, 13507–13512 (2004)
188. Sacco, A., Doyonnas, R., LaBarge, M.A., Hammer, M.M., Kraft, P., Blau, H.M.: IGF-I increases bone marrow contribution to adult skeletal muscle and enhances the fusion of myelomonocytic precursors. J. Cell Biol. 171, 483–492 (2005)
189. Sampaolesi, M., Blot, S., D'Antona, G., Granger, N., Tonlorenzi, R., Innocenzi, A., Mognol, P., Thibaud, J.L., Galvez, B.G., Barthelemy, I., Perani, L., Mantero, S., Guttinger, M., Pansarasa, O., Rinaldi, C., Cusella De Angelis, M.G., Torrente, Y., Bordignon, C., Bottinelli, R., Cossu, G.: Mesoangioblast stem cells ameliorate muscle function in dystrophic dogs. Nature **444**, 574–579 (2006)
190. Dellavalle, A., Sampaolesi, M., Tonlorenzi, R., Tagliafico, E., Sacchetti, B., Perani, L., Innocenzi, A., Galvez, B.G., Messina, G., Morosetti, R., Li, S., Belicchi, M., Peretti, G., Chamberlain, J.S., Wright, W.E., Torrente, Y., Ferrari, S., Bianco, P., Cossu, G.: Pericytes of human skeletal muscle are myogenic precursors distinct from satellite cells. Nat. Cell Biol. **9**, 255–267 (2007)
191. Beauchamp, J.R., Heslop, L., Yu, D.S., Tajbakhsh, S., Kelly, R.G., Wernig, A., Buckingham, M.E., Partridge, T.A., Zammit, P.S.: Expression of CD34 and Myf5 defines the majority of quiescent adult skeletal muscle satellite cells. J. Cell Biol. **151**, 1221–1234 (2000)
192. Halevy, O., Piestun, Y., Allouh, M.Z., Rosser, B.W., Rinkevich, Y., Reshef, R., Rozenboim, I., Wleklinski-Lee, M., Yablonka-Reuveni, Z.: Pattern of Pax7 expression during myogenesis in the posthatch chicken establishes a model for satellite cell differentiation and renewal. Dev. Dyn. **231**, 489–502 (2004)
193. Schultz, E.: Satellite cell proliferative compartments in growing skeletal muscles. Dev. Biol. **175**, 84–94 (1996)

194. Zammit, P.S., Golding, J.P., Nagata, Y., Hudon, V., Partridge, T.A., Beauchamp, J.R.: Muscle satellite cells adopt divergent fates: a mechanism for self-renewal? J. Cell Biol. **166**, 347–357 (2004)
195. Seale, P., Sabourin, L.A., Girgis-Gabardo, A., Mansouri, A., Gruss, P., Rudnicki, M.A.: Pax7 is required for the specification of myogenic satellite cells. Cell **102**, 777–786 (2000)
196. Cerletti, M., Jurga, S., Witczak, C.A., Hirshman, M.F., Shadrach, J.L., Goodyear, L.J., Wagers, A.J.: Highly efficient, functional engraftment of skeletal muscle stem cells in dystrophic muscles. Cell **134**, 37–47 (2008)
197. Cornelison, D.D., Filla, M.S., Stanley, H.M., Rapraeger, A.C., Olwin, B.B.: Syndecan-3 and syndecan-4 specifically mark skeletal muscle satellite cells and are implicated in satellite cell maintenance and muscle regeneration. Dev. Biol. **239**, 79–94 (2001)
198. Cornelison, D.D., Olwin, B.B., Rudnicki, M.A., Wold, B.J.: MyoD(−/−) satellite cells in single-fiber culture are differentiation defective and MRF4 deficient. Dev. Biol. **224**, 122–137 (2000)
199. Brack, A.S., Bildsoe, H., Hughes, S.M.: Evidence that satellite cell decrement contributes to preferential decline in nuclear number from large fibres during murine age-related muscle atrophy. J. Cell Sci. **118**, 4813–4821 (2005)
200. Cornelison, D.D., Wilcox-Adelman, S.A., Goetinck, P.F., Rauvala, H., Rapraeger, A.C., Olwin, B.B.: Essential and separable roles for Syndecan-3 and Syndecan-4 in skeletal muscle development and regeneration. Genes Dev. **18**, 2231–2236 (2004)
201. Cornelison, D.D., Wold, B.J.: Single-cell analysis of regulatory gene expression in quiescent and activated mouse skeletal muscle satellite cells. Dev. Biol. **191**, 270–283 (1997)
202. Volonte, D., Liu, Y., Galbiati, F.: The modulation of caveolin-1 expression controls satellite cell activation during muscle repair. FASEB J. **19**, 237–239 (2005)
203. Irintchev, A., Zeschnigk, M., Starzinski-Powitz, A., Wernig, A.: Expression pattern of M-cadherin in normal, denervated, and regenerating mouse muscles. Dev. Dyn. **199**, 326–337 (1994)
204. Sherwood, R.I., Christensen, J.L., Conboy, I.M., Conboy, M.J., Rando, T.A., Weissman, I.L., Wagers, A.J.: Isolation of adult mouse myogenic progenitors: functional heterogeneity of cells within and engrafting skeletal muscle. Cell **119**, 543–554 (2004)
205. Jesse, T.L., LaChance, R., Iademarco, M.F., Dean, D.C.: Interferon regulatory factor-2 is a transcriptional activator in muscle where it regulates expression of vascular cell adhesion molecule-1. J. Cell Biol. **140**, 1265–1276 (1998)
206. Illa, I., Leon-Monzon, M., Dalakas, M.C.: Regenerating and denervated human muscle fibers and satellite cells express neural cell adhesion molecule recognized by monoclonal antibodies to natural killer cells. Ann. Neurol. **31**, 46–52 (1992)
207. Schubert, W., Zimmermann, K., Cramer, M., Starzinski-Powitz, A.: Lymphocyte antigen Leu-19 as a molecular marker of regeneration in human skeletal muscle. Proc. Natl. Acad. Sci. U.S.A. **86**, 307–311 (1989)
208. Tanaka, K.K., Hall, J.K., Troy, A.A., Cornelison, D.D., Majka, S.M., Olwin, B.B.: Syndecan-4-expressing muscle progenitor cells in the SP engraft as satellite cells during muscle regeneration. Cell Stem Cell **4**, 217–225 (2009)
209. Fukada, S., Higuchi, S., Segawa, M., Koda, K., Yamamoto, Y., Tsujikawa, K., Kohama, Y., Uezumi, A., Imamura, M., Miyagoe-Suzuki, Y., Takeda, S., Yamamoto, H.: Purification and cell-surface marker characterization of quiescent satellite cells from murine skeletal muscle by a novel monoclonal antibody. Exp. Cell Res. **296**, 245–255 (2004)
210. Garry, D.J., Yang, Q., Bassel-Duby, R., Williams, R.S.: Persistent expression of MNF identifies myogenic stem cells in postnatal muscles. Dev. Biol. **188**, 280–294 (1997)
211. Kirk, S., Oldham, J., Kambadur, R., Sharma, M., Dobbie, P., Bass, J.: Myostatin regulation during skeletal muscle regeneration. J. Cell. Physiol. **184**, 356–363 (2000)

212. Mendler, L., Zador, E., Ver Heyen, M., Dux, L., Wuytack, F.: Myostatin levels in regenerating rat muscles and in myogenic cell cultures. J. Muscle Res. Cell Motil. **21**, 551–563 (2000)
213. Saleh, O.A., Sohn, L.L.: Direct detection of antibody-antigen binding using an on-chip artificial pore. Proc. Natl. Acad. Sci. U.S.A. **100**, 820–824, 2003
214. He, M., Herr, A.E.: Microfluidic polyacrylamide gel electrophoresis with in situ immunoblotting for native protein analysis. Anal. Chem. **81**, 8177–8184 (2009)

Author Index

A
Ali Khademhosseini, 153
Ameya Phadke, 19
Amir Foudeh, 153
Andres Bratt-Leal, 45
Ankur Singh, 225
Ashley Carson Brown, 1

B
Behnam Zamanian, 153
Benjamin K. K. Teo, 61

C
Cheong Hoon Kwon, 153
Chien-Wen Chang, 19

E
Elizabeth M. Powell, 247
Eric Jabart, 275
Evelyn K. F. Yim, 61

G
Ge Zhang, 173
Gregory Christopherson, 89
Guoming Sun, 195

H
Hai-Quan Mao, 89
Hang Lu, 119
Hojae Bae, 153

I
Irina Conboy, 275

J
Jason W. Nichol, 153
Jennie B. Leach, 247
Johnna S. Temenoff, 119

K
Korey Kam, 89
Krishnendu Roy, 225

L
Laura J. Suggs, 173

R
Richard Carpenedo, 45

S
Sarah E. Stabenfeldt, 1
Sharon Gerecht, 195
Sharon K. Hamilton, 119
Shawn H. Lim, 89
Shuming Zhang, 89
Shyni Varghese, 19
Soneela Ankam, 61
Sravanti Kusuma, 195
Stephen Fischer, 89

T
Thomas H. Barker, 1
Todd McDevitt, 45